Survey Data Harmonization in the Social Sciences

Survey Data Harmonization in the Social Sciences

Edited by

Irina Tomescu-Dubrow
Institute of Philosophy and Sociology
Polish Academy of Sciences, Warsaw, Poland

Christof Wolf
GESIS Leibniz-Institute for the Social Sciences
University Mannheim, Germany

Kazimierz M. Slomczynski
Institute of Philosophy and Sociology
Polish Academy of Sciences, Warsaw, Poland

J. Craig Jenkins
Department of Sociology
The Ohio State University, Ohio, USA

Published by John Wiley & Sons, Inc., Hoboken, New Jersey.
Published simultaneously in Canada.

For general information on our other products and services or for technical support, please contact our Customer Care Department within the United States at (800) 762-2974, outside the United States at (317) 572-3993 or fax (317) 572-4002.

Wiley also publishes its books in a variety of electronic formats. Some content that appears in print may not be available in electronic formats. For more information about Wiley products, visit our web site at www.wiley.com.

Library of Congress Cataloging-in-Publication Data Applied for:

Hardback ISBN: 9781119712176

Cover Design: Wiley
Cover Image: © Simon Zhu/Unsplash

Set in 9.5/12.5pt STIXTwoText by Straive, Chennai, India

Contents

Preface and Acknowledgments *xv*
About the Editors *xvii*
About the Contributors *xviii*

1 **Objectives and Challenges of Survey Data Harmonization** *1*
Kazimierz M. Slomczynski, Irina Tomescu-Dubrow, J. Craig Jenkins, and Christof Wolf
1.1 Introduction *1*
1.2 What is the Harmonization of Survey Data? *2*
1.2.1 Ex-ante, Input and Output, Survey Harmonization *3*
1.3 Why Harmonize Social Survey Data? *5*
1.3.1 Comparison and Equivalence *6*
1.4 Harmonizing Survey Data Across and Within Countries *7*
1.4.1 Harmonizing Across Countries *7*
1.4.2 Harmonizing Within the Country *8*
1.5 Sources of Knowledge for Survey Data Harmonization *8*
1.6 Challenges to Survey Harmonization *9*
1.6.1 Population Representation (Sampling Design) *10*
1.6.2 Instruments and Their Adaptation (Including Translation) *10*
1.6.3 Preparation for Interviewing (Including Pretesting) *11*
1.6.4 Fieldwork (Including Modes of Interviewing) *11*
1.6.5 Data Preparation (Including Building Data Files) *12*
1.6.6 Data Processing, Quality Controls, and Adjustments *12*
1.6.7 Data Dissemination *13*
1.7 Survey Harmonization and Standardization Processes *13*
1.8 Quality of the Input and the End-product of Survey Harmonization *14*
1.9 Relevance of Harmonization Methodology to the FAIR Data Principles *15*
1.10 Ethical and Legal Issues *15*
1.11 How to Read this Volume? *16*
References *17*

2 **The Effects of Data Harmonization on the Survey Research Process** *21*
Ranjit K. Singh, Arnim Bleier, and Peter Granda
2.1 Introduction *21*
2.2 Part 1: Harmonization: Origins and Relation to Standardization *22*
2.2.1 Early Conceptions of Standardization and Harmonization *22*
2.2.2 Foundational Work of International Survey Programs *23*
2.2.3 The Growing Impact of Data Harmonization *23*
2.3 Part 2: Stakeholders and Division of Labor *25*
2.3.1 Stakeholders *26*
2.3.1.1 International Actors and Funding Agencies *26*
2.3.1.2 Data Producers *26*
2.3.1.3 Archives *27*
2.3.1.4 Data Users *27*
2.3.2 Toward an Integrative View on Harmonization *28*
2.3.2.1 Harmonization Cost *29*
2.3.2.2 Harmonization Quality *29*
2.3.2.3 Harmonization Fit *30*
2.3.2.4 Moving Forward *30*
2.4 Part 3: New Data Types, New Challenges *31*
2.4.1 Designed Data and Organic Data *31*
2.4.2 Stakeholders in the Collection of Organic Data *32*
2.4.2.1 Producers *32*
2.4.2.2 Archives *32*
2.4.2.3 Users *33*
2.4.2.4 Harmonization of Organic Data *33*
2.5 Conclusion *33*
References *35*

Part I **Ex-ante harmonization of survey instruments and non-survey data** *39*

3 **Harmonization in the World Values Survey** *41*
Kseniya Kizilova, Jaime Diez-Medrano, Christian Welzel, and Christian Haerpfer
3.1 Introduction *41*
3.2 Applied Harmonization Methods *42*
3.3 Documentation and Quality Assurance *48*
3.4 Challenges to Harmonization *49*
3.5 Software Tools *51*
3.6 Recommendations *52*
References *54*

4 **Harmonization in the Afrobarometer** *57*
Carolyn Logan, Robert Mattes, and Francis Kibirige
4.1 Introduction *57*
4.2 Core Principles *58*
4.3 Applied Harmonization Methods *60*
4.3.1 Sampling *60*

4.3.2   Training *61*

4.3.3   Fieldwork and Data Collection *62*

4.3.4   Questionnaire *62*

4.3.5   Translation *64*

4.3.6   Data Management *65*

4.3.7   Documentation *65*

4.4   Harmonization and Country Selection *66*

4.5   Software Tools and Harmonization *66*

4.6   Challenges to Harmonization *67*

4.6.1   Local Knowledge, Flexibility/Adaptability, and the "Dictatorship of Harmonization" *68*

4.6.2   The Quality-Cost Trade-off and Implications for Harmonization *68*

4.6.3   Final Challenge: "Events" *69*

4.7   Recommendations *70*

    References *71*

5   Harmonization in the National Longitudinal Surveys of Youth (NLSY) *73*
    Elizabeth Cooksey, Rosella Gardecki, Carole Lunney, and Amanda Roose

5.1   Introduction *73*

5.2   Cross-Cohort Design *75*

5.3   Applied Harmonization *76*

5.4   Challenges to Harmonization *80*

5.5   Documentation and Quality Assurance *82*

5.6   Software Tools *84*

5.7   Recommendations and Some Concluding Thoughts *86*

    References *87*

6   Harmonization in the Comparative Study of Electoral Systems (CSES) Projects *89*
    Stephen Quinlan, Christian Schimpf, Katharina Blinzler, and Slaven Zivkovic

6.1   Introducing the CSES *89*

6.2   Harmonization Principles and Technical Infrastructure *91*

6.3   Ex-ante Input Harmonization *91*

6.3.1   Module Questionnaire *92*

6.3.2   Macro Data *94*

6.4   Ex-ante Output Harmonization *97*

6.4.1   Demographic Variables in CSES Modules *97*

6.4.2   Harmonizing Party Data in Modules *98*

6.4.3   Derivative Variables *99*

6.5   Exploring Interplay Between Ex-ante and *Ex-post* Harmonization *101*

6.5.1   Demographic Variables in CSES IMD *101*

6.5.2   Harmonizing Party Data in CSES IMD *102*

6.6   Taking Stock and New Frontiers in Harmonization *104*

    References *105*

7   Harmonization in the East Asian Social Survey *107*
    Noriko Iwai, Tetsuo Mo, Jibum Kim, Chyi-In Wu, and Weidong Wang

7.1   Introduction *107*

7.2   Characteristics of the EASS and its Harmonization Process   *108*

7.2.1   Outline of the East Asian Social Survey   *108*

7.2.2   Harmonization Process of the EASS   *111*

7.2.2.1   Establishing the Module Theme   *111*

7.2.2.2   Selecting Subtopics and Questions   *112*

7.2.2.3   Harmonization of Standard Background Variables   *113*

7.2.2.4   Harmonization of Answer Choices and Scales   *114*

7.2.2.5   Translation of Questions and Answer Choices   *115*

7.3   Documentation and Quality Assurance   *115*

7.3.1   Five Steps to Harmonize the EASS Integrated Data   *115*

7.3.2   Documentation of the EASS Integrated Data   *117*

7.4   Challenges to Harmonization   *118*

7.4.1   How to Translate "Fair" and Restriction by Copyright   *118*

7.4.2   Difficulty in Synchronizing the Data Collection Phase   *121*

7.5   Software Tools   *122*

7.6   Recommendations   *122*

     Acknowledgment   *123*

     References   *123*

8   **Ex-ante Harmonization of Official Statistics in Africa (SHaSA)**   *125*

     Dossina Yeo

     Abbreviations   *125*

8.1   Introduction   *127*

8.2   Applied Harmonization Methods   *128*

8.2.1   Examples of Ex-ante Harmonization Methods: The Cases of GPS Data and CRVS   *131*

8.2.1.1   Governance, Peace and Security (GPS) Statistics Initiative   *131*

8.2.1.2   Development of Civil Registration and Vital Statistics (CRVS)   *132*

8.2.2   Examples of Ex-post Harmonization: The Cases of Labor Statistics, ATSY, ASY and KeyStats, and ICP-Africa Program   *132*

8.3   Quality Assurance Framework   *134*

8.4   Challenges to Statistical Harmonization in Africa   *136*

8.4.1   Challenges to the Implementation of NSDS   *137*

8.4.2   Challenges with Ex-ante Harmonization: Examples of GPS and ICP Initiatives   *138*

8.4.3   Challenges with Ex-post Harmonization: Examples of KeyStats and ATSY   *139*

8.5   Common Software Tools Used   *139*

8.6   Conclusion and Recommendations   *140*

     References   *142*

     Part II   **Ex-post harmonization of national social surveys**   *145*

9   **Harmonization for Cross-National Secondary Analysis: Survey Data Recycling**   *147*

     Irina Tomescu-Dubrow, Kazimierz M. Slomczynski, Ilona Wysmulek, Przemek Powałko, Olga Li, Yamei Tu, Marcin Slarzynski, Marcin W. Zielinski, and Denys Lavryk

9.1   Introduction   *147*

9.2   Harmonization Methods in the SDR Project   *149*

9.2.1   Building the Harmonized SDR2 Database   *150*
9.3   Documentation and Quality Assurance   *155*
9.4   Challenges to Harmonization   *156*
9.5   Software Tools of the SDR Project   *161*
9.5.1   The SDR Portal   *161*
9.5.2   The SDR2 COTTON FILE   *162*
9.6   Recommendations   *162*
9.6.1   Recommendations for Researchers Interested in Harmonizing Survey
         Data Ex-Post   *162*
9.6.2   Recommendations for SDR2 Users   *163*
         Acknowledgments   *164*
         References   *164*
9.A   Data Quality Indicators in SDR2   *166*

10   Harmonization of Panel Surveys: The Cross-National Equivalent File   *169*
         Dean R. Lillard
10.1   Introduction   *169*
10.2   Applied Harmonization Methods   *170*
10.2.1   CNEF Country Data Sources, Current and Planned   *176*
10.3   Current CNEF Partners   *176*
10.3.1   The HILDA Survey <https://melbourneinstitute.unimelb.edu.au/hilda>   *176*
10.3.2   The SLID <http://www.statcan.ca/start.html>   *176*
10.3.3   The CFPS <https://www.isss.pku.edu.cn/cfps/en>   *177*
10.3.4   The SOEP <https://www.diw.de/en/soep>   *177*
10.3.4.1 The BHPS <https://www.iser.essex.ac.uk/bhps>   *177*
10.3.4.2 Understanding Society, UKHLS <https://www.understandingsociety.ac.uk/>   *178*
10.3.5   The ITA.LI   *178*
10.3.6   The JHPS <https://www.pdrc.keio.ac.jp/en/paneldata/datasets/jhpskhps>   *178*
10.3.7   The RLMS-HSE <https://www.cpc.unc.edu/projects/rlms-hse>   *178*
10.3.8   The KLIPS <https://www.kli.re.kr/klips_eng/contents.do?key=251>   *179*
10.3.9   The Swedish Pseudo-Panel   *179*
10.3.10 The SHP <https://forscenter.ch/projects/swiss-household-panel/>   *179*
10.3.11 The PSID <https://psidonline.isr.umich.edu/>   *179*
10.4   Planned CNEF Partners   *180*
10.4.1   The ASEP   *180*
10.4.2   LISA <https://www.statcan.gc.ca/eng/survey/household/5144>   *180*
10.4.3   The ILS   *180*
10.4.4   The MxFLS <http://www.ennvih-mxfls.org/english/index.html>   *180*
10.4.5   The NIDS <http://nids.uct.ac.za>   *181*
10.4.6   The PSFD <https://psfd.sinica.edu.tw/V2/?page_id=966&lang=en>   *181*
10.5   Documentation and Quality Assurance   *181*
10.6   Challenges to Harmonization   *183*
10.7   Recommendations for Researchers Interested in Harmonizing Panel Survey Data   *185*
10.8   Conclusion   *186*
         References   *187*

11 Harmonization of Survey Data from UK Longitudinal Studies: CLOSER *189*
Dara O'Neill and Rebecca Hardy
11.1 Introduction *189*
11.2 Applied Harmonization Methods *191*
11.2.1 Occupational Social Class *191*
11.2.2 Body Size/Anthropometric Data *193*
11.2.3 Mental Health *194*
11.2.4 Harmonization Methods: Divergence and Convergence *195*
11.3 Documentation and Quality Assurance *196*
11.4 Challenges to Harmonization *198*
11.5 Software Tools *199*
11.6 Recommendations *200*
Acknowledgments *202*
References *202*

12 Harmonization of Census Data: IPUMS – International *207*
Steven Ruggles, Lara Cleveland, and Matthew Sobek
12.1 Introduction *207*
12.2 Project History *208*
12.2.1 Evolution of the Web Dissemination System *210*
12.3 Applied Harmonization Methods *210*
12.4 Documentation and Quality Assurance *215*
12.5 Challenges to Harmonization *217*
12.6 Software Tools *221*
12.6.1 Metadata Tools *221*
12.6.2 Data Reformatting *221*
12.6.3 Data Harmonization *221*
12.6.4 Dissemination System *222*
12.7 Team Organization and Project Management *222*
12.8 Lessons and Recommendations *223*
References *225*

Part III Domain-driven ex-post harmonization *227*

13 Maelstrom Research Approaches to Retrospective Harmonization of Cohort Data for Epidemiological Research *229*
Tina W. Wey and Isabel Fortier
13.1 Introduction *229*
13.2 Applied Harmonization Methods *230*
13.2.1 Implementing the Project *233*
13.2.1.1 Initiating Activities and Organizing the Operational Framework *233*
13.2.1.2 Assembling Study Information and Selecting Final Participating Studies (Guidelines Step 1) *234*

13.2.1.3 Defining Target Variables to be Harmonized (the DataSchema) and Evaluating Harmonization Potential across Studies (Guidelines Step 2) *235*

13.2.2 Producing the Harmonized Datasets *236*

13.2.2.1 Processing Data (Guidelines Step 3a) *236*

13.2.2.2 Processing Study-Specific Data to Generate Harmonized Datasets (Guidelines Step 3b) *237*

13.3 Documentation and Quality Assurance *238*

13.4 Challenges to Harmonization *240*

13.5 Software Tools *241*

13.6 Recommendations *243*

Acknowledgments *244*

References *245*

14 Harmonizing and Synthesizing Partnership Histories from Different German Survey Infrastructures *249*

Bernd Weiß, Sonja Schulz, Lisa Schmid, Sebastian Sterl, and Anna-Carolina Haensch

14.1 Introduction *249*

14.2 Applied Harmonization Methods *250*

14.2.1 Data Search Strategy and Data Access *250*

14.2.2 Processing and Harmonizing Data *253*

14.2.2.1 Harmonizing Partnership Biography Data *253*

14.2.2.2 Harmonizing Additional Variables on Respondents' or Couples' Characteristics *254*

14.3 Documentation and Quality Assurance *255*

14.3.1 Documentation *255*

14.3.2 Quality Assurance *256*

14.3.2.1 Process-Related Quality Assurance *256*

14.3.2.2 Benchmarking the Harmonized HaSpaD Data Set with Official Statistics *256*

14.4 Challenges to Harmonization *258*

14.4.1 Analyzing Harmonized Complex Survey Data *258*

14.4.2 Sporadically and Systematically Missing Data *259*

14.5 Software Tools *260*

14.6 Recommendations *262*

14.6.1 Harmonizing Biographical Data *262*

14.6.1.1 Methodological Recommendations *262*

14.6.1.2 Procedural Recommendations *263*

14.6.1.3 Technical Recommendations *263*

14.6.2 Getting Started with the Cumulative HaSpaD Data Set *263*

Acknowledgments *264*

References *264*

15 Harmonization and Quality Assurance of Income and Wealth Data: The Case of LIS *269*

Jörg Neugschwender, Teresa Munzi, and Piotr R. Paradowski

15.1 Introduction *269*

15.2 Applied Harmonization Methods *271*

15.3   Documentation and Quality Assurance   *275*
15.3.1   Quality Assurance   *275*
          Selection of Source Datasets   *276*
          Harmonization   *276*
          Validation – "Green Light" Check   *276*
15.3.2   Documentation   *278*
15.4   Challenges to Harmonization   *278*
15.5   Software Tools   *281*
15.6   Conclusion   *282*
          References   *283*

16   ***Ex-Post* Harmonization of Time Use Data: Current Practices and
          Challenges in the Field**   *285*
          Ewa Jarosz, Sarah Flood, and Margarita Vega-Rapun
16.1   Introduction   *285*
16.2   Applied Harmonization Methods   *289*
16.2.1   Harmonizing the Matrix of the Diary   *289*
16.2.2   Variable Harmonization   *291*
16.2.3   Other Variables   *293*
16.2.4   Other Types of Time Use Data   *294*
16.3   Documentation and Quality Assurance   *294*
16.3.1   Documentation   *294*
16.3.2   Quality Checks   *296*
16.4   Challenges to Harmonization   *297*
16.5   Software Tools   *300*
16.6   Recommendations   *301*
          References   *302*

          **Part IV   Further Issues: Dealing with Methodological Issues in Harmonized
          Survey Data**   *305*

17   **Assessing and Improving the Comparability of Latent Construct
          Measurements in *Ex-Post* Harmonization**   *307*
          Ranjit K. Singh and Markus Quandt
17.1   Introduction   *307*
17.2   Measurement and Reality   *307*
17.3   Construct Match   *308*
17.3.1   Consequences of a Mismatch   *309*
17.3.2   Assessment   *309*
17.3.2.1 Qualitative Research Methods   *309*
17.3.2.2 Construct and Criterion Validity   *309*
17.3.2.3 Techniques for Multi-Item Instruments   *310*
17.3.2.4 Improving Construct Comparability   *311*

17.4   Reliability Differences   *311*
17.4.1   Consequences of Reliability Differences   *311*
17.4.2   Assessment   *312*
17.4.3   Improving Reliability Comparability   *312*
17.5   Units of Measurement   *312*
17.5.1   Consequences of Unit Differences   *313*
17.5.2   Improving Unit Comparability   *313*
17.5.3   Controlling for Instrument Characteristics   *314*
17.5.4   Harmonizing Units Based on Repeated Measurements   *315*
17.5.5   Harmonizing Units Based on Measurements Obtained from the
          Same Population   *315*
17.6   Cross-Cultural Comparability   *316*
17.6.1   Construct Match   *316*
17.6.1.1 Translation and Cognitive Probing   *317*
17.6.2   Reliability   *317*
17.6.3   Units of Measurement   *318*
17.6.3.1 Harmonizing Units of Localized Versions of the Same Instrument   *318*
17.6.3.2 Harmonizing Units Across Cultures and Instruments   *318*
17.6.4   Cross-Cultural Comparability of Multi-Item Instruments   *318*
17.7   Discussion and Outlook   *319*
          References   *320*

18   **Comparability and Measurement Invariance**   *323*
          Artur Pokropek
18.1   Latent Variable Framework for Testing and Accounting for Measurement
          Non-Invariance   *324*
18.2   Approaches to Empirical Assessment of Measurement
          Equivalence   *325*
18.2.1   Classical Invariance Analysis (MG-CFA)   *326*
18.2.2   Partial Invariance (MG-CFA)   *327*
18.2.3   Approximate Invariance   *327*
18.2.4   Approximate Partial Invariance (Alignment, BSEM Alignment, Partial BSEM)   *328*
18.3   Beyond Multiple Indicators   *329*
18.4   Conclusions   *329*
          References   *330*

19   **On the Creation, Documentation, and Sensible Use of Weights in the Context of
          Comparative Surveys**   *333*
          Dominique Joye, Marlène Sapin, and Christof Wolf
19.1   Introduction   *333*
19.2   Design Weights   *335*
19.2.1   What to do?   *336*
19.3   Post-stratification Weights   *337*

19.3.1   What Should be Done?   *340*
19.4   Population Weights   *341*
19.4.1   What Should be Done?   *342*
19.5   Conclusion   *342*
          References   *344*

20   On Using Harmonized Data in Statistical Analysis: Notes of Caution   *347*
          Claire Durand
20.1   Introduction   *347*
20.2   Challenges in the Combination of Data Sets   *347*
20.2.1   A First Principle: A No Censorship Inclusive Approach   *348*
20.2.2   A Second Principle: Using Multilevel Analysis and Introducing a Measurement Level   *349*
20.2.3   A Third Principle: Assessing the Equivalence of Survey Projects   *351*
20.3   Challenges in the Analysis of Combined Data Sets   *353*
20.3.1   Dealing with Time   *354*
20.3.2   Dealing with Missing Values   *358*
20.3.2.1 Missing Values at the Respondent and Measurement Level   *358*
20.3.2.2 Missing Values at the Survey Level   *359*
20.3.3   Dealing with Weights   *361*
20.4   Recommendations   *362*
          References   *363*

21   On the Future of Survey Data Harmonization   *367*
          Kazimierz M. Slomczynski, Christof Wolf, Irina Tomescu-Dubrow, and J. Craig Jenkins
21.1   What We Have Learned from Contributions on Survey Data Harmonization in this
          Volume   *368*
21.2   New Opportunities and Challenges   *370*
21.2.1   Reorientation of Survey Research in the Era of New Technology   *370*
21.2.2   Advances in Technical Aspects of Data Management   *370*
21.2.3   Harmonizing Survey Data with Other Types of Data   *371*
21.3   Developing a New Methodology of Harmonizing
          Non-Survey Data   *372*
21.3.1   Emerging Legal and Ethical Issues   *372*
21.4   Globalization of Science and Harmonizing Scientific Practice   *373*
          References   *373*

          Index   *377*

Preface and Acknowledgments

This edited volume is about the broad spectrum of harmonization methods that scholars use to create survey data infrastructure for comparative social science research. Contributors from a variety of disciplines, including sociology, political science, demography, and economics, among others, discuss practical applications of harmonization in major social science projects, both at the data collection and processing stages and after data release. They also discuss methodological challenges inherent in harmonization as well as statistical issues linked to the use of harmonized survey data. We thank them all for their valuable input.

An important impetus for preparing this book was the US National Science Foundation grant for the project *Survey Data Recycling: New Analytic Framework, Integrated Database, and Tools for Cross-national Social, Behavioral and Economic Research* (hereafter SDR Project; NSF 1738502). Work within the SDR Project highlighted the need for shared knowledge about harmonization methodology. This volume provides a platform where experts in scientific disciplines such as demography and public health directly communicate with sociology, political science, economics, organizational research, and survey methodology.

The support from the Institute of Philosophy and Sociology of the Polish Academy of Sciences, the Ohio State University's Sociology Department, the Mershon Center for International Security Studies, and the GESIS – Leibniz Institute for the Social Sciences was key to completing this volume. We warmly thank Margit Bäck, GESIS, for her help in proofreading many chapters and communicating with authors.

About the Editors

Irina Tomescu-Dubrow is Professor of Sociology at the Institute of Philosophy and Sociology at the Polish Academy of Sciences (PAN), and director of the Graduate School for Social Research at PAN.

Christof Wolf is President of GESIS Leibniz-Institute for the Social Sciences and professor for sociology at University Mannheim. He has co-authored several papers and co-edited a number of handbooks in the fields of survey methodology and statistics. Aside of his longstanding interest in survey practice and survey research he works on questions of social stratification and health.

Kazimierz M. Slomczynski is Professor of Sociology at the Institute of Philosophy and Sociology, the Polish Academy of Sciences (IFiS PAN) and Academy Professor of Sociology at the Ohio State University (OSU). He co-directs CONSIRT – the Cross-national Studies: Interdisciplinary Research and Training program at OSU and IFiS PAN.

J. Craig Jenkins is Academy Professor of Sociology and Senior Research Scientist at the Mershon Center for International Security at the Ohio State University.

About the Contributors

Chapter 1 – Objectives and Challenges of Survey Data Harmonization

Kazimierz M. Slomczynski is Professor of Sociology at the Institute of Philosophy and Sociology, the Polish Academy of Sciences (IFiS PAN) and Academy Professor of Sociology at the Ohio State University (OSU). He co-directs CONSIRT – the Cross-national Studies: Interdisciplinary Research and Training program at OSU and IFiS PAN.

Irina Tomescu-Dubrow is Professor of Sociology at the Institute of Philosophy and Sociology at the Polish Academy of Sciences (PAN), and director of the Graduate School for Social Research at PAN.

J. Craig Jenkins is Academy Professor of Sociology and Senior Research Scientist at the Mershon Center for International Security at the Ohio State University.

Christof Wolf is President of GESIS Leibniz-Institute for the Social Sciences and professor for sociology at University Mannheim. He has co-authored several papers and co-edited a number of handbooks in the fields of survey methodology and statistics. Aside of his longstanding interest in survey practice and survey research he works on questions of social stratification and health.

Chapter 2 – The Effects of Data Harmonization on the Survey Research Process

Ranjit K. Singh is a post-doctoral scholar at GESIS, the Leibniz Institute for the Social Sciences, where he practices, researches, and consults on the harmonization of substantive measurement instruments in surveys. He has a background both in social sciences and psychology. Research interests include measurement quality of survey instruments as well as assessing and improving survey data comparability with harmonization.

Arnim Bleier is a postdoctoral researcher in the Department of Computational Social Science at GESIS – Leibniz-Institute for the Social Sciences. His research interests are in the field of Computational Social Science, with an emphasis on Reproducible Research. In collaboration with social scientists, he develops models for the content, structure and dynamics of social phenomena.

Peter Granda is Archivist Emeritus at the Inter-university Consortium for Political and Social Research (ICPSR) at the University of Michigan. He maintains a strong interest in international comparative research projects and how data generated from these efforts are archived and made available to the public. He studied the history and cultures of South Asia and spent several years doing archival research in India.

Chapter 3 – Harmonization in the World Value Survey

Kseniya Kizilova, PhD in Sociology, is a Senior Research Fellow at the Institute for Comparative Survey Research (Austria) and Head of Secretariat at the World Values Survey Association (Sweden). Her research focuses on social capital, political culture and political trust, democratization, and political participation. She is a member of the Council of the World Association for Public Opinion Research and an associated researcher at the University of Kharkiv (Ukraine).

Jaime Diez-Medrano is Founding President of JD Systems and Director of the World Values Survey Association's Data archive (Spain). He specializes in telecommunications engineering and has over 20 years of experience in database management software development. Diez-Medrano is actively involved into the data processing and harmonization for a number of large-scale survey research projects such as Afro Barometer, Arab Barometer, Latinobarometro among the others.

Christian Welzel is Professor of Political Culture Research at the Center for the Study of Democracy, Leuphana University (Germany) and Vice-President of the World Values Survey Association WVSA (Sweden). His research focuses on human empowerment, emancipative values, cultural change and democratization. Recipient of multiple large-scale grants, Welzel is the author of more than a hundred-and-fifty scholarly publications and a member of the German Academy of Sciences.

Christian Haerpfer is Research Professor of Political Science at the University of Vienna and Director of the Institute for Comparative Survey Research (Austria). He is President of the World Values Survey Association WVSA (Sweden), Director of the Eurasia Barometer and a member of the European Academy of Sciences and Arts. His research focuses on democratization in Eastern Europe and Eurasia, political trust and regime support, electoral behavior, and political participation.

Chapter 4 – Harmonization in the Afrobarometer

Carolyn Logan is an Associate Professor in the Department of Political Science at Michigan State University, and currently serves as Director of Analysis with Afrobarometer. She has been with Afrobarometer since 2001, including serving as Deputy Director from 2008–2019 and during the network's expansion from 20 to 36 countries in 2011–2013. Her research interests include the role of traditional authorities in democratic governance, and citizen-versus-subject attitudes among African publics.

Robert Mattes is Professor of Government and Public Policy at the University of Strathclyde, and Honorary Professor at the Institute for Democracy, Citizenship and Public Policy in Africa at the University of Cape Town. He is a co-founder of, and Senior Adviser to, Afrobarometer, a ground-breaking regular survey of public opinion in over 30 African countries (www.afrobarometer.org).

Francis Kibirige is co-founder and Managing Director of Hatchile Consult Ltd., a research company based in Uganda. He joined Afrobarometer in 1999, and currently serves as Network Sampling Specialist and co-leads the Afrobarometer team in Uganda. He studied agricultural engineering at Makerere University and has since received four Afrobarometer fellowships to study political research methodology and statistical modeling at University of Cape Town and the Inter-University Consortium for Political and Social Research (ICPSR) at the University of Michigan.

Chapter 5 – Harmonization in the National Longitudinal Survey of Youth (NLSY)

Elizabeth Cooksey (PhD) is Academy Professor Emeritus at The Ohio State University and a senior researcher at CHRR (Center for Human Resource Research) at The Ohio State University. She has worked with the National Longitudinal Survey of Youth data (NLSY) for over 30 years, and has been the Principal Investigator for the NLSY79 Child and Young Adult surveys for the past two decades.

Rosella Gardecki (PhD) is a Research Specialist at CHRR at The Ohio State University. In 1996, she joined CHRR as a data archivist for the National Longitudinal Survey of Youth 1997 (NLSY97). She has contributed to questionnaire design for more than 20 years. With her background in economics, she currently leads the team that creates variables for both the NLSY79 and NLSY97 cohorts.

Carole Lunney (MA) is a data analysis consultant living in Calgary, Alberta, Canada. From 2011 to 2020, Carole worked at CHRR at The Ohio State University on the data archivist team for the National Longitudinal Surveys. In that role, she was involved in survey instrument design and testing, statistical programming, writing study documentation, and data user outreach. She has co-authored publications on posttraumatic stress disorder, music cognition, and communication.

Amanda Roose (MA) has been the NLS Project Manager/Documentation Lead at CHRR at The Ohio State University for two decades, with management responsibilities for all aspects of survey administration including design, fielding, data preparation, documentation, and dissemination. She has experience editing academic publications across a range of disciplines.

Chapter 6 – Harmonization in the Comparative Study of Electoral Systems (CSES) Projects

Stephen Quinlan (PhD, University College Dublin) is Senior Researcher at the GESIS – Leibniz Institute for the Social Sciences in Mannheim and Project Manager of the Comparative Study of Electoral Systems (CSES) project. His research interests are comparative electoral behavior and social media's impact on politics. His work has appeared in journals such as *Information, Communication, and Society, International Journal of Forecasting, Electoral Studies, Party Politics,* and *The European Political Science Review.* E: stephen.quinlan@gesis.org

Christian Schimpf (PhD, University of Mannheim) is a former Data Processing Specialist at the CSES. Previously, he has been Senior Researcher at the University of Alberta and Université du Québec à Montréal. His research interests are comparative electoral behavior and environment/energy policy. His work has appeared in journals such as *The American Political Science Review, Environmental Politics,* and *Political Studies.* E: christianhschimpf@gmail.com

Katharina Blinzler (MA, University of Mannheim) is a Data Processing Specialist and Archivist with the CSES Secretariat at the GESIS – Leibniz Institute for the Social Sciences, Köln. Her research interests are comparative electoral behavior and data harmonization. E: katharina.blinzler@gesis.org

Slaven Zivkovic (PhD, ABD) is a former Data Processing Specialist at the CSES and a PhD candidate at the University of Mainz, Germany. Most recently, he was a Fulbright Fellow at the Florida International University in Miami, USA. His research interests are comparative electoral behavior, especially in post-communist states. His work has appeared in journals such as *The Journal of Contemporary European Studies, Comparative European Politics, and East European Politics and Societies.* E: szivkovi@uni-mainz.de

Chapter 7 – Harmonization in the East Asian Social Survey

Noriko Iwai is Director of the Japanese General Social Survey Research Center and Professor of Faculty of Business Administration, Osaka University of Commerce. She is a PI of JGSS and EASS, a director of the Japanese Association for Social Research, and a member of Science Council of Japan. Her current research project is supporting Japanese researchers in the humanities and social sciences to prepare their data for public usage.

Tetsuo Mo is Research Fellow of the Japanese General Social Survey Research Center, Osaka University of Commerce. His areas of specialty are labor economy, inequality and social exclusion, social survey, and quantitative analysis of survey data. He is responsible for the creation of JGSS question-naires, cleaning of JGSS data, and harmonization of East Asian Social Survey data.

Jibum Kim is a professor in the Department of Sociology and the director of the Survey Research Center at Sungkyunkwan University (SKKU) in Seoul, South Korea. He is also a PI of the Korean General Social Survey (KGSS). He is currently the president of the World Association for Public Opinion Research (WAPOR) Asia Pacific and serves on the editorial board of the International Journal of Public Opinion Research.

Chyi-in Wu is a Research Fellow at the Institute of Sociology, Academia Sinica in Taipei, Taiwan. He is currently the PI of the Taiwan Social Change Survey (TSCS). He currently serves as the editor of The Journal of Information Society (Taiwan) and was the president of the Taiwan Association of Information Society (TAIS).

Weidong Wang is the Director of the Social Psychology Institute, the Executive Deputy Director of the National Survey Research Center (NSRC) , Associate Professor of the Sociology Department at Renmin University of China. He is the PI of the Chinese Religious Life and Survey (CRLS), PI of China Education Panel Survey (CEPS), and the co-founder and Executive Director of the Chinese National Survey Data Archive (CNSDA).

Chapter 8 – Ex-ante Harmonization of Official Statistics in Africa (SHaSA)

Dr. Dossina Yeo is a passionate statistician who led the development and the implementation of the recent major statistical development strategic frameworks in Africa including the African Charter on Statistics and the Strategy for the harmonization of Statistics in Africa (SHaSA 1 and 2. He also led the establishment of the statistical function within the African Union Commission including the creation of the Statistics Unit, its transformation into a Statistics Division and the creation of the African Union Institute for Statistics (STATAFRIC) based in Tunis (Tunisia) and the Pan-African Training Center (PAN-STAT) based in Yamoussoukro (Cote d'Ivoire). Dr. Yeo is currently Acting Director of Economic Development, Integration and Trade at the African Union Commission. He has previously worked for the United Nations Statistics Division (UNSD) and the African Development Bank (AfDB).

Ch 9 – Harmonization for Cross-National Secondary Analysis: Survey Data Recycling

Irina Tomescu-Dubrow is Professor of Sociology at the Institute of Philosophy and Sociology at the Polish Academy of Sciences (PAN), and director of the Graduate School for Social Research at PAN.

Kazimierz M. Slomczynski is Professor of Sociology at the Institute of Philosophy and Sociology, the Polish Academy of Sciences (IFiS PAN) and Academy Professor of Sociology at the Ohio State University (OSU). He co-directs CONSIRT – the Cross-national Studies: Interdisciplinary Research and Training program at OSU and IFiS PAN.

Ilona Wysmulek is an Assistant Professor at the Institute of Philosophy and Sociology, Polish Academy of Sciences in Warsaw. She works in the research group of Professor Kazimierz M. Slomczynski on Comparative Analysis of Social Inequality and is actively involved in the two team's projects: the SDR survey data harmonization project and the Polish Panel Survey POLPAN.

Przemek Powałko is an Assistant Professor at the Institute of Philosophy and Sociology, Polish Academy of Sciences in Warsaw.

Olga Li is a PhD student at the Graduate School for Social Research and is a member of the research unit on Comparative Analyses of Social Inequality at the Institute of Philosophy and Sociology, Polish Academy of Sciences. She has previously worked in Polish Panel Survey (POLPAN) and Survey Data

Recycling (SDR) grant projects. For her PhD thesis, she is conducting a quantitative research on political participation in authoritarian regimes.

Yamei Tu is a Ph.D. student at the Ohio State University, and her research interests include visualization and NLP.

Marcin Slarzynski is an Assistant Professor at the Institute of Philosophy and Sociology at the Polish Academy of Sciences. His research focuses on the national movement of local elites in Poland, 2005–2015.

Marcin W. Zielinski is a Research Assistant, Institute of Philosophy and Sociology, Polish Academy of Sciences, Warsaw, Poland.

Denys Lavryk from Graduate School for Social Research, Polish Academy of Sciences, Warsaw, Poland.

Chapter 10 – Harmonization of Panel Surveys: The Cross-National Equivalent File

Dean Lillard is Professor of consumer sciences in the Department of Human Sciences at The Ohio State University. He received his PhD in economics from the University of Chicago in 1991. From 1991 to 2012, he held appointments at Cornell University. He moved to OSU in 2012. He is a Research Fellow of the German Institute for Economic Research and a Research Associate of the National Bureau of Economic Research.

Chapter 11 – Harmonization of Survey Data from UK Longitudinal Studies: CLOSER

Dara O'Neill was the theme lead in data harmonization at the CLOSER consortium of UK-based longitudinal studies (2018–2022) at University College London (UCL), overseeing diverse cross-study harmonization projects. Previously, Dara held research posts at UCL's Department of Epidemiology and Public Health, at UCL's Institute of Cardiovascular Science and at the University of Surrey's School of Psychology. Dara now works as a statistician/psychometrician in clinical trial research and is an honorary Senior Research Fellow at UCL.

Social Research Institute, University College London, UK

Rebecca Hardy is Professor of Epidemiology and Medical Statistics in the School of Sport, Exercise and Health Sciences at Loughborough University. She previously worked at University College London, where she was CLOSER Director (2019–2022) and Programme Leader in the MRC Unit for Lifelong Health and Aging (2003–2019). Rebecca uses a life course approach to study health and aging, and her interests also include methodological considerations in life course and longitudinal data analysis, as well as cross-study data harmonization.

Chapter 12 – Harmonization of Census Data: IPUMS – International

Lara Cleveland is a sociologist and principal research scientist at the University of Minnesota's Institute for Social Research and Data Innovation (ISRDI) where she directs the IPUMS International census and survey data project. She leads technical workflow development, partner relations, and grant management for the project. Her research interests include data and methods; organizations, occupations, and work; and global standardization practices. She serves on international working groups concerning data dissemination.

Steven Ruggles is Regents Professor of History and Population Studies at the University of Minnesota, He started IPUMS in 1991, and today IPUMS provides billions of records from thousands of censuses and surveys describing individuals and households in over 100 countries from the 18th century to the present. Ruggles has published extensively on historical demography, focusing especially on long-run changes in families and marriage, and on methods for population history.

Matthew Sobek is the IPUMS Director of Data Integration and has overseen numerous projects to harmonize census and survey data collections over the past 30 years. Sobek co-authored the original U.S. IPUMS in the 1990s before shifting focus to international data harmonization. He played a foundational role in the development of IPUMS harmonization and dissemination methods and continues to contribute to their evolution.

Chapter 13 – Maelstrom Research Approaches to Retrospective Harmonization of Cohort Data for Epidemiological Research

Tina W. Wey, PhD, is a research data analyst with the Maelstrom Research team at the Research Institute of the McGill University Health Centre. She has a background in biological research, with experience in data management and statistical analysis in behavior, ecology, and epidemiology.

Isabel Fortier, PhD, is a researcher at the Research Institute of the McGill University Health Centre and Assistant Professor in the Department of Medicine at McGill University. She has extensive experience in collaborative epidemiological and methodological research and leads the Maelstrom Research program, which aims to provide the international research community with resources (expertise, methods, and software) to leverage and support data harmonization and integration across studies.

Chapter 14 – Harmonizing and Synthesizing Partnership Histories from Different German Survey Infrastructures

Dr. Bernd Weiß is head of the GESIS Panel, a probabilistic mixed-mode panel. He also serves as Deputy Scientific Director of the Department Survey Design and Methodology at GESIS – Leibniz Institute for the Social Sciences in Mannheim. His research interests range from survey methodology, research synthesis, and open science to family sociology and juvenile delinquency.

Dr. Sonja Schulz is a senior researcher in the Department Survey Data Curation at GESIS – Leibniz Institute for the Social Sciences. Her current research and services focus on survey data harmonization, family research and social inequality, with a special focus on trends in family formation and marriage dissolution. Recent articles were published in Criminal Justice Review, European Journal of Criminology, and Journal of Quantitative Criminology.

Dr. Lisa Schmid is a social scientist with a research interest in intimate relationships and family demography. As a research associate at GESIS – Leibniz-Institute for the Social Science in Mannheim she is part of the team Family Surveys that conducts the Family Research and Demographic Analysis (FReDA) panel in Germany.

Sebastian Sterl is a scientific researcher in the Interdisciplinary Security Research Group at Freie Universität Berlin focused on creating psychosocial situation pictures in crises and disasters using quantitative approaches. Before, he worked at GESIS – Leibniz Institute for the Social Sciences responsible for harmonizing and synthesizing survey data. His interests include quantitative methods of empirical social and economic research, data management, risk perception, protective and coping behavior, and rational choice theory.

Dr. Anna-Carolina Haensch is a postdoctoral researcher at the LMU Munich and an assistant professor at the University of Maryland. Her work focuses on data quality, especially regarding missing data. She has also been part of the COVID-19 Trends and Impact Surveys (CTIS) team since early 2021 and enjoys teaching courses on quantitative methods at the LMU Munich and the Joint Program in Survey Methodology.

Chapter 15 – Harmonization and Quality Assurance of Income and Wealth Data: The Case of LIS

Jörg Neugschwender has a PhD in Sociology from the Graduate School of Economic and Social Sciences (GESS) at the University of Mannheim, Germany. He is Data Team Manager at LIS Cross-National Data Center in Luxembourg, supervising the data team in harmonizing datasets for the LIS databases, developing and maintaining data production and quality assessment applications, and overseeing the consistency of produced datasets. He is the editor of the LIS newsletter Inequality Matters.

Teresa Munzi, an economist by training (University of Rome and London School of Economics), has been with the LIS Cross-National Data Center in Luxembourg for over 20 years. Since 2019, she is the Director of Operations of LIS, where she is responsible for managing and overseeing all operations. She also carries out research on the comparative study of welfare systems and their impact through redistribution on poverty, inequality, and family well-being.

Piotr Paradowski works at LIS Cross-National Data Center in Luxembourg as a Data Expert and Research Associate. He is also affiliated with the Department of Statistics and Econometrics at the Gdansk University of Technology. In addition, he conducts interdisciplinary research focusing on income and wealth distributions as they relate to economic inequality, poverty, and welfare state politics.

Chapter 16 – Ex-Post Harmonization of Time Use Data: Current Practices and Challenges in the Field

Ewa Jarosz is an assistant professor at the Faculty of Economic Sciences, University of Warsaw. She specializes in cross-national studies and uses comparative survey data, time use data and panel data in her work. Her research interest include time use, demographic change, social inequality, health, and wellbeing.

Sarah Flood is the Director of U.S. Surveys at the IPUMS Center for Data Integration and the Associate Director of the Life Course Center, both at the University of Minnesota. Her data infrastructure work on IPUMS Time Use (https://timeuse.ipums.org/) lowers the barriers to accessing time diary data. Her substantive research is at the intersection of gender, work, family, and the life course.

Margarita Vega-Rapun is a research officer at the Joint research of the European Commission. She is mainly involved in topics related to development goals and territories. She also holds an honorary position at the University College London, Centre for Time Use Research, where she worked on the Multinational Time Use project. Her research interests are time poverty, gender and inequalities, and the impact of covid 19 on time use patterns.

Chapter 17 – Assessing and Improving the Comparability of Latent Construct Measurements in *Ex-Post* Harmonization

Ranjit K. Singh is a post-doctoral scholar at GESIS, the Leibniz Institute for the Social Sciences, where he practices, researches, and consults on the harmonization of substantive measurement instruments in surveys. He has a background both in social sciences and psychology. Research interests include measurement quality of survey instruments as well as assessing and improving survey data comparability with harmonization.

Markus Quandt is a senior researcher and team leader at GESIS Leibniz Institute for the Social Sciences in Cologne, Germany. His research is based on quantitative surveys in cross-country comparative settings. Substantive interests are in political and social participation as collective goods problems; methodological interests concern the comparability and validity of measures of attitudes and values, and the closely related problems of harmonizing data from different sources.

Chapter 18 – Comparability and Measurement Invariance

Artur Pokropek is a Professor at the Institute of Philosophy and Sociology of the Polish Academy of Sciences. His main areas of research interests are statistics, psychometrics and machine learning. He has developed several methodological and statistical approaches for analyzing survey data. He gained knowledge as a visiting scholar at the Educational Testing Service (Princeton, USA) and as an associate researcher at the EC Joint Research Centre (Ispra, Italy).

Chapter 19 – On the Creation, Documentation, and Sensible Use of Weights in the Context of Comparative Surveys

Dominique Joye is professor emeritus of sociology at the University of Lausanne and affiliated researcher at FORS. He is one of the co-editors of the SAGE Handbook of Survey Methodology, and was for a long time the Swiss coordinator of ESS, EVS and ISSP as well as participant in the methodological boards of these international comparative projects.

Marlène Sapin is senior researcher at the Swiss Centre of Expertise in Social Sciences (FORS) and at the Swiss Centre of Expertise in life course research (LIVES), University of Lausanne. She is a specialist in population-based surveys and has been involved in the leading developing team of several national and international surveys (e.g. Swiss Federal Survey of Youths, International Social Survey Programme). She has a strong interest in survey methodology, social networks, and health.

Christof Wolf is President of GESIS Leibniz-Institute for the Social Sciences and professor for sociology at University Mannheim. He has co-authored several papers and co-edited a number of handbooks in the fields of survey methodology and statistics. Aside of his longstanding interest in survey practice and survey research he works on questions of social stratification and health.

Chapter 20 – On Using Harmonized Data in Statistical Analysis: Notes of Caution

Claire Durand is Professor of Sociology at the Institute of Philosophy and Sociology, the Polish Academy of Sciences (IFiS PAN) and Academy Professor of Sociology at the Ohio State University (OSU). He co-directs CONSIRT - the Cross-national Studies: Interdisciplinary Research and Training program at OSU and IFiS PAN.

Chapter 21 – On the Future of Survey Data Harmonization

Kazimierz M. Slomczynski is Professor of Sociology at the Institute of Philosophy and Sociology, the Polish Academy of Sciences (IFiS PAN) and Academy Professor of Sociology at the Ohio State University (OSU). He co-directs CONSIRT – the Cross-national Studies: Interdisciplinary Research and Training program at OSU and IFiS PAN.

Christof Wolf is President of GESIS Leibniz-Institute for the Social Sciences and professor for sociology at University Mannheim. He has co-authored several papers and co-edited a number of handbooks in the fields of survey methodology and statistics. Aside of his longstanding interest in survey practice and survey research he works on questions of social stratification and health.

Irina Tomescu-Dubrow is Professor of Sociology at the Institute of Philosophy and Sociology at the Polish Academy of Sciences (PAN), and director of the Graduate School for Social Research at PAN.

J. Craig Jenkins is Academy Professor of Sociology and Senior Research Scientist at the Mershon Center for International Security at the Ohio State University.

1

Objectives and Challenges of Survey Data Harmonization

Kazimierz M. Slomczynski, Irina Tomescu-Dubrow, J. Craig Jenkins, and Christof Wolf

1.1 Introduction

This edited volume is an extensive presentation of survey data harmonization in the social sciences and the first to discuss ex-ante, or propspective and ex-post, or retrospective harmonization concepts and methodologies from a global perspective in the context of specific cross-national and longitudinal survey projects. Survey data harmonization combines survey methods, statistical techniques, and substantive theories to create datasets that facilitate comparative research. Both data producers and secondary users engage in harmonization to achieve or strengthen the comparability of answers that respondents surveyed in different populations or the same population over time provide (Granda et al. 2010; Wolf et al. 2016). Most data producers employ harmonization ex-ante, when designing and implementing comparative studies, for example, the *World Values Survey* (WVS), the *Survey of Health, Ageing, and Retirement in Europe,* the *International Social Survey Programme* (ISSP), or the *European Social Survey* (ESS), among many others. Secondary users, as well as some data producers, apply harmonization methods *ex-post,* to already released files that are not comparable by design to integrate them into datasets suitable for comparative analysis. The Luxembourg Income Study, the Multinational Time Use Study, IPUMS-International, the Cross-national Equivalent File, and more recently, the Survey Data Recycling (SDR) Database are relevant illustrations of large-scale ex-post harmonization projects.

Harmonizing at the data collection and processing stages (i.e. ex-ante) and harmonizing after data release (i.e. ex-post) present obvious differences, including in scope (what can be harmonized), methods (how to harmonize), organization (who is involved), and expenditure (what is the calculated cost). Nonetheless, both approaches – individually and in relation to each other – play important roles in obtaining survey responses that can be compared. Well-documented ex-ante harmonization procedures inform subsequent ex-post harmonization steps, while lessons learned during ex-post harmonization efforts can aid harmonization decisions in the preparation of new surveys. This book covers both perspectives to offer readers a rounded view of survey data harmonization. Although our examples draw on

Survey Data Harmonization in the Social Sciences, First Edition.
Edited by Irina Tomescu-Dubrow, Christof Wolf, Kazimierz M. Slomczynski, and J. Craig Jenkins.
© 2024 John Wiley & Sons Inc. Published 2024 by John Wiley & Sons Inc.

comparative research, whether cross-national or historical, other harmonization work uses many of the same methods that are presented here.

Over the decades, the number of harmonized social science datasets has grown rapidly, responding to the push for greater comparability of concepts and constructs, representation, and measurement in longitudinal and cross-national projects (Granda et al. 2010) and due to incentives to reuse the wealth of already collected data (Slomczynski and Tomescu-Dubrow 2018). However, documentation of the complex process of harmonization decisions is weak (Granda and Blaszcyk 2016), and the body of methodological literature is small (Dubrow and Tomescu-Dubrow 2016). Explicit discussion of ex-ante harmonization, if present, is usually subsumed under comparative survey methods, and hardly any survey methodology textbook features chapters on ex-post harmonization. The consequence is a scattered field where harmonized datasets are readily available, but harmonization assumptions, as well as challenges that researchers face and solutions they chose, and best practice recommendations, are not widely shared.

To foster the diffusion of harmonization knowledge across the social sciences, this book provides a platform where such scientific disciplines as demography and public health directly communicate with sociology, political science, economics, organizational research, and survey methodology. We structure the volume into four parts that, taken together, integrate the discussion of concepts and methods developed around harmonization with practical knowledge accumulated in the process of building longitudinal and cross-national datasets for comparative research and relevant insights for analyzing harmonized survey data. These four parts are preceded not only by this Introduction (Chapter 1) but also by Chapter 2, which considers data harmonization in the overall survey research process.

Part I of this book focuses on six renowned projects from around the world as case studies of *ex-ante* harmonization. Whether these studies are international surveys (Chapters 3, 4, 6, and 7), a single-nation panel (Chapter 5), or official statistics (Chapter 8), they each share their experiences with harmonization, including its documentation and quality controls, challenges and how they were met, and recommendations. The same structure characterizes the eight chapters in Parts II and III, whose core common theme is *ex-post* harmonization. Chapters 9–12 deal with the integration and harmonization of national surveys, while Chapters 13–16 are devoted to the harmonization of surveys on specific substantive issues, such as health, family, income and wealth, and time use. Contributions in Part IV adopt a user's perspective and discuss methodological issues that statistical analysis of harmonized data will likely raise. We conclude the volume with lessons learned from the chapters and an agenda for moving the survey data harmonization field forward.

1.2 What is the Harmonization of Survey Data?

Harmonization of survey data is a process that aims to produce equivalent or comparable measures of a given characteristic across datasets coming from different populations or from the same population but at different time points. If the aim is to harmonize an entire data collection effort, such as cross-national survey programs or multi-culture surveys seek to do, then the scope of harmonization is broad and concerns sampling design, data collection instrument(s), survey mode(s), fieldwork, documentation, data cleaning, and presenting data and meta-data as machine-readable files. If data are harmonized after they are collected, then the focus lies on ways to code the data into equivalent or comparable categories using appropriate scales and developing variables to capture sources of potential bias among the various surveys incorporated.

1.2.1   Ex-ante, Input and Output, Survey Harmonization

Input harmonization pursues equivalent or comparable representation of the populations studied (samples), instruments of the measurement (questionnaires), methods of data collection (modes), and data documentation (metadata) in a bid to reduce as much as possible the share of methodological artifacts in comparative analyses. This generally involves the agreement of Principal Investigators (PIs) and the national data collection teams to use uniform definitions of concepts and indicators, and the same procedures (e.g. survey mode), training (of, e.g. translators), and technical requirements (e.g. sampling method, minimum response rate) (Ehling 2003; Hoffmeyer-Zlotnik 2016; Kallas and Linardis 2010).

Ex-ante output harmonization assumes that for concepts that exist across populations, comparable estimates can be obtained even if survey conditions differ (Hoffmeyer-Zlotnik 2016). For some characteristics, strict input harmonization is not an option, and this approach is the only way to arrive at comparable measures (Schneider et al. 2016). PIs and national teams agree on a target variable and a common measurement schema to construct it, but first, respondents' answers are gathered using country-specific survey questions (Granda et al. 2010; Kallas and Linardis 2010). Once the data are in, the country-specific items are recoded into the common target variable following the agreed-upon harmonized coding schema. For example, in the ESS, the *International Standard Classification of Education*, ISCED, the harmonized measure of education levels, is obtained via "mapping" of national classifications of education. International survey projects frequently use both input and output harmonization as they seek to balance the need for high-quality cross-national measures with the need for valid and reliable national indicators.

Ex-post survey harmonization aims at producing an integrated data file containing information on the units of observations and variables describing these units that stem from surveys that were not originally meant to be put together. For this type of harmonization, the output in the form of an integrated file is essential. The process of harmonizing ex-post focuses on two aspects of comparability: the question content and the answering scale. The end product should contain metadata on all aspects of the survey life cycle, which can be used to assess comparability and as controls for potential bias associated with differences among included surveys.

In Figure 1.1, we present a simplified relationship between different types of harmonization, including stages of the survey life cycle. We take into account only two projects (A and B), containing three (A) or two (B) surveys, respectively, conducted in the same set of countries, or – in the case of longitudinal studies conducted in one country – different time periods. In the figure, within each project, ex-ante input harmonization refers to all pertinent stages of the survey life cycle. Ex-ante output harmonization should (only) be used for characteristics that cannot be measured with the same categories across contexts. This is different from surveys such as, for example, the EU-Labor Force Survey for which Eurostat did not succeed in prescribing a unified questionnaire for all European countries.

For ex-post harmonization, we assume that substantively both projects deal with the same topics, although they could use slightly different question wordings or other elements of survey production. Of course, ex-post harmonization should be driven by a theoretical and/or practical interest in certain topics. In Figure 1.1, ex-post harmonization adjusts for surveys' differences only at the level of projects because within-project variation has already been eliminated by ex-ante output harmonization. In practice, however, ex-post harmonization often looks at survey differences independently of their project origin. An important part of ex-post harmonization is examining survey differences at all stages of the

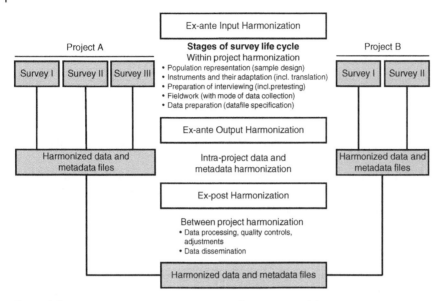

Figure 1.1 Simplified relationships between different types of data harmonization, including stages of survey life cycles.

survey life cycle. Ideally, all differences should be documented as metadata in the form of separate variables.

In Table 1.1, we describe projects included in this volume in terms of the types of survey harmonization they utilize. Among ex-ante output harmonization, we distinguish two situations. The first one follows the ex-ante input harmonization and corrects some survey-specific deviations from the previously agreed-upon solutions. The second situation occurs when some harmonization is needed because collaborating partners bring their own data for which there was no or little prior harmonization. This situation n is similar to ex-post harmonization, when researchers combine surveys on the same topic and harmonize relevant variables. However, ex-post harmonization is conducted within the same project, as in the case of harmonizing data from different waves, either in cross-sectional or panel frameworks. The remaining type of ex-post harmonization refers to the situation when researchers harmonize a number of surveys from different projects, like ISSP and ESS, or Eurobarometer (EB) with its rendition in Asia or Latin America.

Survey harmonization does not mean the simple pooling of data from different international survey projects for analyses performed separately for these projects. For example, in one study, religiosity was re-coded to the unified schema for the World Value Survey and ISSP, but the impact of this created variable on happiness, civic engagement, and health was assessed independently for each project (PEW 2019). Various aspects of pooling survey data and how it is distinct from ex-post harmonization have been examined in the literature (e.g. Kish 1999; Ayadi et al. 2003; Wendt 2007; Roberts and Binder 2009; Malnar and Ryan 2022). We should note, however, that in some projects, survey data integration of different surveys is the first step to harmonization (e.g. in CNEF, Chapter 10).

Table 1.1 Types of survey harmonization in projects included in this volume.

		Ex-ante harmonization			Ex-post harmonization		
			Output				
Chapter #	Project	Input	After input harmo-nization	With no or little input harmo-nization	Within a given survey project	of multi-survey projects	of inde-pendent surveys
3	World Value Survey WVS	X	X		X		
4	Afrobarometer AFB	X	X		X		
5	National Longitudinal Survey of Youth NLSY	X	X	X	X		
6	Comparative Study of Electoral System CSES	X	X	X	X		
7	East Asian Social Survey EASS	X	X		X		
8	Official Statistics in Africa SHaSA	X				X	X
9	Survey Data Recycling SDR					X	
10	Cross-National Equivalent File CNEF						X
11	UK Longitudinal Studies CLOSER					X	X
12	Harmonization of Census Data IPUMS						X
13	Health Survey Data MAELSTROM	X	X	X		X	X
14	Harmonizing Family Biographies HFB					X	
15	Income and Wealth Data LIS/LWS						X
16	Harmonization of Time Use Data					X	X

1.3 Why Harmonize Social Survey Data?

Generally, researchers harmonize survey data to gain new knowledge and solve scientific problems requiring sample sizes that could not be obtained with individual studies. Typically, this requires large geographic and/or temporal coverage as to provide context on the macro-level for explaining individuals' opinions, attitudes, and behaviors. Ideally, in advanced comparative analysis, context should be expressed in terms of specific variables (Przeworski and Teune 1970), dealing with conditions in various relevant dimensions: political (e.g. indexes of democracy), economic (e.g. GDP per capita, foreign debt), social (e.g. indexes of marriage homogamy or social mobility), and cultural (e.g. book readership, religious fractionalization). In addition, as it has been pointed out in the literature, harmonized data "improves the generalizability of results, helps ensure the validity of comparative research, encourages more efficient secondary usage of existing data, and provides opportunities for collaborative and multi-center research." (Doiron et al. 2012, p. 221).

1.3.1 Comparison and Equivalence

How can valid comparisons be made in cross-national research when so many terms and concepts differ in their meanings from country to country?" (Przeworski and Teune 1966, p. 551; see also van Deth 1998). A first naïve answer would be: by using identical measures. The identity of two measures would require that they have the same manifestation in every possible dimension or with respect to every possible aspect. Identity would, for example, require that the language in which the instrument is administered be the same in all instances. Obviously, this is not feasible in comparative research.

Instead, the notion of equivalence of measures has been promoted. It may be suggested that equivalent measures agree with respect to a prespecified dimension. How far the equivalence of measures is empirically achieved can be assessed by analyzing and comparing their relationships with other measures. To the degree to which measures from different data sources show similar or identical relationships with other measures, they can be considered equivalent. A test of these relationships can be made either more qualitatively or more rigorously by statistically analyzing covariances between them (see Chapter 18).

However, as Harkness et al. (2003, p. 14) have noted, the term "equivalence," i.e. "the comparability of measures obtained in different cultural groups," when looked at more closely, has many different meanings. This echoes an earlier claim by Johnson (1998). He lamented already over 25 years ago that the meaning of the term equivalence in the context of comparative research and methodologies to ensure or test for equivalence are underdeveloped. In his review of the literature, Johnson identified over 50 different usages of the term equivalence, ranging from calibration equivalence, over formal equivalence, methodical equivalence, and psychometric equivalence, to verbal and vocabulary equivalence. A further examination of these terms and their meanings led Johnson to distinguish two broad areas in which equivalence might be sought: interpretative equivalence and procedural equivalence. The former aims to describe how abstract concepts can be applied or are valid in different cultural contexts. The latter is concerned with methods and procedures applied in comparative research to ensure equivalence of results. These methods can be further distinguished according to the survey lifecycle into measures in the questionnaire development phase, the questionnaire pretesting phase, the data collection phase, and the data analysis phase (Johnson 1998, p. 12).

This clearly shows that there is not one unified approach to equivalence and that equivalence is not a binary concept. Instead, we have to distinguish different steps in the measurement process and different aspects of the resulting measures. Van Deth (1998, p. 15) and Mohler and Johnson (2010) recommend pragmatic approaches to equivalence. We, therefore, prefer the term comparability, which makes it clear that it is a matter of degree and not a matter of presence or absence of equivalence or identity. It should also be made clear whether the focus is interpretative or procedural equivalence or some combination of the two that is being assessed.

1.4   Harmonizing Survey Data Across and Within Countries

Historically, the first large cross-national survey using ex-ante harmonization, *How Nations See Each Other?* (Buchanan and Cantril 1953), exposed cross-national variation in peoples' opinions with reference to the intercountry experience in World War II. Later on, ex-ante harmonization became a standard procedure in all international survey projects and other projects involving different surveys, including longitudinal ones. Indeed, these projects increase sample sizes and allow researchers to perform more fine-grained contextual analyses.

As described by Dubrow and Tomescu-Dubrow (2016), initial ex-post harmonization efforts began in the 1980s with the "custom" integration of files from similar projects. These efforts were often poorly documented, which hindered the systematic development of ex-post harmonization methodology, and the extensive usage by a larger community of resulting datasets. The rise of specialized ex-post harmonization with extensive documentation is a relatively new development in the past couple of decades. The progress is visible beyond this volume. For example, Biolcati et al. (2020), in their project for the comparative analysis of long-term trends in individual religiosity, used well-documented harmonized data from EB, ESS, *European Values Study* (EVS), ISSP, and WVS.

1.4.1   Harmonizing Across Countries

Chapters 3, 6, and 15 in this volume illustrate the point that harmonization creates greater heterogeneity of theoretically relevant macro-characteristics at the level of countries. In Table 1.2, we show that the harmonization of national surveys from multinational projects increases the countries' coverage in time. In this table, we also included information on the countries' coverage in the SDR project (Chapter 9). As documented in Slomczynski and Tomescu-Dubrow (2006) and Kołczyńska (2014), omitting poor

Table 1.2   Number of countries in the first and last waves of major international survey projects.

International survey projects	First wave		Last wave	
	Year	Countries	Year	Countries
World Value Survey, WVS	1984	11	2022	64
International Social Survey Programme, ISSP	1985	6	2019	29
European Social Survey, ESS	2002	22	2020	40
Comparative Study of Electoral Systems, CSES	2002	33	2022	57
Luxemburg Income and Wealth Study, LIS/LWS	1980	6	2019	53/19
Cross-national Equivalent File, CNEF	1998	4	2020	10
Survey Data Recycling, SDR	2017	142	2022	156

Note: For the first and last waves, we provide the end year of that wave. For some programs, we omitted the current data collection. In the case of SDR, we provide the years of the data release for the SDR1 and SDR2 database versions.

countries from the international survey projects leads to bias in estimating the potential impact of important contextual variables – measured at the level of country-years – on individuals' opinions, attitudes, and behaviors. The inclusion of larger numbers of countries from different multinational projects minimizes this bias since the variability of macro-characteristics is usually increased. For example, the *ISSP* in consecutive waves added countries from Central and Eastern Europe, Asia, Latin America, and Africa (South Africa) to the core of the Western countries, thereby decreasing the potential bias with respect to key macro variables.

1.4.2 Harmonizing Within the Country

Two types of within-country harmonization should be distinguished: (i) harmonizing cross-sectional surveys across time, and (ii) harmonizing panel data. The U.S. *General Social Survey* (GSS) is the longest cross-sectional survey systematically conducted in the world (Marsden et al. 2020). Although the administration of GSS does not use the term "harmonization," some activities reconcile differences between variables across waves and require adjustments to obtain the same meaning of variables through time and hence constitute harmonization. The simplest harmonization takes place when the recoding of the "old" material is performed. For example, as of 2016, "most of job titles and descriptions from prior administration of GSS were . . ., recoded to the 2010 census occupational codes" (Morgan 2017, p. 1). Similarly, there was extensive recoding of the classification of religious denominations and other variables (Marsden et al. 2020). Sometimes, this type of within-country harmonization is built into a larger cross-national project in which all surveys from a given country are subject to constructing variables reflecting well-defined concepts like, for example, disposable income in the *Luxembourg Income Study* (see Chapter 15).

In panel surveys, respondents may be asked repeatedly the same question about the facts that do not change in their biography – for example, about the year of their first job or about their father's occupation when the interviewee was 16 years old. It is likely that the answers to these types of questions for some respondents will differ from one wave to another. Generally, there are two approaches to such a situation in the harmonization process. The first approach, applied mainly to numerical variables, consists of treating the inter-wave discrepancies in the source variables as errors and dealing with them in statistical modeling to produce the target variable. The second approach, applied mainly to nominal variables, tries to reconcile these discrepancies through analyses of the interviewing process. In proceeding, one must take into account the "memory effect," usually resulting in more reliable answers in later waves than in earlier waves (Rettig and Blom 2021). Whatever approach is adopted, we suggest retaining the original responses (source variables) in the file so they are available for further assessment and specific analyses.

1.5 Sources of Knowledge for Survey Data Harmonization

Survey data harmonization in a broad context relies on two main streams of knowledge: the methodology of social surveys as such and data science. Since the book *Scientific Social Surveys and Research: An Introduction to the Background, Content, Methods, and Analysis of Social Studies* was published in the late 1930s (Young and Schmid 1939), the methodology of social surveys has grown and become a rigorous discipline. The World Association of Public Opinion Research (WAPOR), the American Association of

Public Opinion Research (AAPOR), and the European Society of Opinion and Marketing Research (ESOMAR) are leading organizations in promoting a scientific approach to conducting surveys and documenting all activities related to surveys. In addition, many universities offer programs in survey methodology, and there are now numerous textbooks and handbooks on the issue as well as dedicated journals in this field.

The main goal of the International Workshop on Comparative Survey Design and Implementation (CSDI) is to improve comparative survey design, implementation, and related analyses, as evidenced in the periodically organized workshops and conferences. *Advances in Comparative Survey Methodology* (Harkness et al. 2010; Johnson et al. 2019) as well as the *Cross-Cultural Survey Guidelines* (https://ccsg .isr.umich.edu/) provide a solid methodology and technical advice for international survey researchers.

Data science is a buzzword covering several activities of computer specialists in different fields and disciplines. Some of these activities are obviously related to survey data harmonization, database creation, and the management of large data archives, in particular. Ex-post survey harmonization usually requires more attention to the processes where data must be preprepared for analysis through data selection and curation. Data scientists are devoting more research to seeking the most efficient ways to automate this work. Moreover, the analysis of very large numbers of data records can be timely run only by computer algorithms that compress and decompress data.

There are two other areas of data science that are relevant for survey harmonization: visualization and combining harmonized survey data with other digital data (Big Data). For surveys, visualization effectively communicates the structure, patterns, and trends of variables and their relations (Wexler 2022). In addition, visualization can be used to navigate and query harmonized data, enabling users to gain an overview of various data properties. Some recent papers provide recommendations for visualization designers who work with incomplete data, as is often the case with survey harmonized data (Song and Szafir 2019), or for those who use R for visualization of the data from the ESS (Toshkow 2020). Data scientists are also interested in the visualization of regional disparities using data from a harmonized survey project, such as the EVS (Borgoni et al. 2021).

It is interesting in the evolution of the field that in the first volume of the *Data Science and Social Research* (Lauro et al. 2019), the authors of the introduction stated that the availability of Big Data makes surveys obsolete (Amaturo and Aragona 2019, p. 2) while in the second volume under the same title (Mariani and Zenga 2021) the same authors (Amaturo and Aragona 2021, p. 6) advance a revisited epistemological approach in which Big Data is analyzed together with other data within different paradigmatic traditions of the substantive social science disciplines. In our volume, we included chapters describing how harmonized survey data are or can be integrated with other digital data and used for specific analyses (Chapters 1, 3, and 9, in particular). A major advantage of survey data should be knowledge about representativeness relative to a larger universe, while "Big Data" often provides little information about the universe from which samples are drawn.

1.6   Challenges to Survey Harmonization

Many challenges confront survey harmonization. First, we discuss ex-ante input harmonization in terms of the relevant five major stages of the survey life cycle, starting with the question of population representation and ending with that of data preparation (as we listed in Figure 1.1). Ex-ante output harmonization follows these same stages, with the aim of preparing a common database for each survey involved in the harmonization process. After that, we move to the challenges of ex-post harmonization, focusing on the harmonization of surveys from different multi-survey projects or single-survey projects. This discussion encompasses the harmonization of both cross-sectional and longitudinal surveys.

1.6.1 Population Representation (Sampling Design)

In the context of multinational projects, it is often very difficult to apply a uniform sampling design. Sampling frames depend on the national availability of population registers or other sources for representative samples and national legislation governing access to various sampling frames. As described in Chapter 7, the EASS uses a two-stage sample in Japan, three-stage samples in Korea and Taiwan, and a four-stage sample in China. Even the ESS, which applies much stricter harmonization practices, rests on different sampling frames and procedures for the participating countries (The ESS Sampling and Weighting Expert Panel 2018). One point here is that equivalence is not achieved by the mechanical use of identical procedures but must adjust to what is most representative in different contexts.

1.6.2 Instruments and Their Adaptation (Including Translation)

Harmonized survey question formulations sometimes deal with different objects that could be combined into a broader category. An example of this situation is given below:

Original formulation of the questions	Harmonized item
How many cigarettes do you smoke on average every day? How many pipes do you smoke on average every day? How many cigars do you smoke on average every day?	How many cigarettes, pipes, or cigars do you smoke on average every day?

Similarly, identically formulated questions with slightly different response options do not present problems for ex-post harmonization, as in the following example:

Survey 1	Survey 2	Target
How often do you pray?		
Daily	Several times a day daily	Daily
Several times a week	Several times a week	Several times a week
Once a week	Once a week	Once a week
Several times a month	Several times a month	Several times a month
Once a month	Less often	Less often
Less than once a month		
Never	Never	Never

However, sometimes identical question formulations may be associated with quite different meanings. Examples include "political right-left stances" with different meanings in Eastern and Western Europe due to historical reasons (Wojcik et al. 2021), or contraception with different meanings in Africa, Asia, and Europe due to differences in the availability of various contraceptive methods (UN Department of Economic and Social Affairs 2019). The problem is whether specific items formulated in different languages in an identical way, that is being formally equivalent, have the same meanings in the sense of functional equivalency. How does harmonization solve this problem?

In ex-ante input harmonization, it is essential that questionnaires are carefully translated from the language of the original questionnaire – the source language – to the target languages to which they are applied. In the methodological literature, there are disputes over the methods of ensuring the best translations. We recommend the state-of-the-art translation quality assurance provided by a so-called TRAPD model (Translation, Review, Adjudication, Pretesting, and Documentation), described by Harkness (2003; see also Fitzgerald and Zavala-Rojas 2020; Behr and Sha 2018).

Even if strict input harmonization is agreed upon in comparative surveys, researchers often have to resort to using language- or nation-specific terms considered to be functional equivalents. For example, in the East Asia Social Survey (Chapter 7), the 2008 module "Culture and Globalization" contains a question on how much the respondents like "traditional music." In the Japanese Survey, this term was replaced by "Enka," while in Taiwan and China by "traditional opera." In the harmonized file, the variable label remains as "traditional music" since it is assumed that different terms in the national questionnaires assure functional equivalence.

1.6.3 Preparation for Interviewing (Including Pretesting)

Different nations may have very different survey practices regarding interviewing due to cultural and political reasons. As it was pointed out many years ago, the reliability of survey data depends on the degree of the institutionalization (acculturation) of survey studies within a given society (Gostkowski 1974). However, even today's methodological textbooks rarely offer guidance on how to surmount the challenges of complicated cultural and political issues of interviewing in a cross-national setting. Pretesting should not be limited to checking the national questionnaires' properties but should include probing interview situations in respective countries. Such extended pretesting helps to identify possible obstacles in the fieldwork and thus increases the methodological awareness of researchers central to conducting credible cross-national research. The dilemma posed for administrating interviews in different countries consists of choosing between very strict uniform (standardized) interviewing and "allow[ing] interviewers-as-spokesperson to discuss with respondents the intended meaning and purpose of questions, as well as the respondent's answers." (Houtkoop-Seenstra 2000, p. 182). It must be noted, however, that this "spokesperson" strategy increases interviewer effects, which is not desirable. Moreover, it applies only to interviewer-administered surveys, while more and more surveys have switched to self-administered modes, where such interaction is not possible.

1.6.4 Fieldwork (Including Modes of Interviewing)

In well-planned international surveys, obtaining agreements on how to conduct surveys in each country to achieve comparability of results is a routine procedure. The ESS's efforts to coordinate the data gathering during the pandemic are well documented (https://www.europeansocialsurvey.org/about/news/essnews0115.html). In some countries, it was expected to conduct face-to-face fieldwork, supported in this round with video interviews, if respondents did not want to be interviewed face-to-face. But, for the first time in the ESS history (since ESS began in 2002), some countries were allowed to conduct the survey through self-completion methods. All these decisions were taken ex-ante and arguably did not lower comparability of measures.

From the ex-post harmonization standpoint, access to the full documentation of all steps involved in the fieldwork is essential. The ideal situation would be if the information on pretesting, training of interviewers, modes of interviewing, and fieldwork control was recorded in terms of quality control variables and included in the publicly available data file. Some may argue that storing this metadata in the main data file of survey responses is not efficient and should be stored in a separate, sample-level file, which, could be linked to the respondents' answers. We opt for any solution that allows researchers to assess whether the features of the fieldwork could contribute to inter-survey differences in the substantive analyses. Ex-ante, researchers can engage in the same type of assessment. With regard to the ex-post harmonization of the fieldwork, the recording of this kind of information is essential to the SDR approach presented in Chapter 9 and has been shown to affect substantive results (Slomczynski et al. 2022).

1.6.5 Data Preparation (Including Building Data Files)

Sometimes data that should be harmonized come in different formats. For example, in time use surveys (Chapter 16), the data can be either stored in "long format" with time slots as units (e.g. 10 : 01–10 : 10, 10 : 11–20, etc.) or in wide format with persons as units. When combining data sources with different formats, you have to first make sure that the data are stored in the same format in all sources. Also, data preparation may need to make the transformation of numerical scales into a common metric. For income, you may transform original values into a common currency, maybe with purchasing power parity (PPP) correction, or you can use relative measures such as percentiles. In this regard, prominent examples are provided by CNEF (Chapter 10) and LIS/LWS (Chapter 15). If data from different sources are combined in a harmonized file, it is often useful for researchers to also have access to the national version of a variable. For example, ISSP datasets typically include a harmonized variable for educational attainment but also one variable for each country containing nation-specific educational categories. This way, researchers can also attempt to do their own harmonization and assess the effects of different adjustments.

1.6.6 Data Processing, Quality Controls, and Adjustments

Putting the survey data into a digitally readable format consistent with ex-ante harmonization recommendations is a critical but often undervalued step in completing the project. Usually, ex-ante output harmonization involves quality checks and adjustments. Automated data cleaning features can help eliminate inconsistencies and identify suspect data patterns, including nonunique records. In practice, many projects are not free from data processing errors. Some of these processing errors are discovered in the process of ex-post harmonization and have been shown to be significant indicators of bias (Olekyienko et al., 2018; Chapter 9). This underscores the importance of reducing such processing errors in improving the comparability of surveys.

1.6.7 Data Dissemination

Survey harmonization as presented in this volume contributes to an environment of open data. Most projects discussed in this volume distribute data from their own web platforms (see Chapters 3–7, 10–12), but some use data archiving institutions, such as Harvard Dataverse (Chapter 9). The description and documentation of these projects do a lot in terms of the advocacy and promotion of harmonized survey data. They also strengthen the possibility of collaboration between survey researchers and nonacademic organizations, such as government agencies and applied researchers, whose mandate focuses more on data usage and practical evaluations such as social indicator construction and benchmarking.

1.7 Survey Harmonization and Standardization Processes

Harmonization methods have developed in close association with the concept of standardization. Historically, in the quest to understand how results obtained in one country compare with those of other countries, national statistical agencies early on pursued common standards in concept definitions, measurement tools, and data processing – with the goal of collecting data on the social, economic, and political conditions of their populations. Yet, since the 1960s, statistical agencies, as well as international organizations and researchers engaged in cross-national studies, have increasingly understood that complete standardization of data collection and processing was untenable and potentially harmful for comparability. So began the development of harmonization rules for blending cross-national standardized and country-specific survey designs while incorporating continual advances in survey methods and statistical techniques.

What is the relationship between harmonization and standardization? Standardization may be used in ex-ante harmonization: Various marketing and public opinion organizations introduce standards for the questionnaire items on sociodemographic variables (ESOMAR, 1997, is a good example). A standard in one language is assumed and applied in the ex-ante harmonization process in the international project – harmonization includes translation into local languages and pretesting ensuring that the standard is kept. This is sometimes referred to as "standards of procedure."

However, equally important are standards for tools. The International Standard Classification of Occupations (since 1957, prepared and updated by the International Labor Office) and the International Standard Classification of Education (since 1976, prepared and updated by the United Nations Educational, Scientific, and Cultural Organization) are examples of these useful and often used tools in survey research.

Ex-post harmonization is affected by standardization but also influences standardization: finding common grounds for different item formulations sometimes results in proposing a new standard for asking a survey question in a given way. A good example of this is when the World Health Organization, after examining existing formulations on the self-rating of health and attempting to harmonize different questions, proposed a set of simple questionnaire items for use in international research, which has been widely adopted (De Bruin et al. 1996, pp. 51–53). But, as argued above, mechanical standardization without attention to context can undermine functional equivalence, so care is needed in standardization.

1.8 Quality of the Input and the End-product of Survey Harmonization

Ultimately, the harmonization of surveys from different times and different places results in new integrated data files. How can we assess their quality? The general answer is that their quality depends on two components: (i) the quality of the original data and tools used for harmonization, and (ii) the quality of the harmonization process as such. Each of these components differs in ex-ante and ex-post harmonization.

In ex-ante harmonization, researchers and organizations conducting surveys try to achieve the best comparability of the population covered, instruments used, the fieldwork, and documentation. National surveys executed in different countries, or in the same country at different times, could vary considerably in the quality of work on all steps of the survey data life cycle. Cross-national comparability of surveys relies on the degree of fulfilling the standards of dealing with both the representativeness and measurement as explained in the survey methodology textbooks (e.g. Groves et al. 2009) and such documents as Standards and Guidelines for Statistical Surveys (OMB 2006), Standards for the Development, Application, and Evaluation of Measurement Instruments in Social Science Survey Research (RatSWD 2015), or the Task Force Report on Quality in Comparative Surveys (WAPOR/AAPOR 2021).

Survey representativeness is a complex issue since it includes the sampling scheme, the sampling frame, and, if such is the case, specific methods of weighting. In comparative research, the researcher has to decide not only whether a given survey sample is sufficiently representative of a given country at a given time but also about intercountry or inter-time comparability of all the constitutive elements of representativeness. In some cases, the sampling scheme can be deficient (e.g. Kohler 2007), or sampling weights can contain errors (e.g. Zieliński et al. 2018; Chapter 19).

Assessment of the quality of the measurement is not restricted to the instruments – the questionnaires, in particular – but also includes all features of the fieldwork that could potentially influence the validity and reliability of what is supposed to be observed or inferred. Interviewer recruitment, assignment, and training, pilot testing, fieldwork implementation, etc., all influence this.

In ex-post harmonization, it is likely that there will be differences in the quality of the same elements of the survey lifecycle that are relevant in ex-ante harmonization. In the analytic framework of data recycling (Slomczynski and Tomescu-Dubrow 2018; Chapter 9), the quality of surveys should be accounted for via quality-control variables, focusing on the assessment of documentation that reveals how the surveys were conducted, to what extent there were data processing errors, and deficiencies in the data files.

Faniel et al. (2016) studied users of data provided by the Interuniversity Consortium for Political and Social Research and found that the satisfaction of working with data is closely related to the data quality. According to the study on users' satisfaction, the crucial issues are completeness, accessibility, ease of operation, and credibility. The same applies to survey data harmonization. The quality of the final product of ex-post harmonization, the integrated data, depends on the mindfulness and cautiousness of those preparing the data and the extent to which they address these issues. Our main recommendation is that all steps of the harmonization process should be well documented so that users can be confident of completeness, accessibility, ease of operation, and credibility of the final product.

In this volume, chapters on specific survey harmonization projects devote considerable space to documents that could be independently assessed with respect to the quality of described harmonization steps. In addition, as Lillard (Chapter 10) points out, researchers should not only have access to harmonized data and harmonization algorithms but also to the original data. Only then can they check the harmonization algorithms. Also, this allows them to test alternative harmonization approaches and incorporate such considerations in their analyses and presentations of results.

1.9 Relevance of Harmonization Methodology to the FAIR Data Principles

The importance of the four principles of FAIR (Wilkinson et al. 2016) for data harmonization is to ensure the management and stewardship of scientific data in ways that promote broader use. FAIR stands for Findability, Accessibility, Interoperability, and Reusability – which are basically ways to maximize the reuse and effective archiving of scientific data. The FAIR principles respond to the needs of science funders, publishers, and government agencies for better standardization and transparency in the collection, annotation, and archival use of scientific data. In effect, these are principles that make possible the ex-post harmonization of data and, going forward, should guide the construction of data so that future users will be able to harmonize.

A key point for Wilkinson and colleagues is the importance of developing mechanisms for finding and integrating scientific data, especially from archives. Unless data follow some set of standard procedures for data cataloging, users, who are typically using searching devices, will not be able to locate data. Thus, a critical feature is the development of software supporting repositories and code modules that focus on metadata publication and searchability and are compatible with privacy considerations.

Harmonization mostly contributes to interoperability of data in the sense that through harmonized variables, different data sources can be linked and integrated, for example, surveys conducted in different countries or at different time points. Additionally, harmonization may contribute to the reusability of data because these data can be more easily integrated with other data sources. Also, in the same vein, harmonized data contribute more to the cumulation of research findings – in particular, if harmonization leads to standardization of survey measurement.

1.10 Ethical and Legal Issues

Conducting survey research involves specific ethical issues, especially respondents' consent, the use of incentives, privacy, confidentiality, and anonymity, among others. Most of the academic and private organizations conducting surveys follow their own codes of ethics, and those set up by such institutions as WAPOR, AAPOR, and ESOMAR – already mentioned earlier – or follow principles of sound research as set up by national research councils and funders. In Europe, the General Data Protection Regulation (GDPR) requires that survey respondents be informed about what will be done with the data they provide and how long these data will be kept.

Sensitive data need encryption. Encrypted files could be stored, modified, and backed up relatively easily, and such files are often used in the transmission of data from computer-assisted interviewing. Ex-post harmonized panel data may also need encryption if there are reasons to believe that the identification of respondents is possible, such as in the case of patterns of rare respondents' characteristics coupled with their locations with very small population sizes.

Ex-post harmonization is based on the reuse of data, which may involve additional ethical and legal issues. In general terms, these issues are discussed in the literature (e.g. Boté-Vericad and Témens 2019; Lipton 2020). "At first sight, it may appear that copyright regimes do not apply to data and dataset. Research data in its own right is unlikely to meet the originality standards and, therefore, is unlikely to qualify as a protectable subject matter. However, copyright can apply to original compilations of data and thus to databases." (Lipton 2020, Section 7.1). In 1996, the Council of the European Union passed the directive on the legal protection of databases, giving legal rights to certain computer records. These rules may apply to some survey files.

In our opinion, whatever the legal disputes regarding copyright may be, the reuse of survey data for ex-post harmonization purposes should be based on an agreement with the distributor of the original surveys or – as in the case of the large international survey projects – the compiled files. And these distributors have the right to restrict the redistribution of the original data, although they may release the data to make them freely open and disseminated in the new form. Such an arrangement protects the interests of the original data developers and the integrity of the scientific record while also making it possible for ex post harmonization projects that build on these data. A clear chain of agreements and acknowledgements is necessary.

1.11 How to Read this Volume?

Readers of this volume may bring diversified interests. Those who seek information on ex-ante survey harmonization may find Part I most relevant. They would realize that in some projects (WVS, in particular), ex-post harmonization of surveys from consecutive waves is an important element of the broader harmonization endeavor. Although ISSP and ESS have been important for the development of harmonization procedures, this volume has no separate chapters devoted to these important projects. We encourage readers to examine their web platforms regarding recommended procedures for ex-ante input and output harmonization. These projects differ in their organizational structure, with ESS being more driven by national representatives with respect to substance and methodology and ESS having stronger central control by the coordinating administrative body. However, both these projects have contributed greatly to our knowledge of the central issues of ex-ante harmonization.

Ex-post harmonization has become a fast-developing subfield. Readers may appreciate the variety of projects that we included in Parts II and III. An important aspect of these parts consists of methodological similarities and differences in approaching ex-post harmonization. We stress here two similarities: attention given to the quality of data and open access to the harmonized data. We also point out that the differences are driven not only by the substantive goals of the projects but also by organizational conditions. For those readers who are mostly interested in general methodological issues of ex-post harmonization, we recommend Part IV. However, some of these general issues are also raised in chapters devoted to particular projects.

We end with the observation that a minimum prerequisite for all forms of global survey harmonization is dedicated and sustained communication between funding organizations, data producers, and actors who reprocess extant data. Such dialogue should develop a shared understanding of the benefits and drawbacks of increasingly "inter-linkable" datasets, including "organic" or non-survey forms of data. This volume should contribute to such a conversation, by allowing all the main actors to look at the challenges of different harmonization approaches and the outcomes of such efforts.

References

Amaturo, E. and Aragona, B. (2019). Introduction. In: *Data Science and Social Research* (ed. N.C. Lauro, E. Amaturo, M.G. Grassia, et al.), 1–6. Cham, Switzerland: Springer.

Amaturo, E. and Aragona, B. (2021). Digital methods and the evaluation of the epistemology of the social science. In: *Data Science and Social Research II* (ed. P. Mariani and M. Zenga), 1–8. Cham, Switzerland: Springer.

Ayadi, M., Krishnakumar, J., and Matoussi, M. (2003). Pooling surveys in the estimation of income and price elasticities: an application to Tunisian households. *Empirical Economics* 28: 181–201.

Behr, D. and Sha, M. (2018). Introduction: translation of questionnaires in cross-national and cross-cultural studies. *International Journal of Translation and Interpreting Research* 10 (2): 1–3.

Biolcati, F., Molteni, F., Quandt, M., and Vezzoni, C. (2020). Church Attendance and Religious change Pooled European dataset (CARPE): a survey harmonization project for the comparative analysis of long-term trends in individual religiosity. *Quality & Quantity* 56: 1729–1753.

Borgoni, R., Caragini, A., Michelaneli, A., and Zaccagnini, F. (2021). Attitudes toward immigration inclusion: a look at the special disparities across European countries. In: *Data Science and Social Research II* (ed. P. Mariani and M. Zenga), 79–90. Cham, Switzerland: Springer.

Boté-Vericad, J.-J. and Térmens, M. (2019). Reusing data: technical and ethical challenges. *DESIDOC Journal of Library & Information Technology* 39 (6): 329–337.

Buchanan, W. and Cantril, H. (1953). *How Nations See Each Other: A Study in Public Opinion*. Urbana, IL.: University of Illinois Press.

De Bruin, A., Picavet, H.S.J., and Nossikow, A. (ed.) (1996). *Health Interview Surveys: Toward International Harmonization of Methods and Instruments*. Copenhagen: World Health Organization Regional Office for Europe.

van Deth, J.W. (1998). Equivalence in comparative political research. In: *Comparative Politics. The Problem of Equivalence* (ed. J.W. van Deth), 1–19. London: Routledge.

Doiron, D., Raina, P., Ferretti, V. et al. (2012). Facilitating collaborative research – implementing a platform supporting data harmonization and pooling. *Norsk Epidemiologi* 21 (2): 221–224.

Dubrow, J.K. and Tomescu-Dubrow, I. (2016). The rise of cross-national survey data harmonization in the social sciences: emergence of an interdisciplinary methodological field. *Quality and Quantity* 50: 1449–1467.

Ehling, M. (2003). Harmonising data in official statistics. In: *Advances in Cross-National Comparison* (ed. J.H.P. Hoffmeyer-Zlotnik and C. Wolf), 17–31. Boston, MA: Springer.

ESOMAR (1997). *ESOMAR Standard Demographic Classification*. Amsterdam: A System of International Socio-Economic Classification of Respondents to Survey Research.

Faniel, I.M., Kriesberg, A., and Yakel, E. (2016). Social scientists' satisfaction with data reuse. *Journal of the Association for Information Science and Technology* 67 (6): 1404–1416.

Fitzgerald, R. and Zavala-Rojas, D. (2020). A model for cross-national questionnaire design and pretesting. In: *Advances in Questionnaire Design, Development, Evaluation and Testing* (ed. P. Beatty, D. Collins, L. Kaye, et al.), 493–520. Wiley.

Gostkowski, Z. (1974). Toward empirical humanization of mass surveys. *Quality and Quantity* 8: 11–26.

Granda, P. & Blasczyk, E. (2016). *Data harmonization. Guidelines for Best Practice in Cross-Cultural Surveys*. Ann Arbor, MI: Survey Research Center, Institute for Social Research, University of Michigan. Retrieved 2022.09.18, from http://ccsg.isr.umich.edu/.

Granda, P., Wolf, C., and Hadorn, R. (2010). Harmonizing survey data. In: *Methods in Multinational, Multicultural and Multiregional Contexts* (ed. J. Harkness, M. Braun, B. Edwards, et al.), 315–332. Hoboken, NJ: Wiley.

Groves, R.M., Fowler, F.J., Couper, M. et al. (2009). *Survey Methodology*. Hoboken, NJ: Wiley.

Harkness, J. (2003). Questionnaire translation. In: *Cross-Cultural Survey Methods* (ed. J. Harkness, F.J.R. van de Vijver, and P.P. Mohler), 35–56. Hoboken, NJ: Wiley.

Harkness, J.A., Mohler, P.P., and van de Vijver, F.J.R. (2003). Comparative research. In: *Cross-Cultural Survey Methods* (ed. J. Harkness, F.J.R. van de Vijver, and P.P. Mohler), 3–16. Hoboken, NJ: Wiley.

Harkness, J.A., Braun, M., Edwards, B. et al. (ed.) (2010). *Methods in Multinational, Multicultural and Multiregional Contexts*. Hoboken, NJ: Wiley.

Hoffmeyer-Zlotnik, J.H. (2016). *Standardisation and Harmonisation of Socio-Demographic Variables (Version 2.0)*, GESIS Survey Guidelines. Mannheim, Germany: GESIS – Leibniz Institute for the Social Sciences.

Houtkoop-Seenstra, H. (ed.) (2000). *Interaction and the Standardized Survey Interview*, The Living Questionnaire. Cambridge: Cambridge University Press.

Johnson, T.P. (1998). Approaches to equivalence in cross-cultural and cross-national survey research. In: *ZUMA-Nachrichten Spezial 3: Cross-Cultural Survey Equivalence* (ed. J. Harkness), 1–40. Mannheim: ZUMA.

Johnson, T.P., Pennell, B.-E., Stoop, I., and Dorer, B. (ed.) (2019). *Advances in Comparative Survey Methods: Multinational, Multiregional, and Multicultural Contexts*. Hoboken, NJ: Wiley.

Kallas, J. and Linardis, A. (2010). A documentation model for comparative research based on harmonization strategies. *IASSIST Quarterly* 32 (1): 12–25.

Kish, L. (1999). Cumulating/combining population surveys. *Survey Methodology* 25 (2): 129–138.

Kohler, U. (2007). Surveys from inside: an assessment of unit nonresponse bias with internal criteria. *Survey Research Methods* 1 (2): 55–67.

Kołczyńska, M. (2014). Representation of Southeast European countries in international survey projects: assessing data quality. *Ask: Research and Methods* 23 (1): 57–78.

Lauro, N.C., Amaturo, E., Grassia, M.G. et al. (ed.) (2019). *Data Science and Social Research*. Cham, Switzerland: Springer.

Lipton, V.J. (2020). *Open scientific data – why choosing and reusing the RIGHT DATA matters. IntechOpen* https://doi.org/10.5772/intechopen.87201.

Malnar, B. and Ryan, L. (2022). Improving knowledge production in comparative survey research: cross-using data from four international survey programmes. *Czech Sociological Review* 57 (1): 683–706.

Mariani, P. and Zenga, M. (ed.) (2021). *Data Science and Social Research II*. Cham, Switzerland: Springer.

Marsden, P.V., Smith, T.W., and Hout, M. (2020). Tracking US social change over a half-century: the general social survey at fifty. *Annual Review of Sociology* 46: 109–134.

Mohler, P. and Johnson, T.P. (2010). Equivalence, comparability, and methodological progress. In: *Survey Methods in Multinational, Multiregional, and Multicultural Contexts* (ed. J. Harkness, M. Braun, B. Edwards, et al.), 17–29. Hoboken, New Jersey: Wiley.

Morgan, S.L. (2017). A coding of social class in the General Social Survey. GSS Methodological Report No. 125. https://gss.norc.org/Documents/reports/methodological-reports/MR125.pdf

Office of Management and Budget [OMB] (2006). *Standards and Guidelines for Statistical Surveys*. Washington, DC: US Office of Management and Budget https://www.ftc.gov/system/files/attachments/data-quality-act/standards_and_guidelines_for_statistical_surveys_-_omb_-_sept_2006.pdf.

Oleksiyenko, O., Wysmiułek, I., and Vangeli, A. (2018). Identification of processing errors in cross-national surveys. In: *Advances in Comparative Survey Methods: Multinational, Multiregional, and Multicultural Contexts (3MC)* (ed. T.P. Johnson, B.-E. Pennell, I.A.L. Stoop, et al.), 1015–1034. Hoboken, NJ: Wiley.

PEW (2019). *Religion's Relationship to Happiness, Civic Engagement and Health Around the World*. Washington, DC: Pew Research Center.

Przeworski, A. and Teune, H. (1966). Equivalence in cross-national research. *Public Opinion Quarterly* 30: 551–568.

Przeworski, A. and Teune, H. (1970). *The Logic of Comparative Social Inquiry*. New York: Wiley.

RatSWD (Rat für Sozial- und Wirtschaftsdaten) (2015). *Quality Standards for the Development, Application, and Evaluation of Measurement Instruments in Social Science Survey Research*. Prepared and written by the Quality Working Group: (ed. C. Beierlein, E. Brähler, M. Eid, et al.). Berlin: SCIVERO Verlag.

Rettig, T. and Blom, A.G. (2021). Memory effects as a source of bias in repeated survey measurement. In: *Measurement Error in Longitudinal Data* (ed. A. Cernat and J.W. Sakshaug), 3–18. Oxford: Oxford University Press.

Roberts, G., & Binder, D. (2009). Analyses based on combining similar information from multiple surveys. in *Proceedings of the Joint Statistical Meetings*, Section on Survey Methods Research. 2138–2147. http://www.asasrms.org/Proceedings/y2009/Files/303934.pdf

Schneider, S.L., Joye, D., and Wolf, C. (2016). When translation is not enough: background variables in comparative surveys. In: *The SAGE Handbook of Survey Methodology* (ed. C. Wolf, D. Joye, T.W. Smith, and Y. Fu), 288–308. London, UK: SAGE Publications.

Slomczynski, K.M. and Tomescu-Dubrow, I. (2006). Representation of European post-communist countries in cross-national public opinion surveys. *Problems of Post-Communism* 53 (4): 42–52.

Slomczynski, K.M. and Tomescu-Dubrow, I. (2018). Basic principles of survey data recycling. In: *Advances in Comparative Survey Methods* (ed. T.P. Johnson, B.-E. Pennell, I.A. Stoop, and B. Dorer), 937–962. Hoboken: Wiley.

Slomczynski, K.M., Tomescu-Dubrow, I., and Wysmulek, I. (2022). Survey data quality in analyzing harmonized indicators of protest behavior: a survey data recycling approach. *American Behavioral Scientist* 66 (4): 412–433.

Song, H. and Szafir, D.A. (2019). Where's my data? evaluating visualizations with missing data. *IEEE Transactions on Visualization and Computer Graphics* 25 (1): 914–924.

The ESS Sampling and Weighting Expert Panel (2018). *European Social Survey Round 9 Sampling Guidelines: Principles and Implementation*. European Social Survey. https://www.europeansocialsurvey.org/docs/round9/methods/ESS9_sampling_guidelines.pdf.

Toshkov, D. (2020). *Visualizing Data from the European Social Survey with R*. http://dimiter.eu/Visualizations_files/ESS/Visualizing_ESS_data.html (accessed 27 April 2022).

UN Department of Economic and Social Affairs (2019). *Contraceptives by Used Methods*. (ST/ESA/SER.A/435). https://www.un.org/development/desa/pd/sites/www.un.org.development.desa.pd/files/files/documents/2020/Jan/un_2019_contraceptiveusebymethod_databooklet.pdf.

WAPOR/AAPOR (2021). *Task Force Report on Quality in Comparative Surveys*. World Association of Public Opinion Research https://wapor.org/resources/aapor-wapor-task-force-report-on-quality-in-comparative-surveys/.

Wendt, M. (2007). *Consideration before pooling data from two different cycles of the General Social Survey*. CiteSeer http://citeseerx.ist.psu.edu/viewdoc/summary?doi=10.1.1.607.1003.

Wexler, S. (2022). Visualizing Survey Data. Version (2022).1 – February 2022. Published at the DataRevelations.com. /https://www.datarevelations.com/wp-content/uploads/2022/02/DataRevleations_whitepaper_surveydata_2022.1-1.pdf (Accessed 2 May 2022).

Wilkinson, M., Dumontier, M., Aalbersberg, I. et al. (2016). The FAIR guiding principles for scientific data management and stewardship. *Scientific Data* 3: 160018.

Wojcik, A.D., Cislak, A., and Schmidt, P. (2021). 'The left is right': left and right political orientation across Eastern and Western Europe. *The Social Science Journal* https://doi.org/10.1080/03623319.2021.1986320.

Wolf, C., Schneider, S.L., Behr, D., and Joye, D. (2016). Harmonizing survey questions between cultures and over time. In: *The SAGE Handbook of Survey Methodology* (ed. C. Wolf, D. Joye, T.W. Smith, and Y. Fu), 502–524. London: Sage.

Young, P.V. and Schmid, C.F. (1939). *Scientific Social Surveys and Research: An Introduction to the Background, Content, Methods, and Analysis of Social Studies*. New York: Prentice-Hall.

Zieliński, M.W., Powałko, P., and Kołczyńska, M. (2018). The past, present, and future of statistical weights in international survey projects. In: *Advances in Comparative Survey Methods: Multinational, Multiregional, and Multicultural Contexts (3MC)* (ed. T.P. Johnson, B.-E. Pennell, I.A.L. Stoop, et al.), 1035–1052. Hoboken, NJ: Wiley.

2

The Effects of Data Harmonization on the Survey Research Process

Ranjit K. Singh[1], Arnim Bleier[2], and Peter Granda[3]

[1]*GESIS – Leibniz Institute for the Social Sciences, Department of Survey Design and Methodology, Mannheim, Germany*
[2]*GESIS – Leibniz Institute for the Social Sciences, Department of Computational Social science, Cologne, Germany*
[3]*University of Michigan, Institute for Social Research, Ann Arbor, MI, USA*

2.1 Introduction

The historical evolution of the survey research process, often gradual but occasionally abrupt, involves more than complex methodological advances and wondrous new technologies. It also reveals how once novel but now established procedures can evolve and affect the entire survey data lifecycle, a term which refers to all aspects of the production of data from initial project planning to its distribution and reuse. Such is the case with data harmonization. Long viewed as of principal interest to large-scale data collection projects, harmonization techniques and strategies are now recognized by all stakeholders (international organizations, funding agencies, data producers and users in government and academic institutions, archivists, and methodologists) as critical for the effective planning of survey projects and in producing valid and useful outputs for policymakers and the research community. Data harmonization has assumed even greater importance as social science research expands into areas of inquiry using non-traditional resources not previously available, such as data collected from social media organizations and from sensors that routinely monitor various energy and transportation systems. These advances suggest that it is not premature to state that the harmonization of survey data constitutes a key development in the history of social science.

This chapter will describe the effects of data harmonization through its evolution from a tool to standardize measures over time to a comprehensive conceptual framework that now affects all aspects of the survey research process. It recognizes the widespread uses of the term in cross-national survey research, which extend well beyond simply recoding individual variables, and highlights the early work of national statistical agencies and international survey programs. Recent data collection efforts have built on the work of these organizations so that harmonization now affects how these projects are planned, conducted, checked for quality, and how the data they generate are made available to end users.

The initial discussion will focus on comparative research projects that gained widespread analytical use and affected the work of both those who produced the data and repositories that focused on preservation and dissemination. Secondly, attention will be given to the current uses of data harmonization among the growing number of stakeholders who produce and analyze comparative data. Additional descriptions of the complex challenges and trade-offs involved with various harmonization strategies will follow along with an analysis of the contributions made by the growing number of experts and organizations who perform harmonization tasks. The aim is to explore a more integrative view on the complex division of labor in survey research and harmonization. Finally, how will the growing use of "organic" data, defined as self-generated data that accumulate as a natural by-product of the current digital ecosystem (Groves 2011), affect data harmonization? While work on the harmonization of new big and organic data is still in an early stage, the chapter hopes to contribute to the discussion regarding the challenges and opportunities that arise from this new type of data.

2.2   Part 1: Harmonization: Origins and Relation to Standardization

2.2.1   Early Conceptions of Standardization and Harmonization

Harmonization developed through its close association with the concept of standardization. Standards can take many different forms, such as specifications, classifications, clearly defined processes, or validated instruments that can be applied across many different survey programs. The need for standards became a fundamental component of survey research practice at an early stage because of the emphasis placed on comparability. Many national statistical agencies, almost from the time of their inception, collected basic measurements on the social, economic, and political conditions and beliefs of their populations and wanted to understand how the results they obtained compared with those of other nations (Joseph 1973). In order to produce accurate assessments over time, those who collected social and economic measurements had to develop precise definitions of the objects they intended to measure, a clear understanding of how they would collect the information, and specific rules and procedures to convert the resulting data into quantitative values that would reveal valid changes from year to year (National Institute on Drug Abuse [NIDA] 1975).

But rules and procedures change in response to the need to accommodate new objectives and goals. Statistical agencies understood that imposing a set of standards developed for collecting and reporting their own statistics was neither realistic nor intellectually sound when seeking national comparisons (Franchet 1991). A more flexible structure was required, which soon focused on the development of various harmonization principles that would act as "linking mechanisms" to facilitate the development of agreed classifications and definitions. Ideally, such structures would produce comparative statistics across countries on similar economic and social issues. Standardization was the goal, but specific harmonization rules would be the means to achieve this goal (Tay and Parker 1990).

With the ongoing development of data harmonization tools and strategies, these national statistical agencies continued to seek more standardized measurements over time as conditions changed in each country. When combined with the efforts of governments to report new measurements in conjunction with continual advances in data collection methods and statistical theory, data harmonization became an indispensable component of the survey research process and one without which any international comparisons of overall societal well-being would be meaningless.

2.2.2 Foundational Work of International Survey Programs

It was not only the national statistical agencies that emphasized the collection of comparative data over long periods of time. Many international organizations and academic researchers have equally long-standing histories in surveying populations across nations, regions, and cultures. Whether it is the many "barometer" surveys conducted in multiple countries of which the Eurobarometer (EB) is the oldest, the International Social Survey Programme (ISSP), World Values Survey (WVS), or, more recently, the European Social Survey (ESS) among others, all sought to create datasets that would facilitate cross-national comparisons. These organizations began to strengthen comparability through additional processing steps that produced common coding schemes across countries for many key variables. This allowed them to produce single "integrated" public-use data files in which they combined and harmonized responses of those interviewed in many countries into a single dataset to facilitate use by other researchers (https://ec.europa.eu/commfrontoffice/publicopinion/index.cfm).

In addition to harmonizing the meaning of select variables, the oldest survey programs developed appropriate weighting schemes for the national populations they surveyed and began to provide secondary analysts with basic information about their efforts to create datasets that facilitated comparability. At this early stage, the social science community had not yet developed standards and agreed best practices for producing public-use data and documentation files. The overall role that harmonization played in project planning decisions was not always fully described to users. In particular, detailed information about how responses from different countries on similar questions were harmonized into single variables varied from project to project, or sometimes was completely absent from the documentation. While having access to the integrated files, researchers often had no access to the individual country files from which they derived. The integrated files made it easier for users to run comparative analyses, but they depended on the quality of all of the harmonization efforts employed by the producers at all stages of the survey design, data collection, and final production processes. As the volume of comparative surveys rose exponentially, the focus on developing overall standards for documenting such studies became increasingly important for both data producers and data users (Burkhauser and Lillard 2005).

Over time, the efforts of both national statistical agencies and international survey programs, in conjunction with the broader survey research community, produced a set of specific and widely acknowledged data harmonization methods. Data harmonization strategies are now broadly categorized as ex-ante (harmonization intended before data collection) and ex-post (harmonization not intended before data collection). There is a distinction within ex-ante harmonization between strategies employed at the stages of survey design and implementation, which seek to use a set of rules that apply to collecting data in multiple countries (input harmonization), and strategies employed after data collection by individual countries to enhance comparability before the data are shared with others (ex-ante output harmonization). The effects of these various strategies are discussed in more detail in the remaining sections of this chapter.

2.2.3 The Growing Impact of Data Harmonization

The efforts of the statistical agencies and international survey programs produced advances in methods and practices that advanced all aspects of the survey research process. Data harmonization is now viewed as a distinct field within this domain. Comparative survey projects regularly employ harmonization tools and strategies in how they conceive and administer their workflows. The expectation is that a body of theory and practical guidelines for implementing projects from start to finish will follow (Dubrow and Tomescu-Dubrow 2016; Fortier et al. 2017). Based on these long-term developments in survey design and implementation, all those involved in the production and dissemination of comparative data view harmonization as playing a key role in their efforts to produce data of high quality and to make these resources available to users as quickly as possible.

Harmonization affects how data producers design and implement surveys, including how they pose questions to respondents, the values for the answers they collect, and how they and data repositories present the processed data to other users. At the same time, harmonization advances have had direct benefits for researchers. They have encouraged the creation of additional resources for the public good: new datasets produced from existing ones, as funders realized that the value of the individual data collections that they spent considerable money to collect is significantly enriched if it is combined with other, related, datasets. These new resources, of which the Integrated Fertility Survey Series (IFSS; Smock et al. 2015) is a notable example, derived from the use of ex-post data harmonization techniques on individual surveys, often completed over many years, which focused on the same research topics but were not specifically intended to be merged into composite datasets. In this case, ex-post harmonization was not only used to create comparable variables over time but also drove the entire data production process, from investigating the contents of the original surveys to select which topics and questions would best address a wide variety of research questions, to constructing appropriate imputation variables and weights across the entire data series. These new datasets provided many more cases to study and enabled meta-analyses with the potential of uncovering new insights into specific social and health issues.

Both the increased concern with comparability and the increased use of harmonization have created what might be termed as an overall harmonization "mindset" throughout the data lifecycle, from initial project planning to such survey-related issues as: structuring the content of questions posed to respondents; how to translate them into different languages (see discussion on "Shared Language Harmonization" (Harkness et al. 2016)); what harmonization strategies to use to collect and report responses; and how to monitor overall fieldwork operations. Projects with sufficient resources often establish central coordinating centers that provide overall guidelines to field offices, receive completed interviews into a central database, execute standardized checking procedures, provide specific guidance when questions arise, and make adjustments, when necessary, in future field operations.

Harmonization practices also influenced parts of the survey lifecycle that did not always receive the attention they deserved (e.g. University of Reading Research Services 2021; for its specific application to survey data, see: Data Documentation Initiative 2021). Often neglected in the past, activities such as preservation, sharing, and replication have attracted new interest because of the ongoing increase in the number and size of data collections produced. Repositories face new pressures to decide what to keep permanently, what to hold for a limited time period, and what to reject. In the age of open data, government granting agencies in many countries require producers to share their collections as an obligation when they receive grants, not only to allow other researchers to generate new analyses but also to allow others to replicate published findings (SPARC 2015).

These practices reflect a concerted effort to address more systematically principles embedded in the Total Survey Error (TSE) model as they apply to comparative surveys (Lyberg and Stukel 2010). The TSE paradigm attempts to describe all possible sources of error regarding representation and measurement in survey estimates (Groves and Lyberg 2010). Input harmonization provides strategies for reducing error by implementing procedures in multiple countries before interviews are conducted, through such techniques as coordinated sample designs, common question wording, standardized translation methods, and common interviewing practices. After data collection, ex-ante output harmonization can further reduce error through the creation of unified target variables when the data are processed. Finally, ex-post harmonization, in addition to its importance in the production of new data resources from existing ones, has more recently led to advances in the archiving practices of institutions, which preserve data resources and create public-use data files and their associated documentation. Social science data archives and similar repositories have now begun to create harmonized datasets from their own holdings or work directly with data producers on harmonization projects. Many of these institutions also employ specific harmonization strategies and decision rules regarding the ingest and evaluation of new data collections, such as uniform processing steps to check for inconsistencies, a specific set of editing options for making necessary corrections, and a standardized format for the structure and content of the documentation they produce for secondary analysts.

For data repositories, the primary goal is to generate public-use data files, which seek to minimize error and improve overall quality. They strive to meet principles associated with the dimensions of the total quality framework with regard to relevance (the data will satisfy the needs of users), accessibility (users will find the data easy to use), timeliness/punctuality (the production of data adheres to announcements about its release and serves the needs of secondary analysts), and usability/interpretability (that the documentation is clear and complete) (Biemer 2010). The last dimension of this survey quality framework is crucial for researchers who want to use secondary data. Since they did not collect it themselves, they require comprehensive documentation to allow them to take full advantage of these sources. They need to know pertinent details about sampling, questionnaire design and implementation, translation, and data collection operations, and all of the work performed on the data records from the time they were collected until they are ready for use. Data producers and repositories increasingly recognize this need, including documenting and even providing data files about the collection process itself, commonly known as paradata (Kreuter 2013).

Where either ex-ante or ex-post data harmonization efforts are employed, documentation challenges and expectations from users are magnified (AAPOR/WAPOR Task Force Report 2021). Ideally, analysts need to understand the rules used to harmonize specific variables, the decision process for invoking these rules, and even the code and programs, which generated the final results. They may also want to reuse or edit these programs to check the outputs of the harmonized variables or create new versions. There is as yet no complete agreement among social scientists about what the final documentation files should include. But new attempts show great promise. These include the insertion, directly into ex-post harmonized datasets, of quality control variables, which describe the completeness of the original survey documentation, combined with extensive reports and syntax files, which document the entire ex-post data harmonization process (Slomczynski et al. 2016).

2.3 Part 2: Stakeholders and Division of Labor

As we have seen, standardization, harmonization, and the survey data life cycle are closely entwined, and many stakeholders involved in harmonization or survey research are embedded in a complex web of interdependencies. In this part, we want to untangle this web. Caught up in the everyday business of survey research, different stakeholders may lose sight of the whole ecosystem. This, in turn, might mean making decisions that negatively impact other stakeholders needlessly, or letting opportunities for greater reusability and comparability pass by. In this part, we explore the possibilities of developing harmonization further into a more integrated ecosystem. Stakeholders and actors may find it useful to think of the bigger picture to form more optimal decisions and to lobby other actors (e.g. data producers) more effectively.

2.3.1 Stakeholders

First, we look at different stakeholders, starting with those whose decisions impact many eventual research projects and shape the survey data life cycle early on. Thus, we will begin with international actors and large funding agencies, then data producers, archives, and finally harmonization projects and data users. However, this structure should not let us lose sight of the various interactions between these stakeholders and the inherently cyclical nature of the survey data life cycle.

2.3.1.1 International Actors and Funding Agencies

We first look at the large entities that shape the financial and political environment in which survey research projects compete for resources. This entails funding agencies (e.g. national research councils; Gulbrandsen 2005) as well as large national and international public entities with stakes in the data collection and harmonization process (e.g. National Ministries and Statistical Offices, Eurostat, the OECD, or UNESCO). The decisions of such stakeholders often impact many research projects. This grants them tremendous leverage, which can reduce costs and increase the quality of harmonization projects. The downside, however, can be a reduced fit for particular projects, especially those beside the mainstream.

Many standards useful for ex-ante or ex-post harmonization are within the purview of such large entities (e.g. UNESCO et al. 2015), which have the necessary sway to encourage widespread adoption of such standards. After all, harmonization standards only unfold their full potential if a critical number of projects adhere to them. Funding criteria can play a role too, where they drive applicants to promise adherence to harmonization standards to signal their criteria compliance, such as promising DDI compliance (Data Documentation Initiative 2021), to signal data transparency and reusability.

The financial and political environment also sets the boundaries for ex-ante and ex-post projects by choosing which projects to fund (Gulbrandsen 2005). This point might seem trivial, but the practice of harmonization will certainly benefit from the fact that many funding agencies now actively support data reuse. This is driven by desiderata of scientific policy such as FAIR Data, Open Science, and replicability (Link et al. 2017). As a consequence, funding agencies now increasingly weigh the necessity of collecting new data against the potential for data reuse to answer research questions. This increases the pressure of ex-ante harmonization projects to justify new data collection. At the same time, ex-post harmonization studies might benefit in the competition for grants.

2.3.1.2 Data Producers

Data producers, meanwhile, are often the most influential stakeholders for both ex-ante and ex-post harmonization. Ex-ante harmonization is predominantly governed by data producers (Granda et al. 2010). However, this does not mean that data producers themselves are not bound to complex trade-offs. Projects involving extensive ex-ante harmonization (e.g. international comparative surveys) are usually subject to a complex multilevel game between local institutions and the international governance structure. Such coordination is necessary but can also shape the survey data landscape in unintentional ways. For example, many international survey programs strive for cross-national comparability by first designing a source questionnaire that is then localized into national questionnaires. This often creates tension between cross-national comparability and the desire to truthfully represent national idiosyncrasies. This example illustrates a more general pattern: comparability in one direction (here between the different countries) might come at the expense of comparability in another direction (here within each country) (Lynn et al. 2006; Granda et al. 2010). Harmonization standards and conventions play an important role here, as they reduce cost and complexity and facilitate coordination across different survey programs. However, harmonization standards can also reduce the fit of harmonized data for specific research agendas.

Ex-post harmonization, meanwhile, occurs within the boundaries of decisions made during data production, since it relies on the survey data pool in relevant societies (Dubrow and Tomescu-Dubrow 2016). While some decisions during data production can be compensated by ex-post harmonization practitioners (e.g. different scale formats), others cannot (e.g. missing or too coarsely grained information). However, while data producers cannot provide information that has not been measured, they can and should more frequently choose to provide raw measurements alongside their harmonized and aggregated data. For example, why not provide ISCED codes (Ortmanns 2020) alongside the raw national measurements for education? After all, advances in technology and archival techniques make providing tailored datasets easier than ever. This would give data users a choice of whether to use the harmonization work done for them or to find their own harmonization approach, which is then tailored to their specific needs.

2.3.1.3 Archives

Archives are complex actors in that they are the interface between many of the stakeholders described here. For data producers, archives are the gateway to eventual users, and for the users, archives and their services are often the face of the survey program. Via the metrics archives collect, user behavior also influences the survey programs. Archives are also remarkable in that they tend to network more and support each other's efforts; often across discipline, research topic, and national boundaries (Dekker et al. 2019). An example of such network activities is the Consortium of European Social Science Data Archives (CESSDA).

Archives thus often become the drivers of many standardization efforts, especially technological or metascientific harmonization standards. Data producers profit from delegating harmonization work to the archives, for example by asking archives to perform some data cleaning and transformation in line with harmonization standards or by having archives establish compliance with certain technological harmonization standards. Archives, in turn, achieve a critical mass of expertise and experience with harmonization as to be more efficient than separate survey programs themselves. Similarly, neither ex-ante nor ex-post harmonization projects would be possible without the work done by archives. At the same time, many decisions have already been made once data reaches the archive. Consequently, getting archival input early on, ideally even during the conception of grant proposals, can be crucial.

2.3.1.4 Data Users

Data users are a very diverse group of stakeholders, but also arguably the most important. After all, data users transform data into scientific and societal value. At the same time, individual users interact with data comparatively late in the survey data life cycle and are thus subject to many previously made decisions in data production and ex-ante harmonization. Of course, users can and do influence data production and harmonization, for example via advisory boards or simply via data usage metrics. The status quo for most users is, however, that they are presented with the result of many choices and thus have fewer remaining degrees of freedom. This can have advantages, since harmonization work done by data producers and archives saves users much effort. Also, such actors often have harmonization expertise that not all individual users can match. On the other hand, harmonization is seldom neutral or agnostic with regard to the domain matter. If surveys applied specific classification systems, for example, then this often has far-reaching substantive implications. ISCED 2011, for example, groups very different qualifications into the same ISCED categories; different both within and across nations (Schneider 2013). Clad in the authority of a widely used standard, these implications are then not always reflected by users in the necessary depth.

Ex-post harmonization is a way of regaining agency for data users. In essence, ex-post harmonization moves some choices back into the hands of users, but at the cost of substantial extra work (Dubrow and Tomescu-Dubrow 2016; Haensch and Weiß 2020). Of course, ex-post harmonization is limited by the available data, but the choices of specific survey programs may be compensated to some degree by the collective landscape of available data from other surveys and also non-survey data sources. Just because one survey omitted a necessary construct or did not span some years of interest does not mean that all available survey programs did. Ex-post harmonization thus has two major advantages. First, users can compensate for at least some of the substantive, temporal, geographical, or population limitations set by individual surveys. Second, users can increase the rigor of their research. Drawing data from many sources paves the way for replications via meta-analyses using individual respondent data, and the very act of harmonizing multisource data entails what we think of as comparative methodology. Just as comparative research across countries reveals patterns invisible in national surveys, ex-post harmonization across different surveys can reveal methodological patterns invisible to users of only a single survey program (Slomczynski et al. 2016).

2.3.2 Toward an Integrative View on Harmonization

Next, we discuss the trade-offs and dynamics that we implicitly touched upon while discussing different stakeholders. Harmonization work can be understood as a form of division of labor across many diverse stakeholders. By explicitly recognizing and perhaps restructuring this division of labor, much can be gained in terms of efficiency, quality, and fit for the needs of eventual data users. To better understand these dynamics, we will consider three broad and deliberately simplified ideas.

First, division of labor means that stakeholders faced with harmonization tasks often must decide whether to use existing harmonization solutions, commissioning harmonization specialists, create their own harmonization solutions, or pass harmonization work down the line to other stakeholders. Imagine an ex-post harmonization project looking to create an internationally comparable harmonized variable for educational attainment. The project could try to adapt an existing solution, such as ISCED. They seek specialist council or support. They could create their own educational coding scheme that best fits their project's goals. Or they could pass on raw education data to the users of their harmonized dataset, passing harmonization work onward. Of course, not every option is available for every task and for every stakeholder. And sometimes the options are not mutually exclusive.

Second, these decisions impact three entangled harmonization aspects: harmonization cost, harmonization quality, and harmonization fit. Harmonization cost means time, personnel, and materials necessary to perform the harmonization work. Harmonization quality is the degree of comparability established for a specific dimension of comparability, e.g. across countries, across time, or across subpopulations. Harmonization fit means how well a harmonized dataset fits the specific goals of a stakeholder. This means there can be high-quality but poor-fit scenarios, where comparability is only established across a dimension that is not of interest to a research project. Often, division of labor decisions in harmonization entail a trade-off along these dimensions. For example, using an existing harmonization solution might increase quality and decrease cost but might also decrease the fit for a stakeholder's needs.

Third, division of labor in harmonization is often hierarchical. For one, the relationship between different stakeholders is often asymmetrical. Decisions made by data producers cannot always be compensated by data users, for example. Furthermore, stakeholders differ in how many eventual research projects their decisions impact. If an international organization, such as the UNESCO, changes a harmonization standard, such as ISCED (Ortmanns 2020), then these changes impact countless research projects.

These three interlinked ideas obviously do not capture the full complexity of the matter. Nonetheless, they provide us with a heuristic with which we can examine the current division of labor in harmonization and perhaps to rethink it. In the following, we will thus look at the three harmonization aspects cost, quality, and fit and break down how division of labor decisions shape these outcomes.

2.3.2.1 Harmonization Cost

Division of harmonization labor between different stakeholders can reduce costs by realizing economies of scale. For example, we can outsource harmonization to harmonization in existing academic organizations or perhaps to organizations directly specializing in harmonizing data (e.g. Maelstrom Research 2020; Singh 2020). In any case, we increase efficiency through standardized procedures, innovation, and learning effects. While such outsourcing is not yet a widespread practice, it might well become more frequent due to increasing specialization and professionalization in the maturing field of harmonization. The potential drawbacks are increased transaction costs and principal-agent problems: both the fit and the quality of the harmonization might suffer as a result if harmonization partners are not carefully chosen.

Another approach is to reuse existing harmonization solutions. This might mean using existing standards, databases, scripts, or infrastructures instead of reinventing the wheel. Such a push for standardization is more effective when established early on in the survey data life cycle, since it then touches upon many different eventual use cases and saves data users much redundant work. However, we can only expect large efficiency gains in areas where the needs of many data users align.

2.3.2.2 Harmonization Quality

By harmonization quality, we mean the level of comparability across a certain comparability dimension, such as comparability across time, countries, languages, survey modes, measurement instruments, or sociodemographic groups. The overall quality of a harmonized data product is then a multiplicative composite of harmonization quality and the data qualities of the separate source surveys: high-quality surveys can still be hard to compare. Establishing comparability and thus ensuring harmonization quality is a complex and costly endeavor. Thus, ex-ante and ex-post harmonization quality can benefit from similar division of labor decisions as we discussed above for cost reduction. Costly but high-quality harmonization earlier in the survey data life cycle can still be more efficient than separate efforts because many data users benefit. Similarly, outsourcing complex harmonization tasks can mean benefitting from specialized expertise and a wealth of harmonization experience that individual projects cannot always match. Furthermore, this density of harmonization activities means acquiring specialized expertise has greater returns through economies of scale. Such centers of specialized expertise can thus also become incubators for methodological innovation.

However, this does not mean that ex-post harmonization projects performing their own harmonization work are always inferior in quality. In fact, ex-post harmonization can often reveal new insights not only into the comparability but also the source quality of survey data (Slomczynski et al. 2016). Combining and analyzing data in configurations not envisioned by the data producers often shines a new light on quality aspects. Furthermore, ex-post harmonization projects often focus on a specific topic. Ex-post harmonization practitioners are thus often domain experts with unique insights into the subject matter, and in many harmonization efforts, such domain expertise is crucial to establish comparability.

2.3.2.3 Harmonization Fit

Lastly, harmonization fit is the fit of a harmonized dataset for a stakeholder's specific goals. For ex-post harmonization practitioners, that might mean how well the data fits a specific research agenda. Harmonization fit has two main components: harmonization dimensions and coverage. Harmonization dimensions refer to the problem of harmonization quality we mentioned earlier. Harmonization efforts often focus on specific dimensions, such as time or countries. And improving one dimension does not necessarily guarantee comparability over time, for example. The important thing to note is that which comparability dimensions data producers tackle are an implicit division of labor decision. After all, much of the work done in ex-post harmonization projects focuses on establishing comparability across dimensions not intended by data producers (Granda et al. 2010). At the same time, this means that ex-post harmonization is an effective tool to increase harmonization and is fit for specific research agendas.

The second aspect of harmonization fit is data coverage across countries, points in time, subpopulations, and concepts. Much of harmonization work is done for the specific purpose of increasing coverage, such as through ex-ante harmonized cross-national surveys aiming for coverage across many countries. At the same time, coverage is perhaps the arena in which stakeholders' interests collide most often. Large-scale survey programs must make hard choices of which topics to include, which populations to survey, and in which intervals. The resulting gaps in coverage hindering specific research agendas then become the raison d'être for ex-post harmonization projects. Such projects can compensate for gaps in coverage of specific survey programs if other parts of the survey landscape provide the needed data coverage. However, ex-post harmonization efforts are hampered if survey programs omit raw, detailed data or if many programs systematically omit important concepts of interest. For example, we could increase the statistical resolution for infrequent subpopulations, such as LGBTIQ people, by pooling the data of many surveys. However, if many surveys omit identifiers for such subpopulations (Vries et al. 2020), ex-post harmonization can no longer compensate for gaps in coverage.

2.3.2.4 Moving Forward

Lastly, we stress again that our take on stakeholders and their interdependencies is a deliberate simplification. Nonetheless, our heuristic framework makes it salient that every seemingly isolated action in survey data collection or harmonization is both shaped by previous decisions and, in turn, shapes the degrees of freedom for other actors. Given that harmonization in general and ex-post harmonization in particular have not reached full maturity as a practice and a methodological field, there are certainly gains in harmonization efficiency, quality, and fit that we can achieve if we rethink division of labor in harmonization. Much harmonization work still features needless redundant work done by isolated practitioners. At the same time, other harmonization decisions are made too early in the survey data life cycle, needlessly removing degrees of freedom for data users and ex-post harmonization practitioners. And lastly, different stakeholders may simply not be aware of the needs of other stakeholders and the consequences their actions have for their research goals.

2.4 Part 3: New Data Types, New Challenges

In the final part, we move our focus from the complexities of current survey data harmonization to future harmonization challenges posed by new and nontraditional data types and sources. Modern society features a host of technological innovations that, increasingly, allow the collection of data on social attitudes and behaviors as well as their analysis on a large scale. This collection of data may happen via designed instruments, or dedicated sensors, such as on entrance gateways to public transport systems, or on social media platforms, that were not originally intended for later use in research. In the following, we look at the effects this changing technological environment has on data-driven research in the social sciences and the role that harmonization may play in providing innovative processing strategies to enable reproducible and valid analysis pipelines. The goal is to discuss how changes in data types relate to the stakeholders of harmonization, the challenges they are likely to face in pursuit of data comparability, as well as an outlook on the interplay between traditional survey data and this new form of data.

2.4.1 Designed Data and Organic Data

To understand the effects that technological changes have on data harmonization in social science research, we will distinguish between the effects in the context of the well-established data collected via designed survey instruments and the opportunities and pitfalls arising in the emerging research areas enabled by found organic data.

Forty years ago, mail-in surveys and interviewers conducting face-to-face surveys were complemented by telephone surveys. With rising access to the Internet in large segments of society, these types of surveys were soon joined by web-based surveys, prompting Witte (2009) to assert that this type of data collection "has reached a level of maturity such that it can be considered an essential part of the sociological tool kit." Since web-surveys are self-administered, data about the process of responding to the survey stimuli itself can be collected (Kaczmirek 2009). These process data are frequently referred to as paradata, with response time measurements (see, e.g. Yan and Tourangeau 2008 or Posch et al. 2019) being typical examples. Finally, microtask platforms such as MTurk (https://www.mturk.com) and Appen (https://appen.com) allow the crowdsourcing of participants. Recruitment of participants via crowdsourcing is increasingly used in social psychology research and offers considerable scale (Posch et al. 2020) and cost advantages over student and convenience samples (e.g. Paolacci and Chandler 2014).

Our ability to collect data about the behavior of individuals and groups in large quantities can also be a by-product of the increasingly technology-mediated social environment we live in. Following Groves (2011), we refer to this new type of data as "organic data." This terminology distinguishes this new type of data from "designed data" collected via survey instruments that are designed specifically to answer research questions. Organic data are nothing fundamentally new, of course. Social scientists have long made use of public records or other data, which public or private organizations had collected for purposes other than research. However, organic data has now gained popularity not only for providing more direct and nonreactive empirical observations but also for its seemingly widespread availability in large quantities (i.e. Big Data) at relatively low cost in times where response rates to traditional surveys are continually declining (Stedman et al. 2019; Japec and Lyberg 2020). Besides providing the empirical basis for numerous research papers analyzing phone records (Blumenstock 2012), web browsing histories (Araujo et al. 2017), or social media posts (Stier et al. 2018), the advent of this new type of organic data arguably contributed to the emergence of a new subdiscipline: computational social science (Lazer et al. 2009).

While technological changes in data generation and processing pose substantial challenges for designed data as well as organic data (Savage and Burrows 2007), we follow a similar strategy as in part two: unraveling the complexity by better understanding the involved stakeholders and their roles associated with the different types of harmonization in the area of designed data. This is in part due to the data collection process being controlled by the scientists involved, and in part due to the progressive but continuous and constructive nature of the development in the area of designed survey data. However, in the case of organic data, the situation is different. Unlike traditional survey data, organic data are by definition not intentionally produced to answer research questions. Furthermore, the process of data production is typically controlled by private companies driven by economic goals. This has considerable effects on the roles of stakeholders in the field of data harmonization.

2.4.2 Stakeholders in the Collection of Organic Data

2.4.2.1 Producers

The private companies that collect organic data of their users are clearly aware of considerations of data harmonization and data integration (Halevy et al. 2005). Yet, these harmonization and integration efforts are typically guided by the respective strategic business objectives and either aligned with the commercial ecosystem these companies operate their platforms in or even aimed at designing a data ecosystem in its own right (Evans and Basole 2016). International web-based data-centered companies such as Twitter and Facebook frequently develop their own standards for interoperability. They also decide, as part of their platform design, what is collected, in which way, what is shared, and under which conditions. They are bound by some legal constraints; however, those are for the most part still in their early stages of development (Schneble et al. 2018). In this environment, much of the harmonization work and the decision of what aspects of the available data can be harvested and used for research is done long after the original data were produced, making ex-post harmonization the primary mode to work with organic data in social science research.

2.4.2.2 Archives

The traditional role of archives, as the interface between data producers and users as well as important coordinating actors in the area of data harmonization, is challenged in a number of ways by this new paradigm of data gathering. While producers of organic data market their data increasingly for-profit directly to potential data users, data users are often bound by restrictive data-sharing policies asymmetrically imposed on them by large companies with little to no influence on the data producers (Acker and Kreisberg 2020). This leads, in the extreme, to the situation where even those subsets of organic data on which research has been carried out can no longer be archived in line with the respective data sharing policies of the producers, such as is the case with large Twitter data sets (Puschmann and Burgess 2013; Lietz et al. 2014). Legal frameworks governing the use and access to Big Data for research are in their infancy, and traditional anonymization techniques are frequently hard to apply for this new form of data (Japec and Lyberg 2020). However, lawmakers are starting to react, and some recent proposals draw not only "inspiration from the principles for data management and reuse developed for research data" (European Union 2020) but also aim at (re-) enabling the much-needed access to this increasingly important form of organic data. On the technological side, the volume, variety, and variability of big organic data (Couper 2013) are unlike the well-structured data, collected via traditional instruments. In particular, variety and variability constitute a messiness that poses an additional challenge to data archives.

2.4.2.3 Users
While data users can be considered the most important stakeholder group in the work with designed data, the role of social scientists is even more pronounced in the work with found data (Johnson and Smith 2017). It no longer suffices to transform well-formed survey data into works of scientific value. Social scientists are now faced with data that originally were not intended to answer scientific research questions (Savage and Burrows 2007; Chang et al. 2014), frequently provided directly by the producers, and in many cases without the benefits of any ex-ante consideration of data harmonization for scientific use.

2.4.2.4 Harmonization of Organic Data
In our minds, these modified roles of different stakeholders, and in particular the elevated status of the data users, are a transitional period made necessary by the rapid development in the area of nontraditional data. In this transitional period of exploring the potential of found data, users cannot rely on the luxury of established best practices and research data institutions. Hence, it is not surprising that technical aspects of ex-post harmonizing different data sources and formats as well as extracting conceptual relevant features are currently dominating much of the work in this emerging field. In fact, it is the often complex and high-dimensional nature of this nontraditional type of data, along with its volume and variety, that enables the exploration of new substantive questions while pioneering necessary harmonization and standardization efforts.

2.5 Conclusion

Data harmonization concepts and strategies were developed in close association with advances in survey research seeking to produce comparable data over time and space. As a consequence, harmonization began to affect the entire data lifecycle of survey projects. All the many stakeholders and actors involved in the planning, production, creation, and dissemination of survey data introduced and continually expanded the use of harmonization processes in their workflows. The number of actors and processes has reached a point that there now exists an integrated harmonization ecosystem in the survey research space. But it is an ecosystem that requires more integration so that actors do not work at cross-purposes. It is this challenge that affects survey researchers today but has even greater implications as social research itself continues to evolve.

We hope that the integrative view we proposed stresses the need for even more mutual awareness, communication, and cooperation between the different stakeholders in survey research and harmonization. It also reminds us that the current division of labor is not set in stone. Of course, few can unilaterally change longstanding practices, especially those currently in the purview of other stakeholders and actors. However, even separate research groups have to make such division of labor choices: for example whether to develop their own harmonization solution, using established standards, or perhaps pooling efforts with other ex-post harmonization projects to find a mutual and ideally reusable harmonization solution. After all, the trend toward open science, data convergence, FAIR data, and thus more harmonization shows no signs of slowing down.

Lastly, the emergence of new data types, measurement paradigms, and analysis approaches begs the question of how this will impact survey research and the harmonization of survey data. Reactions to the newly emerging methods range from fundamental rejection to euphoric embracing of the new purported possibilities. To our mind, these new possibilities will obviously not make survey research obsolete. There will always be a place in the social sciences for asking members of society questions and considering their answers. Instead, the rich and diverse toolbox of social science simply became a bit richer and more diverse.

Of course, as with any large methodological innovation, new data types and digital organic data also come with discussions about epistemology and about the prerogative of interpretation of old and new social questions. Just think of the controversial discussion if social science should be mainly theory-driven (e.g. Groves 2011) or if data-driven approaches can offer valuable new insights. And on the business side of academia, new methods and research paradigms also mean that another contender has entered the ring in the fight for funding, career merits, and access to decision-makers.

Still, those growing pains of our evolving and expanding discipline should not cloud the view for the potential of collaboration and synergy between survey research and the new paradigms. After all, the new data sources and research paradigms can complement survey research both in terms of substantive research and in terms of increased quality and efficiency (Japec and Lyberg 2020). In fact, this is already happening. Consider the growing role of digitally collected paradata in surveys, which give us new insights into respondent behavior. Do respondents hesitate? Do they change their mind? Do they take long breaks? And while web-surveys are the most well-known example, such paradata are not limited to digital, self-administered surveys. After all, most interviews (personal and via telephone) are conducted using a computer. Furthermore, surveys can ask respondents for their social media account handles. This allows researchers to supplement the survey responses with data from their Twitter posts or Facebook likes (Stier et al. 2020). The new possibilities of collecting behavioral data, such as geospatial movement and proximity profiles (Génois et al. 2019), mean survey researchers have new ways of validating their survey instruments. With a look toward ex-post harmonization projects, these new data sources and research paradigms mean that we have new contextual data available to enrich our harmonized datasets. Adding public record data, such as regional unemployment rates, to enrich comparative data sets is common practice. We have to be aware that an all too naive use of Google search trends, or of aggregated social media data, as a replacement for more traditional data is problematic (Lazer et al. 2009;

Jungherr et al. 2012). However, this should not preclude us from the careful analysis and use of these types of data in general. We feel that being at least aware of the newly emerging data sources and paradigms is beneficial to all people involved in the survey data lifecycle. Otherwise, promising opportunities to enrich our data and research might remain unrealized.

References

AAPOR/WAPOR Task Force Report on Quality in Comparative Surveys (2021). https://wapor.org/resources/aapor-wapor-task-force-report-on-quality-in-comparative-surveys

Acker, A. and Kreisberg, A. (2020). Social media data archives in an API-driven world. *Archival Science* 20 (2): 105–123. https://doi.org/10.1007/s10502-019-09325-9.

Araujo, T., Wonneberger, A., Neijens, P., and de Vreese, C. (2017). How much time do you spend online? Understanding and improving the accuracy of self-reported measures of internet use. *Communication Methods and Measures* 11 (3): 173–190. https://doi.org/10.1080/19312458.2017.1317337.

Biemer, P.P. (2010, 2010). Total survey error design, implementation, and evaluation. *Public Opinion Quarterly* 74 (5): 817–848. https://doi.org/10.1093/poq/nfq058.

Blumenstock, J.E. (2012). Inferring patterns of internal migration from mobile phone call records: evidence from Rwanda. *Information Technology for Development* 18 (2): 107–125. https://doi.org/10.1080/0268110 2.2011.643209.

Burkhauser, R.V. and Lillard, D.R. (2005). The contribution and potential of data harmonization for cross-national comparative research. *Journal of Comparative Policy Analysis* 7 (4): 313–330. https://doi.org/10.1080/13876980500319436.

Chang, R.M., Kauffman, R.J., and Kwon, Y. (2014). Understanding the paradigm shift to computational social science in the presence of big data. *Decision Support Systems* 63: 67–80. http://dx.doi.org/10.1016/j.dss.2013.08.008.

Couper, M.P. (2013). Is the sky falling? New technology, changing media, and the future of surveys. *Survey Research Methods* 7 (3): 145–156. https://doi.org/10.18148/srm/2013.v7i3.5751.

Dekker, R., Smith, E., Veršić, I. I., et al. (2019). CESSDA Annual Report 2018 (Final). Zenodo. https://doi.org/10.5281/ZENODO.3492085.

Dubrow, J.K. and Tomescu-Dubrow, I. (2016). The rise of cross-national survey data harmonization in the social sciences: emergence of an interdisciplinary methodological field. *Quality and Quantity* 50 (4): 1449–1467. https://doi.org/10.1007/s11135-015-0215-z.

European Union (2020). Proposal for a regulation of the European parliament and of the Council on European data governance (Data Governance Act) Electronic copy. https://eur-lex.europa.eu/legal-content/EN/TXT/?uri=CELEX%3A52020PC0767 (accessed 11 May 2021).

Evans, P.C. and Basole, R.C. (2016). Revealing the API ecosystem and enterprise strategy via visual analytics. *Communications of the ACM* 59 (2): 26–28. https://doi.org/10.1145/2856447.

Fortier, I., Parminder, R., Van den Heuvel, E.R. et al. (2017). Maelstrom research guidelines for rigorous retrospective data harmonization. *International Journal of Epidemiology* 46 (1): 103–105. https://doi.org/10.1093/ije/dyw075.

Franchet, Y. (1991). International comparability of statistics: background, harmonization principles and present issues. *Journal of the Royal Statistical Society: Series A (Statistics in Society)* 154 (1): 19–22. https://www.jstor.org/stable/2982691.

Génois, M., Zens, M., Lechner, C. et al. (2019). Building connections: How scientists meet each other during a conference. arXiv preprint arXiv:1901.01182.

Granda, P., Wolf, C., and Hadorn, R. (2010). Harmonizing survey data. In: *Survey Methods in Multinational, Multiregional, and Multicultural Contexts* (ed. J.A. Harkness, M. Braun, B. Edwards, et al.), 315–332. Wiley https://doi.org/10.1002/9780470609927.ch17.

Groves, R.M. (2011). Three eras of survey research. *Public Opinion Quarterly* 75 (5): 861–871. https://doi .org/10.1093/poq/nfr057.

Groves, R.M. and Lyberg, L. (2010). Total survey error: past, present, and future. *Public Opinion Quarterly* 74 (5): 849–879. https://doi.org/10.1093/poq/nfq065.

Gulbrandsen, M. (2005). Tensions in the research council-research community relationship. *Science and Public Policy* 32 (3): 199–209. https://doi.org/10.3152/147154305781779524.

Haensch, A.-C. and Weiß, B. (2020). Better together? Regression analysis of complex survey data after ex-post harmonization [preprint]. *SocArXiv* https://doi.org/10.31235/osf.io/edm3v.

Halevy, A.Y., Ashish, N., Bitton, D. et al. (2005). Enterprise information integration: successes, challenges and controversies. In: *Proceedings of the 2005 ACM SIGMOD international conference on Management of data*, 778–787. https://doi.org/10.1145/1066157.1066246.

Harkness, J.A., Dorer, B., and Mohler, P. (2016). *Shared Language Harmonization. Guidelines for Best Practice in Cross-Cultural Surveys*. Ann Arbor, MI: Survey Research Center, Institute for Social Research, University of Michigan https://ccsg.isr.umich.edu/chapters/translation/harmonization (accessed 24 April 2021).

Japec, L. and Lyberg, L. (2020). Big data initiatives in official statistics. *Big Data Meets Survey Science: A Collection of Innovative Methods* 273–302. https://doi.org/10.1002/9781118976357.ch9.

Johnson, T.P. and Smith, T.W. (2017). Big data and survey research: Supplement or Substitute? In: Thakuriah, P., Tilahun, N., Zellner, M. (eds)., *Seeing Cities through Big Data*, (ed. P. Thakuriah, N., Tilahun, M. Zellner) 113–125. Cham: Springer https://doi.org/10.1007/978-3-319-40902-3_7.

Joseph, A.W. (1973). Harmonizing statistical data. *Journal of the Royal Statistical Society. Series A (General)* 136 (2): 248–254. https://doi.org/10.2307/2345111.

Jungherr, A., Jürgens, P., and Schoen, H. (2012). Why the pirate party won the German election of 2009 or the trouble with predictions: a response to Tumasjan, A., Sprenger, T. O., Sander, P. G., & Welpe, I. M. "Predicting Elections With Twitter: What 140 Characters Reveal About Political Sentiment". *Social Science Computer Review* 30 (2): 229–234. https://doi.org/10.1177/0894439311404119.

Kaczmirek, L. (2009). *Human-Survey Interaction. Usability and Nonresponse in Online Surveys*. Cologne, Germany: Herbert von Halem Verlag.

Kreuter, F. (ed.) (2013). *Improving Surveys with Paradata*. Hoboken, NJ: Wiley.

Lazer, D., Pentland, A., Adamic, L. et al. (2009). *Computational Social ScienceScience* 323 (5915): 721–723. https://doi.org/10.1126/science.1167742.

Lietz, H., Wagner, C., Bleier, A., and Strohmaier, M. (2014). When politicians talk: assessing online conversational practices of political parties on Twitter. In: *Proceedings of the International AAAI Conference on Web and Social Media*, vol. 8. (1). https://ojs.aaai.org/index.php/ICWSM/article/view/14521.

Link, G., Lumbard, K., Germonprez, M. et al. (2017). Contemporary issues of open data in information systems research: considerations and recommendations. *Communications of the Association for Information Systems* 41 (1): 587–610. https://doi.org/10.17705/1CAIS.04125.

Lyberg, L. and Stukel, D.M. (2010). Quality assurance and quality control in cross-national comparative studies. In: *Survey Methods in Multinational, Multiregional, and Multicultural Contexts* (ed. J.A. Harkness, M. Braun, B. Edwards, et al.), 227–249. Hoboken, NJ: Wiley https://doi .org/10.1002/9780470609927.ch13.

Lynn, P., Japec, L., and Lyberg, L. (2006). What's so special about cross-national surveys? In: *Conducting Cross-National and Cross-Cultural Surveys: Papers from the 2005 Meeting of the International Workshop on Comparative Survey Design and Implementation (CSDI)*, Madrid: 7–20.

Maelstrom Research (2020). *Harmonization Platforms. Maelstrom Research* https://www.maelstrom-research .org/page/harmonization-platforms (accessed 24 April 2021).

National Institute on Drug Abuse (1975). *Operational Definitions in Socio-Behavioral Drug Use Research*, Research Monograph Series 2. Rockville, MD: National Institute on Drug Abuse.

Ortmanns, V. (2020). Explaining inconsistencies in the education distributions of ten Cross-National Surveys – the role of methodological survey characteristics. *Journal of Official Statistics* 36 (2): 379–409. https://doi.org/10.2478/jos-2020-0020.

Paolacci, G. and Chandler, J. (2014). Inside the Turk: understanding mechanical Turk as a participant pool. *Current Directions in Psychological Science* 23 (3): 184–188. https://doi.org/10.1177/0963721414531598.

Posch, L., Bleier, A., Lechner, C.M. et al. (2019). Measuring motivations of Crowdworkers: the multidimensional Crowdworker motivation scale. *ACM Transactions on Social Computing* 2 (2): 1–34. https://doi.org/10.1145/3335081.

Posch, L., Bleier, A., Flöck, F., Lechner, C. M., Kinder-Kurlanda, K., Helic, D., and Strohmaier, M. (2020). Characterizing the global crowd workforce: a cross-country comparison of crowdworker demographics. *Human Computation* 9(1), 22–57. https://doi.org/10.15346/hc.v9i1.106.

Puschmann, C. and Burgess, J. (2013). The politics of Twitter data. Electronic copy http://ssrn.com/ abstract=2206225 (accessed 11 May 2021).

Savage, M. and Burrows, R. (2007). The coming crisis of empirical sociology. *Sociology* 41 (5): 885–899. https://doi.org/10.1177/0038038507080443.

Schneble, C.O., Elger, B.S., and Shaw, D. (2018). The Cambridge Analytica affair and internet-mediated research. *EMBO Reports* 19 (8): e46579. https://doi.org/10.15252/embr.201846579.

Schneider, S.L. (2013). The International Standard Classification of Education 2011. In: *Class and Stratification Analysis*, (Comparative Social Research), vol. 30 (ed. G.E. Birkelund), 365–379. Bingley: Emerald Group Publishing Limited https://doi.org/10.1108/S0195-6310(2013)0000030018.

Scholarly Publishing and Academic Resources Coalition (SPARC) (2015). Summary of Open Data Policy Harmonization Workshop. https://sparcopen.org/policy-harmonization-statement (accessed 24 April 2021).

Singh, R.K. (2020). GESIS: Harmonizing substantive instruments. https://www.gesis.org/en/services/ data-analysis/data-harmonization/harmonizing-substantive-instruments (accessed 24 April 2021).

Slomczynski, K., Tomescu-Dubrow, I., Jenkins, J. et al. (2016). Democratic Values and Protest Behavior. Harmonization of Data from International Survey Projects.

Smock, P., Granda, P., and Hoelter, L. (2015). *Integrated Fertility Survey Series, Release 7, 1955–2002 (United States)*. Ann Arbor, MI: Inter-University Consortium for Political and Social Research (distributor) https:// doi.org/10.3886/ICPSR26344.v7.

Stedman, R.C., Connelly, N.A., Heberlein, T.A. et al. (2019). The end of the (research) world as we know it? Understanding and coping with declining response rates to mail surveys. *Society & Natural Resources* 32 (10): 1139–1154. https://doi.org/10.1080/08941920.2019.1587127.

Stier, S., Bleier, A., Lietz, H., and Strohmaier, M. (2018). Election campaigning on social media: politicians, audiences, and the mediation of political communication on Facebook and Twitter. *Political Communication* 35 (1): 50–74. https://doi.org/10.1080/10584609.2017.1334728.

Stier, S., Breuer, J., Siegers, P., and Thorson, K. (2020). Integrating survey data and digital trace data: key issues in developing an emerging field. *Social Science Computer Review* 38 (5): 503–516. https://doi .org/10.1177/0894439319843669.

Tay, J.S.W. and Parker, R.H. (1990). Measuring international harmonization and standardization. *Abacus* 26: 71–88. https://doi.org/10.1111/j.1467-6281.1990.tb00233.x.

UNESCO, OECD, Eurostat, & UNESCO Institute for Statistics (2015). *ISCED 2011 Operational Manual*. OECD.

University of Reading Research Services (2021). The Research Data Lifecycle. www.reading.ac.uk/en/ research-services/research-data-management/about-research-data-management/the-research-data-lifecycle (accessed 24 April 2021).

Vries, L.D., Fischer, M., Kasprowski, D. et al. (2020). LGBTQI*-Menschen am Arbeitsmarkt: Hoch gebildet und oftmals diskriminiert. *DIW Wochenbericht* https://doi.org/10.18723/DIW_WB:2020-36-1.

Welcome to the Data Documentation Initiative (DDI) (2021). https://ddialliance.org (accessed 24 April 2021).

Witte, J.C. (2009). Introduction to the special issue on web surveys. *Sociological Methods & Research* 37 (3): 283–290. https://doi.org/10.1177/0049124108328896.

Yan, T. and Tourangeau, R. (2008). Fast times and easy questions: the effects of age, experience and question complexity on web survey response times. *Applied Cognitive Psychology: The Official Journal of the Society for Applied Research in Memory and Cognition* 22 (1): 51–68. https://doi.org/10.1002/acp.1331.

Part I

Ex-ante harmonization of survey instruments and non-survey data

3

Harmonization in the World Values Survey

Kseniya Kizilova[1,2,3], Jaime Diez-Medrano[4,5], Christian Welzel[6,7,8], and Christian Haerpfer[7,9]

[1]*Department of Methods of Sociological Studies, Kharkiv National University, Kharkiv, Ukraine*
[2]*Institute for Comparative Survey Research, Vienna, Austria*
[3]*World Values Survey Association, Vienna, Austria*
[4]*JD Systems, Madrid, Spain*
[5]*World Values Survey Association, Madrid, Spain*
[6]*Political Culture Research at the Center for the Study of Democracy, Leuphana University, Lüneburg, Germany*
[7]*World Values Survey Association (WVSA), Stockholm, Sweden*
[8]*Ronald Inglehart Laboratory for Comparative Social Research at the National Research University - Higher School of Economics, Moscow & St. Petersburg, Russia*
[9]*Political Science at the University of Vienna, Austria and Director of the Institute for Comparative Survey Research, Vienna, Austria*

3.1   Introduction

The World Values Survey (WVS) is one of the world's oldest and largest survey research programs dedicated to the study of people's values and beliefs. With a time series spanning 40 years and a cross-sectional coverage of over 120 countries on the globe, the WVS's key product is its wave-after-wave updated and pooled dataset covering countries and years (Table 3.1). Against this backdrop, the WVS requires its survey data to be comparable both over time and across nations, which makes data harmonization – a set of procedures aimed at improving the comparability of different surveys – a pivotal element in the WVS research effort (Ehling 2003; Granda et al. 2010; Mohler and Johnson 2010; Wolf et al. 2016; Smith 2019).

The overall choice of the data harmonization paradigm in the WVS has not changed over time. Since its emergence back in 1981 in partnership with the European Values Study (EVS), the WVS's initial goal was to produce comparable cross-country survey data. Due to this aim, *input and ex-ante output harmonization* (hereafter, *ex-ante harmonization*) have largely prevailed as the main strategy throughout the whole WVS's history. Like some other cross-cultural comparative survey research programs discussed in this volume, the WVS follows the "ask-the-same question" design format with all WVS waves being based on a single master questionnaire with the logic of the questionnaire, the sequence of questions and answer options being carefully followed by all national teams.

The early WVS rounds have witnessed the formation of national schools of survey research in many countries. Data collected for the 1981–1984 wave thus was crucial not only as a part of a comparative

Survey Data Harmonization in the Social Sciences, First Edition.
Edited by Irina Tomescu-Dubrow, Christof Wolf, Kazimierz M. Slomczynski, and J. Craig Jenkins.

Table 3.1 Overview of WVS survey waves coverage.

Round	Years	Countries/territories	Respondents
1	1981–1983	11	14,840
2	1989–1992	21	28,998
3	1995–1998	56	77,818
4	2000–2004	41	60,041
5	2005–2008	58	83,975
6	2011–2014	60	89,565
7	2017–2022	66[a]	98,867[a]
1–7	**1981–2022**	**122**	**454,104[a]**

[a] WVS-7 still ongoing at the time of writing.

international research program, but also as an important national inquiry into the state of values, perceptions, and attitudes in a given society. This implied a leading role of the local survey teams and their principal investigators (PI) in sample design, translation, and localization of the survey tools among other efforts. Therefore, techniques of *ex-post harmonization* have been applied to some socioeconomic and demographic variables in the early waves of the WVS. With the expansion of its membership and geographic coverage, and particularly the inclusion of developing countries, the WVS was challenged with the growing concern about the data comparability. This resulted in a critical revision of the WVS survey procedures and data harmonization approach in favor of *ex-ante* harmonization (both input and output) at all stages of the survey cycle, including the choice of globally valid indicators for the core items, elaboration of centralized sampling, and data collection and data cleaning provisions.

In the most recent period of its history, the WVS witnessed a growing use of its data by international organizations and development agencies, whose researchers frequently combine public opinion research findings with macroeconomic and other statistical indicators. As these indicators are based on categories of international classifications (such as ISCED-2011 scale for education or the ISO 3166-2 classification for regions), this actualized the employment of these standards for demographic and background variables. Correspondingly, this led to the further revision and improvement of the WVS harmonization strategy.

The chapter is organized into several sections in which we will address (i) the overall approach and related harmonization methods in the WVS, (ii) the software and documentation style used, and finally (iii) the main conclusions and recommendations derived from the challenges that the WVS has faced.

3.2 Applied Harmonization Methods

Being designed as a comparative cross-cultural time-series study, the WVS employs an *ex-ante* harmonization approach, combining both *input* harmonization and *ex-ante output* harmonization (Granda et al. 2010; Dubrow and Tomescu-Dubrow 2016; Wolf et al. 2016). Exceptions, however, occurred in early waves when relevance of the classification (characteristics such as education, ethnic belonging, or income) in the particular national and cultural context was sometimes given priority over cross-country comparability. Departing from the WVS common standards for this purpose was perceived acceptable in waves 1 and 2. As a result, for example, WVS-1 (1981–1983) does not have a harmonized education variable (though some countries, like Mexico, coded it using a country-specific scale). In wave 1 dataset, the age when the respondent finished their studies was used instead as a proxy to compare levels of education across countries.

In the 1990s, with the geographical expansion of the WVS, *ex-post* harmonization of demographic and socioeconomic variables became more and more challenging. Project growth eventually facilitated the establishment of the WVS Association (WVSA) in 1994 and the program's coordination group (Executive and Scientific Committees), which turned the survey into a much more centrally coordinated effort with a more elaborated survey planning stage. More efforts have also been put into harmonization. As a result, the WVS core group made a first separation of the country-specific variables that cannot be harmonized (region, political parties, and national income scales) and a set of variables where an agreed classification can serve as a basis of input harmonization (language, education, employment status, religion, occupation, and town size). Unlike other international studies, it was decided the WVS datasets will feature only harmonized target variables without all national source variables being included as that would mean having too many variables. That standard has been maintained until today (except for education). As for sampling, the WVS set up already in wave 3 (1995–1998) a methodological working group with the task of preapproving sample designs and supervising enforcement of the sampling rules. Multistage random sampling with or without substitution was approved as a desired model at that time (as we discuss later in this chapter, alternations from this model and use of quota sampling techniques took place occasionally and rarely, but still occurs at present).

In the early 2000s, the WVSA Data archive was moved from University of Michigan (United States) to Madrid (Spain), and its first task was to undergo a retrospective harmonization of older waves while also filling the gaps in the documentation by trying to obtain the missing parts from the PIs and team members by mail. The first ex-ante harmonization rules were defined for some sociodemographic variables, which, however, were not based on standard international scales, mainly because of the difficulty to "export" them to countries outside the OECD. Instead, the WVS tried to enforce compliance with their own standards of coding. The regular exchange with the national teams and the identification of typical problems led to the development of more elaborated ex-ante output harmonization methodologies in wave 5 (2005–2008). At the same time, a joint working group of the WVS and the EVS was established to build the joint EVS/WVS aggregate in 2008. This led to the new procedures to be agreed upon for the future waves in both the WVS and the EVS, such as the need to cover rural areas in all countries representatively, tightening rules for the respondent substitution, stressing the necessity to move to pure random samples, and use of international standards for sociodemographic variables and regions.

Further development of survey research and dissemination of professional standards worldwide (guidelines of AAPOR, WAPOR, and ESRA) allowed the WVS to standardize requirements to sample designs and interview procedures. The main mode of data collection remains a face-to-face interview, which constitutes 88% of cases in WVS-7 divided between 51% computer-assisted personal interviews (CAPI) and 37% paper-and-pencil interview (PAPI). The spread of Internet and growing costs of in-person data collection have given the inspiration to employ online methods. Computer-assisted web-interview (CAWI) was first used in WVS-6 (Netherlands) and most recently in WVS-7 (solely or as a part of a mixed-mode design in Australia, Canada, Hong Kong SAR, and Netherlands). The precondition for applying online self-administered interview mode is high Internet penetration (over 90%). A few countries are conducting postal surveys when a randomly selected sample of respondents receives an invitation to complete the survey by post. All WVS surveys are required to employ the noninstitutionalized adult population 18 years and older, citizens or long-term residents of the country, as their target population. Furthermore, all WVS sample designs are required to be representative of at least 95% of the adult population, which means no major social groups or regions, with the exception of very remote, hard-to-reach areas, can be excluded. Reaching greater standardization and comparability oversampling approaches remains challenging, foremost due to financial, security, or administrative reasons.

The main subject of harmonization in the WVS remains its measurement instrument – the WVS survey questionnaire. *Input harmonization* is applied to all core and majority demographic and socioeconomic variables, including income and occupation group; *ex-ante output harmonization* methods are used for variables such as religious and ethnic belonging, education level, and spoken language. The WVS follows the "ask-the-same question" design format (Harkness et al. 2010a; Jong et al. 2019) with all WVS waves being based on a single master questionnaire with the logic of the questionnaire, the sequence of questions and answer options being carefully followed by all national survey teams. With the WVS geographical coverage reaching 120 countries spread across all continents, the WVS questionnaire design procedures aim to ensure that questions can be meaningfully asked in all countries and in all cultural contexts and languages – which is the basic requirement for the survey data comparability (Harkness et al. 2003a, 2003b; Wolf et al. 2016).

The WVS employs the so-called "parallel" approach to the master questionnaire development which, opposed to "sequential" development, addresses cross-cultural considerations at the stage of the master questionnaire design rather than at the translation stage only (Harkness et al. 2010b). This strategy helps to minimize the cultural and linguistic bias in the WVS questionnaire by involving different modes of cross-national collaboration, such as the international questionnaire drafting and review committees, regular exchanges with the national survey teams worldwide, and input from the international academic community. In every wave, the WVS international questionnaire drafting group is established from the members of the executive and scientific committees and includes representatives of all world's main cultural zones, as well as scholars working in different domains (Anthropology, Economics, Social Psychology, Sociology, Political Science, Religious studies). Every new wave questionnaire is drafted over a period of around 12–24 months, when both the critical revision of the relevance and reliability of the previously used survey measures and the pilot and assessment of the cross-cultural validity of newly designed measures via several rounds of translation, backtranslation, and qualitative analysis with the involvement of national and language experts are conducted (Schaffer and Riordan 2003; Huang and Wong 2014). Cognitive pretests (Willis 2016; Jong et al. 2019) of the new questions are conducted in a face-to-face interview mode by at least two WVS survey teams from every major cultural zone.

While the master questionnaire is drafted in English, advanced translations into several widely spoken languages are produced at the questionnaire design stage and the translation reports are used to amend and finalize the English master questionnaire. Questionnaire translation is organized as a multistep process with several cycles of translation, testing, and adjustment being made. All WVS teams are advised to prepare more than one independent translation with the subsequent comparison and mediation between those to finalize the main translation to be used in the survey fieldwork. To increase data comparability over time for repeated items, all national teams are encouraged to employ the same translation of the question, which has already been asked in the country in one or several of the previous survey rounds. This rule becomes particularly important as continuity in the national survey team composition might not be possible, and sometimes every new wave is conducted by a new team of researchers, who are tempted to prepare their own – supposedly better – version of the questionnaire translation. Maintaining the same translation, therefore, reduces the total survey error and allows to increase the overtime data comparability. Exceptions from this rule are possible in case if the older translation has been recognized as inappropriate, that is wrongly transmitting the meaning of the question or answer options.

Demographic variables (ethnic belonging, religious denomination, language spoken at home, and education level attained) that use country-specific scales are subject to *ex-ante output harmonization*. Substantial harmonization efforts in the most recent WVS survey round (2017–2022) are dedicated to the utilization of the ISCED-2011 international classification of education levels, where we can distinguish between the mappings between the country-specific and unified scales being made either before (ex-ante) or after (ex-post) the data collection phase. The first approach – i.e. mapping done prior to data collection – is used in those countries where the survey teams have chosen to develop a country-specific classification schema of completed educational levels fully compatible with the ISCED-2011 1-digit education codes and at the same time meaningful for the respondents in the current national context (see Table 3.2 for examples). Such coding schemes require extensive preparatory work and consultancy between the survey teams, the WVSA Secretariat/Data archive, and in some cases experts in the field of education. This approach was more frequent among university-based teams, countries where a corresponding ISCED mapping for 2011 (or at least 1997) has been proposed by UNESCO (2013) (see: http://

Table 3.2 Target variable and source variables on education level in WVS-7 (2017–2021).

Target variable: Respondent's education_ISCED-2011	Source variable: Respondent's education in Russia	Source variable: Respondent's education in the United States
0) No/early childhood (ISCED 0)	1) No/early childhood education	1) Early childhood education and kindergarten
1) Primary (ISCED 1)	2) Primary school	
2) Lower secondary (ISCED 2)	3) Incomplete secondary 4) Primary professional education without secondary general	2) Elementary or middle school
3) Upper secondary (ISCED 3)	5) Complete secondary school	3) High school graduation or GED 4) Some college or nondegree programs
4) Postsecondary non-tertiary (ISCED 4)	6) Primary professional education – with secondary general	5) Vocational or certificate programs
5) Short-cycle tertiary (ISCED 5)	7) Secondary professional education	6) Associate degree
6) Bachelor or equivalent (ISCED 6)	8) Bachelor's degree	7) Baccalaureate degree
7) Master or equivalent (ISCED 7)	9) Master's degree 10) Higher by single-tier system	8) Master's degree
8) Doctoral or equivalent (ISCED 8)	11) Academic degree	9) Doctorate or professional degree

uis.unesco.org/en/isced-mappings), and countries where data collection was fully or partly covered from the WVS central budget and thus greater compliance was achieved.

The second approach features the country-specific measure of education levels used at the fieldwork stage, with the subsequent re-coding of national education codes into the ISCED-2011 unified coding scheme. This approach has been used primarily in surveys conducted using self-administered data collection modes (e.g. postal surveys in Japan and New Zealand), but has been recognized as less efficient as post-coding was likely to contain a greater number of inconsistencies between the country-specific and ISCED-2011 education codes. This approach was more widespread among survey teams with less staff and those who had no capacity or knowledge to introduce the mapping. Consequently, beyond the ISCED-2011 harmonized education variable, the WVS has developed another one with the differentiation between the three basic education levels (functional equivalents of primary, secondary, and tertiary). Mapping for the three-categories' education variable is applicable to all countries surveyed in WVS-7 as well as being used to link the new and old WVS education variables (see Table 3.3). This three-categories' variable is also included in the next version of the EVS/WVS 2017 aggregate.

In the latest round, the WVS has introduced a new common target variable for the respondents' religious belonging, which allowed to create a more detailed classification of religions as opposed to the seven-items list of major denominations employed in the past. The problem with the previous scale was twofold. First, Christianity was represented through three categories (Catholicism, Protestantism, and Orthodoxy), which was useful in Europe, but proven complicated in other countries where the share of

Table 3.3 Education variables in WVS (1990–2021).

Target variable: Education 3 groups	Source variable (2017–2021): Education ISCED-2011	Source variable (1990–2011): Education WVS
1 Lower	0 No/early childhood (ISCED 0)	1 No formal education
	1 Primary (ISCED 1)	2 Incomplete primary school
	2 Lower secondary (ISCED 2)	3 Complete primary school
		4 Incomplete secondary: technical/vocational
		5 Incomplete secondary: university-preparatory
2 Medium	3 Upper secondary (ISCED 3)	6 Complete secondary: university-preparatory
	4 Postsecondary non-tertiary (ISCED 4)	7 Complete secondary: technical/vocational
3 Higher	5 Short-cycle tertiary (ISCED 5)	8 Some university-level education, no degree
	6 Bachelor or equivalent (ISCED 6)	9 University – level education, with degree
	7 Master or equivalent (ISCED 7)	
	8 Doctoral or equivalent (ISCED 8)	

Christians was very small, or the respondents belonged to specific religious groups, not always knowing which bunch this corresponds to. For example, in wave 7 in Australia, where self-administered interview mode was used, 17.3% of the respondents as their religious affiliation indicated just "Christianity." In Ukraine, the opposite, the respondents provide more detailed answers specifying that they belong to "Byzantine Rite" or "Evangelical Baptist Union of Ukraine." At the same time, other world major religions were represented with just one category each, with no specification of brunches or subgroups.

Recognizing the importance of the data on religious belonging and upon reflection on the issues stated above, in the seventh survey round, the WVS introduced a new common target variable for the respondents' religious belonging, which employs a hierarchical and multidigit coding approach based on the "tree of religions" (Table 3.4). WVS "tree of religions" comprises 1492 categories at four levels. The first digit signifies the broadest category of the unified coding scheme, with subsequent digits indicating

Table 3.4 Fragment of the hierarchical coding scheme used for the target variable "Religion_detailed list."

Level 1	Level 2	Level 3	Level 4
Catholic, nfd (10000000)	Roman Catholic; Latin Church (10100000)	–	–
	Eastern Catholic Churches nfd (10200000)	Alexandrian Rite (10201000)	Coptic Catholic Church (10201010) Eritrean Catholic Church (10201020) Ethiopian Catholic Church (10201030)
		Armenian Rite (10202000)	Armenian Catholic Church (10202010)
		Byzantine Rite (10203000)	Albanian Greek Catholic Church (10203010) Belarusian Greek Catholic Church (10203020) Bulgarian Greek Catholic Church (10203030) [. . .] Slovak Greek Catholic Church (10203130) Ukrainian Greek Catholic Church (10203140)
		East Syriac Rite (10204000)	Chaldean Catholic Church (10204010) Syro-Malabar Catholic Church (10204020)
		West Syriac Rite (10205000)	Maronite Church (10205010) Syriac Catholic Church (10205020) Syro-Malankara Catholic Church (10205030)

increasingly detailed subgroups. Levels 1, 2, and 3 are comprehensive in covering all the world's main religions and their sub-denominations. Exhaustion of the categories list at level 4 will be constantly challenged as new religious groups and churches continue to emerge. Therefore, depending on the coding level selected by the respondent or the national survey team, every respondent receives a relevant code from levels 1, 2, 3, or 4. The advantage of this approach is that it both largely preserves the original, detailed, and country-specific information and yet enables comparisons across countries or across religious groups, since same and overlapping religions obtain identical codes within different national datasets.

3.3   Documentation and Quality Assurance

Maintaining a comprehensive and detailed documentation is an important part of data harmonization in any cross-cultural project (Granda and Blasczyk 2016; Behr et al. 2019), but especially in a time-series study. However, early WVS waves surveyed in the 1980s featured quite a few exceptions for a number of reasons. From a technological point of view, it was almost out of question to transmit data electronically and to have regular exchanges, with the local teams in charge of data archiving in their countries. Both data on an electronic diskette and printed survey documentation were often submitted by post. Interaction, if any, was made by postal letters, sometimes phone calls and fax. This resulted in the originally surveyed (localized and translated) questionnaires for several surveys from 1981 to 1990 being still absent in the WVS documentation collection. Furthermore, production of the wave aggregates was done in a less elaborate way, in research departments at Michigan University, a pioneer in this task, as the survey data archives were only starting to emerge at those times. Harmonization was mainly based on strict compliance of the source variables in the national datasets with the target variable. Divergences in coding could be left without attention or simply excluded if no codebook was provided alongside the data file and thus no means existed to apply recoding. Cases, when data archiving documentation was incomplete, made it impossible to estimate later the validity of some harmonization decisions made.

To improve the quality of the WVS data and to ease coordination of the harmonization efforts, presurvey validation was introduced in WVS wave 5 (2005–2008). At present, the WVS integrated dataset includes only studies for which the questionnaire translation and the survey design have been verified and validated by the WVSA prior to the data collection. Validation is run by the Scientific Committee, the Secretariat, and the Data archive and includes checks of the questionnaire translation, compatibility between the proposed country-specific scales and agreed target variables, and verification of the sampling model. Decisions on the well-reasoned requests to omit one or several questions from the core questionnaire are made by the Executive Committee. While minor alterations of the national questionnaires from the core questionnaire logic sometimes occur, the input harmonization of the questionnaire – that is, asking the same questionnaire items across all countries in order to maximize standardization and comparability – is emphasized as the basic methodological foundation of the WVS project.

The assessment and approval of the sampling approach are made by the Scientific Committee, foremost against compliance with the WVS basic requirements regarding the target population, coverage, and random sample design. While in terms of sampling the WVS relies heavily on random probability designs and generally no quota or convenience sampling is accepted, divergencies sometimes occur. For example, in wave 7 in Venezuela, following the random selection of survey locations, quota sampling was employed at the last stage of selection (selection of the respondent in the household) as due to the difficult security situation the interviewers had no possibility to do the regular random selection with several revisits. Similar to this, in wave 7 in Mongolia because of the very narrow fieldwork window due to the covid-pandemic situation, quota sampling was used for the selection of the respondent in the household upon the random choice of the primary sampling units. Control quotas (primarily by gender) are sometimes employed in the Middle East (e.g. Morocco in wave 7), where females are often less keen to participate in surveys. To summarize, about 10% of surveys in wave 7 included elements of nonrandom selection of the respondent. Other alterations from the WVS standard methodology might include limiting the target population to nationals only – in case the existing national laws forbid surveying noncitizens. While all national survey companies involved in the data collection are obliged to follow the laws, discrepancies in the target population definition certainly affect the findings, especially in countries with high shares of foreign workers or migrants, who hold a different citizenship (e.g. Qatar in wave 6). The cross-national WVS datasets, therefore, include both samples of residents and citizens, and a variable inquiring about the respondent's citizenship status can be included in the analysis by the data-users if necessary. Adjustments to the lower age limit can be made if the voting age commences earlier in a country or if there is a long-standing research tradition for surveying individuals (usually) in the age 16 and over (e.g. Nicaragua in wave 7 and South Africa in wave 6). WVS methodology forbids introducing upper age limits in the sampling plans. Given the great length and complexity of the WVS questionnaire, however, rare exceptions are sometimes made (e.g. Iran and Vietnam in wave 7, where due to particularly fragile condition of the elderly respondents' upper age limit was set up to 65 years). Any alteration from the overall WVS methodology must be described in the survey plan alongside with an extensive justification; it is subject to approval by the WVS Scientific Committee prior to the survey. For example, introducing an upper age limit in the sample would be possible only if, apart from an extensive justification, the share of the national population in the age 65+ does not exceed 5%, to comply with the WVS's overall requirement for the sample to be representative of a minimum of 95% of the population. Generally, alterations from the WVS sampling requirements are relatively rare: they are applicable to 10–15% of surveys and as a rule are justified by financial or administrative, security, and political reasons, which make it impossible to conduct a survey unless an amendment is accepted.

3.4 Challenges to Harmonization

Challenges to harmonization occur at all stages of the survey cycle, where *ex-ante* harmonization input is introduced in the WVS. While the WVS surveys mostly employ various modes of multistage stratified random sample and rely on the common definition of the target population, input harmonization of the survey procedures remains quite limited. The preferred data collection mode in the WVS remains the face-to-face interview; the vast majority of data in the WVS across all waves has been collected in a face-to-face mode. Growing competition in the field of survey research, not lastly over-limited funds, declining response rates, and most recently the outbreak of the corona pandemic, have forced some WVS teams to employ alternative data collection methods. Postal surveys have been conducted in the WVS in the past waves in countries where high costs of data collection (Japan) or population dispersed across a great territory (Canada and Australia) made the application of face-to-face interviews challenging. Use of online survey (solely or in a mixed-mode design) in the WVS must be justified by a high Internet penetration rate to meet the basic requirement of the sample representativity (for example, online survey proposal for Uruguay in WVS-7 was rejected by the WVS). Data from the few conducted online surveys in the WVS still shows an overrepresentation of social groups such as those with tertiary education and urban residents; for such cases, weighting was applied. Furthermore, comparability of interviewer-administered and self-administered surveys (those conducted without the interviewer where the respondent completes an online or a printed questionnaire) is challenging due to the different traditions of treating nonresponse options ("don't know," "refuse to answer," and similar).

As in other surveys, WVS interviewers are instructed not to read nonresponse options and code those only if volunteered by the respondent; this model does not work in a self-administered printed questionnaire design, where nonresponse options are to be either printed and visible to the respondent or not. Given the very limited number of self-administered surveys in the WVS, country-specific decisions based on the expert advice of the local team and justified by the national cultural peculiarities have been accepted in WVS-7. For example, the team in Australia opted for not printing the nonresponse options, explaining that doing otherwise might increase the temptation of the respondent to select this option every time a question requires a bit of thinking. In contrast, the team in Japan suggested that not having nonresponse options might make the respondent feel that they are forced into an answer and thus a possibility to evade an answer must be explicit. Both approaches and respective experiences will be analyzed to elaborate a common preferred methodology for the next WVS wave in 2023–2026 when we anticipate a higher number of self-administered surveys.

To further improve data comparability and in line with the "Total Survey Error" perspective (Lyberg and Weisberg 2016; Smith 2019), in the recent (2018–2022) in wave 7, the WVS has pursued some input harmonization of sampling and survey methodology by requiring all national surveys to follow common basic requirements on the sample design and interview procedures. This included such provisions as nonacceptance of non-probability samples, standardization of requirements to the sample size, cluster size limit, respondent selection procedures, number of revisits, elaboration of a common set of instructions, and training materials for the interviewers. Introducing the same, identical data collection protocols in all WVS surveys so far have not been possible, due to the different situations with the availability of sampling frames, existing national standards and practices in data collection, dissimilarities in the security situation, and response rates across countries among the many reasons.

A central challenge to harmonization in multicultural survey projects is the issue of concept equivalence. Therefore, major input harmonization in the WVS is made at the stage of international master questionnaire development, where the aim is to ensure that questions can be meaningfully asked in all countries and in all cultural contexts. The few shortcomings in this respect have been linked to somewhat limited input of scholars from Africa, South Asia, and the Middle East into the questionnaire design process. This resulted in some items being later identified as culturally biased. For example, the WVS-7 questionnaire is asking to estimate the scale of corruption in the country, additionally specifying that corruption occurs *"when people pay a bribe, give a gift or do a favor to other people in order to get the things they need done or the services they need."* While the definition seems to be appropriate and legally relevant for many countries, teams in a few countries pointed out that giving a gift or doing a favor in exchange for something is an inherent part of their culture and is not perceived as corruption as such. And thus, referring to the practices of gift-giving in the context of the question on corruption perceptions might lead to misinterpretation of the measured concept and subsequently data incomparability with other countries. Based on the experience with this and similar cases, the WVS plans for the greater involvement of scholars from the previously underrepresented regions and further intensification of communication and exchange at the stage of questionnaire design in the next survey round in 2023–2026.

Ex-ante output harmonization methods play a crucial role in maximizing the standardization and comparability of variables such as education level, religious and ethnic belonging, or languages spoken at home. Yet, using existing standard classifications brings also challenges, as the ISCED-2011 shows. Thanks to its international, global coverage and the availability of mapping between ISCED-2011 levels and national equivalents, this standard classification of education is of great use in WVS. At the same time, while UNESCO made available a set of Excel spreadsheets with the local adaptations of the ISCED-2011 education scale (ISCED national mappings), the experience of WVS, similar to ESS (Schneider 2016), has shown that survey teams, who have no substantial past experience in education research faced significant difficulties with using the proposed adaptations. In addition, particularly challenging implementation of the ISCED-2011 scale for education has been in those territories, where the mapping has not been yet identified by UNESCO (in WVS wave 7: Mongolia, Myanmar, Guatemala, Nicaragua, and Iran).

3.5 Software Tools

Data harmonization in the WVS is powered by its own software "JDSurvey" developed at the WVS Data archive in 2004 and is being used and constantly improved since then. Datasets and available documentation for all previous WVS rounds have also been uploaded into this system. JDSurvey system is based on the Data Documentation Initiative (DDI) (2014) standards for codebook (see: https://ddialliance.org/Specification/DDI-Codebook/2.5). The architecture of the JDSurvey can be described as a set of coordinated databases combining the survey datasets for all countries and all waves into one system, with the integrated textual fields that can be used for remarks, questions wording and translations, codes, and labels as well as command line to run the harmonization syntax. Additional elements include thesaurus dictionaries, thematic indexes, geographical scope structure, and country-specific variables dictionaries. The software has been developed with the overall aim to automate all procedures (where technical assistance is possible and meaningful) at all stages of the survey cycle.

The JDSurvey software is first used at the beginning of each new wave when the new master questionnaire is uploaded into the software database. The system splits the questionnaire into variables with relevant codes and labels, which thus generates the structural matrix for the subsequent data file. The system checks for an overlap between the new items and the earlier versions of the WVS questionnaires used in the previous waves, identifying repetitive (time series) and unique questions. While every WVS wave dataset features its own sequence and numbering of the variables, the system allocates identical variable names to items repeated across several WVS waves, which is later used to automatically produce the time-series dataset.

The JDSurvey is also useful at the first stage of questionnaire translation validation. While the automatic checks, run by the software, do not substitute the assessment and feedback by the local language experts, JDSurvey's basic validation procedures include checking the number of questions and answer options and comparing them to the master English version; identifying and recording any additional questions introduced by the survey teams and thus not included into the master questionnaire (such variables are released later as a part of a national dataset). JDSurvey also includes a function for comparative overview of the master English questionnaire and an automated English backtranslation of the national questionnaires (backtranslation first needs to be run independently). The system marks any potential items where discrepancies in the meaning of the question might be present, and which are subsequently checked by language experts.

All national WVS survey datasets upon their submission are uploaded into the JDSurvey. The software runs algorithms to perform the data quality checks. The technical aspects of data cleaning and data harmonization are also run automatically by JDSurvey, though person-made decisions are required on the exact recodings to be made and steps to follow. Development of new harmonization procedures is usually done by the WVSA Data archive and Secretariat, with the expert advice of the national teams if required. The most typical use of JDSurvey is automatic detection of variables featuring "irrelevant" labels and codes – those which are not included in the structure of the target variables. The program allows both assigning correct codes and storing the original codes. Thus, original data is never lost, and the recoding or harmonization decision can be revisited any time. Computing rules are developed in such a way that applying syntax to a variable that participates in another variable's syntax results in cascade recalculation. For example, if mapping between country-specific and ISCED-2011 education scales has proven to have an error, once the correction is introduced, the three-category education variable is also updated automatically. The second group of quality checks includes the filter verification procedures, noncomplying cases are signaled, and the variable can be adjusted optionally using recodes or syntax to match the filter rule.

Once the syntax with the harmonization decision is written, a note on the decisions made and the applied mapping is added into the particular variable for the particular national dataset (by mapping here, we mean a correspondence schema between the source and target variables). Such notes are stored and can be accessed any time in the future to check or amend the decision. The program documents every cleaning and harmonization step being undertaken for each deposited national dataset. JDSurvey is also used to store the national survey technical documentation as well as to produce textual (pdf) and tables (excel) outputs for different stages of the survey cycle. The software has built-in functions for checking the consistency, completeness, and comparability of the harmonization procedures and outputs. Additional functions can be added to JDSurvey once a code is written by the WVS Data archive programmers.

3.6 Recommendations

Based on four decades (1981–2021) of cross-national survey experience, the WVS strongly supports input harmonization as the main approach in producing the country-and-time pooled datasets. This approach presumes that survey planning and coordination is organized as a complex multistage process that foremost requires all project members to carefully follow the agreements and decisions made by the survey design team with respect to sampling, survey mode, interview procedures, and the questionnaire logic. Decisions on the harmonization of the survey protocols and sample procedures should be made with the best practices in mind, as promoted in the survey research literature and the codes of the international professional associations (such as WAPOR, AAPOR, and ESRA). At the same time, elaboration of the project's survey methodology must be pursued in close collaboration with all national teams to ascertain that the final accepted survey design complies with the national laws and other legal provisions and is appropriate to all cultural settings, also given the country- or region-specific political, economic, or security conditions. General survey provisions in a cross-national survey effort must be supplemented with a detailed guidance on their implementation, scope of acceptable alterations, and situations or conditions when alterations are justifiable. The WVSA's past experience suggests that many of the survey provisions are in fact far from being self-evident, and information and training sessions to clarify the agreed common standards are highly desirable.

Given that input harmonization of sampling designs remains a challenge, especially on a global scale, where the data collection spans across societies that differ dramatically in the level of income, availability of statistical and census data, and security and transportation conditions, the measurement tool – the questionnaire – often becomes the main subject of input harmonization. While following the "ask-the-same-question" rule allows reaching a high level of standardization, another important aspect is the choice of cross-culturally applicable indicators, which have equal validity for the measurement of the studied concept. Both the choice of indicators and the development of unambiguous question wordings require establishing an international questionnaire design team with the equal participation of researchers from all major studied regions. Further steps such as regular parallel exchanges with all participating national survey teams and seeking advice of the international academic community allow addressing the cross-cultural considerations to minimize cultural and linguistic bias at the stage of the master questionnaire design when changes are still feasible. In a time-series project, like the WVS, to ascertain the over-time comparability of the data, maintaining the wording of questions and answer options is essential, provided these measures retain their validity for the measurement of the studied concept. In large-scale surveys, which involve many countries, data harmonization can benefit from detailed guidance and descriptions of survey protocols and from requirements as to the translation and interview procedures to be developed by the coordination team.

The role of international classifications for data comparability and standardization is crucial. Introduction of the internationally established official classifications into the survey datasets has proven to increase their utility for both academic research, developmental analysis, and policymaking. Yet utilization of these classifications requires substantial mapping effort to ensure that equivalences between the country-specific categories and the international coding scheme have been unequivocally established and validated. Often the development of such mapping requires deep knowledge of the concept in question. And here, further input from the international organizations who authored the respective classifications would be highly welcome. Depending on the available budget and survey team capacity, international codes can be applied both during the data collection phase and at the post-fieldwork stage. Regardless of the selected approach and to avoid coding errors, training sessions on using international classifications should ideally involve national team members participating at all stages of the survey cycle.

The WVS coordination team is constantly working on the improvement of all survey aspects to ensure greater data comparability. In this, the WVS relies both on the quality control measures run by the Data archive and review of the analysis published by the WVS data-users. With the WVS dataset being an open resource, the program became one of the most widely used public opinion data sources (for instance, Google Scholar indicates more than 60,000 citations of the WVS). This gives an outstanding opportunity to learn about the strong and weak points of many indicators included in the WVS dataset as well as other survey-related aspects – an analysis that the WVS group could never conduct on their own at such a comprehensive, pervasive, and all-encompassing level. In particular, the issues of measurement equivalence (or invariance) in cross-national research are frequently discussed drawing on WVS data, including such indicators as democracy–autocracy preference and democratic performance evaluation scales (Ariely and Davidov 2011); trust (Freitag and Bauer 2013; Breustedt 2018); the post-materialism scale (Davis and Davenport 1999; Ippel et al. 2014, among others); traditional-secular-rational and survival-self-expression values (Alemán and Woods 2016; Sokolov 2018); gender role attitudes (Constantin and Voicu 2015; Lomazzi 2018); and emancipative and secular values indexes (Alemán and Woods 2016; Sokolov 2018). These and similar studies, while reaching both positive and negative conclusions about the measurement invariance in the WVS, rely on multigroup confirmatory factor analysis (MGCFA) as a tool of measurement validation that tests psychological constructs for group-to-group invariance in the respective items' interconnections and concludes that cross-cultural dissimilarity in a construct's inter-item connections is an infallible sign of encultured differences in the semantic understanding of the items.

Yet the most recent analyses conducted by WVS scholars prove that claims of cross-cultural incomparability of the WVS indicators are inconclusive and that it is safe to use such WVS measures in cross-cultural comparison (Welzel et al. 2021). Using the index of the emancipative value as an example, the authors demonstrate that the arithmetic of closed-ended scales in the presence of sample mean disparity is what becomes the main source of non-invariance. And since arithmetic principles are culture-unspecific, the non-invariance that these principles enforce in statistical terms is inconclusive of encultured in-equivalences in semantic terms. The applied invariance testing methodology, therefore, operates only within a restricted sphere of "reflective" constructs, which are based on the "similarity" criterion of index formation, while "formative" constructs tailored to the "complementarity" criterion are "out of its judgmental authority" (Welzel et al. 2021). Due to the nomological notion of equivalence, a construct's external linkages should be applied instead as a validity benchmark: numerically similar averages of the construct are equivalent in a substantive sense if they map in significantly corresponding fashion on the construct's supposed antecedents, outcomes, or correlates (Welzel and Inglehart 2016). The data-users are recommended to employ single confirmatory factor analysis (CFA) over the pooled data, using a two-level design to examine correspondence in dimensionality simultaneously (Duelmer et al. 2015). At the same time, following a mono-dimensional approach in index construction should not be slavishly followed as an ironclad law: constructs do not have to be mono-dimensional in order to be meaningful theoretically or to be consequential empirically. Multidimensional constructs with mutually complementary components created as a result of conceptually thoughtful work can form a meaningful higher-ordered construct with important consequences (Welzel et al. 2021).

References

Alemán, J. and Woods, D. (2016). Value orientations from the world values survey. *Comparative Political Studies* 49 (8): 1039–1067.

Ariely, G. and Davidov, E. (2011). Can we rate public support for democracy in a comparable way? Cross-national equivalence of democratic attitudes in the world value survey. *Social Indicators Research* 104 (2): 271–286.

Behr, D., Dept, S., and Krajčeva, E. (2019). Documenting the survey translation and monitoring process. In: *Advances in Comparative Survey Methods. Multinational, Multiregional, and Multicultural Contexts (3MC)* (ed. T. Johnson, B.-E. Pennell, I. Stoop, and B. Dorer), 341–355. Hoboken, NJ: Wiley.

Breustedt, W. (2018). Testing the measurement invariance of political trust across the globe. A multiple group confirmatory factor. Analysis. *Methoden, Daten, Analysen* 12 (1): 39.

Constantin, A. and Voicu, M. (2015). Attitudes towards gender roles in cross-cultural surveys: content validity and cross-cultural measurement invariance. *Social Indicators Research* 123 (3): 733–751.

Data Documentation Initiative (2014). www.ddialliance.org

Davis, D.W. and Davenport, C. (1999). Assessing the validity of post-materialism index. *American Political Science Review* 93 (3): 649–664.

Dubrow, J. and Tomescu-Dubrow, I. (2016). The rise of cross-national survey data harmonization in the social sciences: emergence of an interdisciplinary methodological field. *Quality & Quantity* 50 (4): 1449–1467.

Duelmer, H., Inglehart, R., and Welzel, C. (2015). Testing the revised theory of modernization: measurement and explanatory aspects. *World Values Research* 8: 68–100.

Ehling, M. (2003). Harmonising data in official statistics: development, procedures, and data quality. In: *Advances in Cross-National Comparison. A European Working Book for Demographic and Socio-Economic Variables* (ed. J.H.P. Hoffmeyer-Zlotnik and C. Wolf), 17–31. New York: Kluwer Academic/ Plenum Publishers.

Freitag, M. and Bauer, P.C. (2013). Testing for measurement equivalence in surveys: dimensions of social trust across cultural contexts. *Public Opinion Quarterly* 77: 24–44.

Granda, P. and Blasczyk, E. (2016). Data harmonization. In: *Guidelines for Best Practice in Cross-Cultural Surveys* (ed. T.P. Johnson, B.-E. Pennell, I.A.L. Stoop, and B. Dorer), 617–636. Ann Arbor, MI: Survey Research Center, Institute for Social Research, University of Michigan http://ccsg.isr.umich.edu/images/ PDFs/CCSG_Full_Guidelines_2016_Version.pdf.

Granda, P., Wolf, C., and Hadorn, R. (2010). Harmonizing survey data. In: *Survey Methods in Multinational, Multicultural and Multiregional Contexts* (ed. J.A. Harkness, M. Braun, B. Edwards, et al.), 315–334. Hoboken, NJ: Wiley.

Harkness, J., Mohler, P.P., and van de Vijver, F.J.R. (2003a). Comparative research. In: *Cross-Cultural Survey Methods* (ed. J. Harkness, F.J.R. van de Vijver, and P.P. Mohler), 3–16. Hoboken, NJ: Wiley.

Harkness, J., van de Vijver, F.J.R., and Johnson, T.P. (2003b). Questionnaire design in comparative research. In: *Cross-Cultural Survey Methods* (ed. J. Harkness, F.J.R. van de Vijver, and P.P. Mohler), 19–34. Hoboken, NJ: Wiley.

Harkness, J.A., Edwards, B., Hansen, S.E. et al. (2010a). Designing questionnaires for multipopulation research. In: *Survey Methods in Multicultural, Multinational, and Multiregional Contexts* (ed. J.A. Harkness, M. Braun, B. Edwards, et al.), 33–58. Hoboken, NJ: Wiley.

Harkness, J.A., Villar, A., and Edwards, B. (2010b). Translation, adaptation, and design. In: *Survey Methods in Multicultural, Multinational, and Multiregional Contexts* (ed. J.A. Harkness, M. Braun, B. Edwards, et al.), 117–140. Hoboken, NJ: Wiley.

Huang, W.Y. and Wong, S.H. (2014). Cross-cultural validation. In: *Encyclopedia of Quality of Life and Well-Being Research* (ed. A.C. Michalos). Dordrecht: Springer.

Ippel, L., Gelissen, J., and Moors, G. (2014). Investigating longitudinal and cross-cultural measurement invariance of Inglehart's short post-materialism scale. *Social Indicators Research* 115 (3): 919–932.

Jong, J., Dorer, B., Lee, L. et al. (2019). Overview of questionnaire design and testing. In: *Advances in Comparative Survey Methods. Multinational, Multiregional, and Multicultural Contexts (3MC)* (ed. T. Johnson, B.-E. Pennell, I. Stoop, and B. Dorer), 115–138. Hoboken, NJ: Wiley.

Lomazzi, V. (2018). Using alignment optimization to test the measurement invariance of gender role attitudes in 59 countries. *Methoden, Daten, Analysen* 12 (1): 77–103.

Lyberg, L. and Weisberg, H. (2016). Total survey error: a paradigm for survey methodology. In: *The SAGE Handbook of Survey Methodology* (ed. C. Wolf, C.D. Joye, T. Smith, and Y. Fu), 27–40. London: SAGE Publications.

Mohler, P. and Johnson, T. (2010). Equivalence, comparability, and methodological progress. In: *Survey Methods in Multinational, Multicultural and Multiregional Contexts* (ed. J.A. Harkness, M. Braun, B. Edwards, et al.), 17–32. Hoboken, NJ: Wiley.

Schaffer, B.S. and Riordan, C.M. (2003). Review of cross-cultural methodologies for organizational research: a best-practices approach. *Organizational Research Methods* 6: 169–215.

Schneider, S.L. (2016). The conceptualisation, measurement, and coding of education in German and Cross-National Surveys. GESIS survey guidelines. *Mannheim, Germany: GESIS – Leibniz Institute for the Social Sciences* https://doi.org/10.15465/gesis-sg_en_020.

Smith, T. (2019). Improving multinational, multiregional, and multicultural (3MC) comparability using the total survey error (TSE) paradigm. In: *Advances in Comparative Survey Methods. Multinational, Multiregional, and Multicultural Contexts (3MC)* (ed. T. Johnson, B.-E. Pennell, I. Stoop, and B. Dorer), 13–44. Hoboken, NJ: Wiley.

Sokolov, B. (2018). The index of emancipative values: measurement model misspecifications. *The American Political Science Review* 112 (2): 395–408. https://doi.org/10.1017/S0003055417000624.

UNESCO (2013). ISCED-2011 International Standard Classification of Education and Country-specific Mappings Overview. http://uis.unesco.org/en/isced-mappings

Welzel, C. and Inglehart, R. (2016). Misconceptions of measurement equivalence: time for a paradigm shift. *Comparative Political Studies* 49: 1068–1094.

Welzel, C., Brunkert, L., Kruse, S., and Inglehart, R.F. (2021). Non-invariance? An overstated problem with misconceived causes. *Sociological Methods & Research* 1–33.

Willis, G. (2016). Questionnaire pretesting. In: *The SAGE Handbook of Survey Methodology* (ed. C. Wolf, D. Joye, T. Smith, and Y. Fu), 359–381. London: SAGE Publications.

Wolf, C., Schneider, S., Behr, D., and Joye, D. (2016). Harmonizing survey questions between cultures and over time. In: *The SAGE Handbook of Survey Methodology* (ed. C. Wolf, D. Joye, T. Smith, and Y. Fu), 502–524. London: SAGE Publications.

4

Harmonization in the Afrobarometer

Carolyn Logan[1,2], Robert Mattes[3,4], and Francis Kibirige[5]

[1]*Department of Political Science, Michigan State University, East Lansing, MI, USA*
[2]*Director of Analysis, Afrobarometer, Accra, Ghana*
[3]*School of Government and Public Policy, University of Strathclyde, Glasgow, Scotland*
[4]*Institute for Democracy, Citizenship and Public Policy in Africa, University of Cape Town, Cape Town, South Africa*
[5]*Hatchile Consult Ltd, Kampala, Uganda*

4.1 Introduction

Afrobarometer (AB) is a comparative series of public opinion surveys on democracy, governance, development, quality of life, and civil society in Africa. Like other cross-national survey enterprises, it is a social science project committed to using state-of-the-art methods to produce valid and reliable measures of citizens' values, preferences, and experiences. Unlike many other projects, Afrobarometer's donors support us because they see it as a political intervention that advances democracy by providing a voice to ordinary Africans, who have too often been ignored by elites and policymakers in their own countries. Thus, while the scholarly community is an important recipient of our work, our primary target audience consists of elected officials, policymakers, and donors, with whom we communicate through a dedicated program of dissemination and direct engagement, and by releasing results to news media and via social media. More than 20 years after its inception and in the midst of its ninth round of surveys, Afrobarometer has earned a reputation as Africa's most reliable source of public opinion data and analysis, having by now interviewed more than 330,000 respondents across 39 countries.[1]

Afrobarometer has made several major contributions to the study of politics in Africa. First, we have demonstrated that collecting high-quality and reliable public attitude data is possible even in some of the most challenging physical, social, and political environments. Second, we have made Africa less exotic, demonstrating that the region can be compared to other regions across common indicators, even if it sometimes occupies a unique position. Third, we have produced evidence that has challenged many

---

1 Afrobarometer covered 12 countries in Round 1 (1999–2001), and expanded to 16 in Round 2 (2002–2003), 18 in Round 3 (2004–2006), 20 in Round 4 (2008–2010), 35 in Round 5 (2011–2013), 36 in Round 6 (2014–2015), and 34 in Round 7 (2016–2018) and Round 8 (2019–2021). The network hopes to cover 40 countries in Round 10 (2021–2022).

Survey Data Harmonization in the Social Sciences, First Edition.
Edited by Irina Tomescu-Dubrow, Christof Wolf, Kazimierz M. Slomczynski, and J. Craig Jenkins.
© 2024 John Wiley & Sons Inc. Published 2024 by John Wiley & Sons Inc.

long-standing scholarly assumptions about the values, opinions, and behaviors of ordinary Africans, almost all of which were based on anthropological observation and expressed in qualitative narratives. Given the difficulties that Africa presents to survey researchers through its geographic, environmental, economic, social–cultural, and political diversity, these achievements would not have been possible without a strong emphasis on data harmonization, especially *ex ante* harmonization achieved not just by using a harmonized questionnaire, but also by devising and applying standard methods for drawing samples, conducting interviews, and managing data. In short, harmonization has been a key pillar of Afrobarometer's success.

4.2 Core Principles

We have based the work of Afrobarometer on five fundamental principles. First, because we were conscious not to replicate colonial modes of production – i.e. employing Africans to mine raw data, while shipping it outside the continent to be analyzed, interpreted, and turned into a finished product by outsiders – we designed the project as a *research network* rather than a hierarchy. The network consists of both core partners,[2] as well as national partner organizations within each country based at universities, nongovernmental organizations, or private sector research companies. The national partners not only carry out the actual surveys but also have primary responsibility for reporting results and disseminating them within their own countries. The commitment to run Afrobarometer as a network rather than a hierarchy means that, as much as possible, we treat national partners as coinvestigators rather than contractors.

Second, in virtually everything, we do we follow a process of *context-sensitive standardization.* That is, we design standard procedures to the greatest extent possible, but we also try to anticipate situations where the survey environment differs so radically that the common rules cannot be followed in any reasonable way, and devise alternatives that will achieve the same overall purpose.

This is necessary because partners in different countries, and even different interview teams within the same country, encounter radically different religious, cultural, ethnic, linguistic, political, and economic environments in which to draw representative samples, contact respondents, and carry out interviews. In some places, fieldwork teams encounter compact and densely populated settlements, while in others they must travel vast expanses to conduct their interviews, sometimes exiting their vehicle to cross lakes and rivers by boat, or navigate mountainous terrain by foot, horseback, or mule. Some teams weave their way through informal shacks, while others must negotiate their way past security guards to access households in gated communities or apartments in wealthy high-rise apartment buildings. Our partners conduct some interviews one-on-one in private homes, but others occur in a crowded home or in the open air, in the presence of family or even neighbors. And in some countries, national statistical offices provide detailed household lists and maps, while in others, a selected enumeration area (EA) map may be nothing more than a rough outline with a few key landmarks noted.

---

2 Initially these were the Centre for Democratic Development-Ghana (CDD-Ghana), the Institute for Democracy in South Africa (IDASA), and Michigan State University (MSU). Now, CDD-Ghana is joined by the Institute for Justice and Reconciliation (IJR) in South Africa and the Institute for Development Studies (IDS) at the University of Nairobi in Kenya.

In the face of this extreme heterogeneity, Afrobarometer argues that both the instrument and the data collection process must be standardized as much as possible to ensure the quality, validity, and comparability of our data. These two aspects are inextricably linked. Harmonization is not only about making sure that individual questions and response options are widely applicable and generate comparable responses in many different contexts. It is also about the rigor and consistency of the entire data collection process, from sampling to fieldworker training to fieldwork management.

Obviously, this requires a common set of procedures and *written* protocols so that everyone in the organization is working from the same set of rules. We capture these rules in the *Afrobarometer Survey Manual,* sometimes referred to as the "AB Bible." This is a public document, updated in each round, and available on the Afrobarometer website,[3] that contains extensive discussion of sampling, training, data collection and management, and overall survey management (including even budgeting).

The necessity of such an obsession with standardization was driven home quite early in the project because Afrobarometer began as an amalgamation of two different survey initiatives. The Southern African Democracy Barometer was a seven-country (Botswana, Lesotho, Malawi, Namibia, South Africa, Zambia, and Zimbabwe) study conducted by Robert Mattes at the Institute for Democracy in South Africa (IDASA) in 1999–2000. At the same time, a separate set of three surveys was conducted in Ghana (1999), Nigeria, and Uganda (2000) by Michael Bratton of Michigan State University (MSU) and E. Gyimah-Boadi at the Center for Democratic Development in Ghana (CDD-Ghana). After joining these initiatives under the new name Afrobarometer, two additional surveys were conducted in Mali and Tanzania (2001). Merging the "Afrobarometer Round 1" data from these 12 surveys, based on three quite distinct questionnaires, required countless decisions about which questions were functionally equivalent and could be included in the merged data set, and painstaking effort to match and recode equivalent response options while still capturing important distinctions and thresholds where necessary (e.g. education and occupation).[4] While this process enabled us to produce the first cross-national analysis of public opinion in Africa (Bratton et al. 2005), the time and effort required for *ex post* harmonization in Round 1 clearly pointed to the necessity of *ex ante* harmonization in future rounds.

At the same time, tremendous geographic and social diversity means that there is rarely a "one-size-fits-all" solution to many sampling and fieldwork issues. Afrobarometer therefore, seeks to elaborate rules and protocols that are both broad and sufficiently detailed to guide partners when they encounter substantially different or difficult circumstances. To take one example, the process of identifying start points and then proceeding along random walk patterns to select households for interviews has to be designed quite differently depending on whether the community is a densely spaced urban township in South Africa, a sparsely populated rural settlement in Mali, a high rise in Tunis, or a commercial farm in Kenya.

Yet with the forces of urbanization, population growth, and the spread of communications technology, Africa is changing rapidly, and while some challenges fade, new ones emerge. Thus, a third principle of the project is *constant learning*. In preparation for each new round of surveys, all partners gather to intensively review lessons learned in the previous round in order to update each section of the "AB Bible" for the future.

---

3 https://www.afrobarometer.org/surveys-and-methods
4 For a brief overview of the formation and early findings of Afrobarometer, see Mattes (2012).

Finally, the fourth and fifth principles might be termed *constrained flexibility* and *maximum support*. While standardizing and harmonizing is an essential goal, we also have to maintain the flexibility to accommodate unique circumstances, and be ready to respond to unforeseen events. The protocols laid out in the Afrobarometer Survey Manual can never account for every circumstance that partners may encounter in the field. Moreover, as important as it is, the Afrobarometer Survey Manual cannot stand alone – it must be complemented by other critical inputs, most notably training, technical advice, and problem-solving support. In every round, we work with new partners, go to new countries and new EAs, and encounter new challenges. Thus, within the broad emphasis on standard procedures, the diversity of conditions, challenges, and unanticipated problems require a flexible and adaptable approach to problem-solving. And especially when partners are new to Afrobarometer and our rigorous methods, or to nationally representative surveys altogether, they may need the network's support to understand the purpose of the protocols, why they matter, and how to implement them effectively.

We will discuss all of these principles further throughout the course of this chapter.

4.3   Applied Harmonization Methods

4.3.1   Sampling

Harmonization in the Afrobarometer begins with drawing a nationally representative, multistage, stratified sample that employs random and systematic selection at all stages using probability proportionate to (adult) population size. Our protocols require that only official census and population data are used to draw a fresh national sample for each survey round. Where census data are more than five years old, partners are required to obtain a recent projection of the study population.

The first phase begins with stratification of primary sampling units (PSUs) – usually census EAs – both along the topmost geographic census administration level (province, region, state, etc.) and by urban or rural location. EAs or PSUs are then selected from these stratified lists with probability proportionate to population size (PPPS). Then they record the final list of selected PSUs, complete with full location details and measurements of size, in a sampling reporting template. National partners can only make substitutions of selected PSUs in the case of insecurity or true inaccessibility (e.g. flooded rivers); remoteness and difficulty of access alone are not accepted as justifications for PSU substitutions. We limit substitutions to a maximum of 5% of the selected PSUs. If more substitutions are likely to be required for any reason, the sample must be redrawn.

However, drawing a stratified PPPS sample may be easier said than done. African census systems vary widely with regard to factors such as which population counts (adults versus total population) are available and at what levels of disaggregation, and in the availability and specificity of updates and projections. In addition, while in some cases Afrobarometer national partners are able to obtain direct access to sample frames and can thus draw samples according to AB protocols, our partners often work with sampling specialists from the national statistics office, who may prefer their own standard procedures. Agreeing on sample parameters can often involve a lengthy process of negotiation and, sometimes, compromise.

In the first few rounds, we relied on national partners to navigate these issues and provide a sample for central review by Afrobarometer, but we eventually created a position for a centralized network sampling specialist. This individual works closely with partners in each country, assisting them to liaise with their national statistics offices, plan and draw the sample, and ensure that consistent practices are followed to the maximum extent possible. In addition to the value of the consistent input, advice, and oversight that the specialist provides, we also find that his experience and affiliation with an international network like Afrobarometer can sometimes facilitate access and accommodation that national partners cannot always achieve on their own.

Fieldwork teams usually implement the second and third stages of sampling – selection of households and of individual respondents. In a small number of countries, national statistical offices are able to provide either detailed household maps or household lists for the selected PSUs enabling national partners to randomly preselect the households to be visited by fieldwork teams. But in most countries, we select a random "sampling start point" (SSP) in each PSU using the map and a grid. The general rule is that all four members of a fieldwork team move out from this SSP in four different directions, and select households using a 5th household/10th household selection pattern. However, vast differences in settlement patterns, map quality, and in how households are structured, defined, identified, and approached have required the network to define alternative protocols for managing these processes under a wide range of field conditions. These are described in extensive detail in the Afrobarometer Survey Manual.

One area where Afrobarometer has experienced major improvements over the last two decades in many countries is the greater availability, higher quality, lower cost, and improved format of the EA maps we can obtain. In many African countries, census enumerator maps are now available in digital formats, making it possible to integrate them into electronic data collection routines, and making deployment of field teams and infield sampling much easier and faster.

Standard procedures also apply to respondent selection. In order to ensure gender parity in our samples, fieldworkers alternate interviews between male and female respondents. Within each selected household, they list the first names of all adult citizens of the required gender who are usually members of the selected household, and the final respondent is randomly selected. In the past, this was done by having a senior member of the household select a playing card or numbered card – an approach that was implemented in place of Kish grids because of its transparency and understandability for household members. But since the advent of electronic data capture (EDC), we use a random selection script on the tablets. The Afrobarometer Survey Manual provides rules for household substitution under different circumstances as well as call back procedures to follow if the selected respondent is temporarily unavailable. Afrobarometer only substitutes households, not respondents within households.

4.3.2 Training

During Round 8, Afrobarometer national partners employed nearly 1000 fieldworkers, so training fieldworkers to the same consistent standard is another essential aspect of harmonization. During training interviewers develop interview skills, learn how to establish rapport and an effective interview atmosphere, review research ethics and informed consent, are thoroughly familiarized with the questionnaire, conduct pilot interviews (in multiple languages where necessary), and learn the interviewer's critical role in sampling. They should also come away with a broader understanding of the survey purpose and the importance of the protocols: To the extent possible, we seek buy-in and commitment from interviewers as members of a research team in the same way that we seek buy-in from national partners. The Afrobarometer Survey Manual lays out a plan for a six-day training course – unusually long by most survey training standards – that covers all of these aspects. Training is also an important opportunity for pretesting and final refinement of the questionnaire, especially the local language translations.

4.3.3 Fieldwork and Data Collection

Afrobarometer's two primary goals in managing fieldwork are (i) consistency, efficiency, and effectiveness of the data collection process, and (ii) supervision for quality control.

Our efforts to create standard rules for fieldwork begin with the very organization of the fieldwork team. We require that each fieldwork team include a driver, a team supervisor, and four fieldworkers. The specific numbers correspond to our policy of conducting eight interviews within each selected PSU (each fieldworker conducts two interviews per PSU) and to our supervision protocols, as well as with the number of people that can comfortably fit inside a typical field vehicle or SUV. The national partner then takes the selected sample and devises routes that each interview team will travel over the data collection period, which averages around three weeks. The composition of each team and the travel routes need to take into account the local language proficiency of the interviewers, as well as attention to local administrative, cultural, or religious norms that may affect data collection.

The fieldwork supervisor is responsible for ensuring the team reaches the correct PSU and start point, introducing the team to local authorities when necessary, gathering information about the availability of key facilities and services (schools, police, mobile service, etc.) in the PSU, and, most importantly, quality control, ensuring that fieldworkers correctly implement both their sampling responsibilities and their interviews. Supervisors sit in on early interviews and back-check a minimum of one in every eight interviews throughout the course of fieldwork.

Fieldwork supervision has changed radically since the introduction of EDC in Round 7 (2016–2018). Prior to this, aside from a daily check-in with the home office when phone service permitted, and periodic visits from roving senior management teams, field supervisors were largely on their own. With EDC, fieldworkers conduct interviews using preprogrammed tablets that ensure that questionnaires are filled out completely (no skips/missing data) and correctly (no incorrect question linkages), while also capturing always accurate date and time stamps as well as GPS markers. Since interviews are uploaded to central servers, usually on the same day they are captured, Afrobarometer's data quality officer can review them for problems (interviews that are too short or long, or that are flagged for other reasons) in near real time and alert supervisors to make necessary adjustments. The software also offers greater opportunities to randomize question or response ordering and other innovative survey methods. But supervisors still bear responsibility for ensuring that fieldworkers correctly implement sampling protocols and that they are handling their interview interactions appropriately and effectively.

4.3.4 Questionnaire

The complex challenges of *ex post* harmonization of questions and responses experienced by Afrobarometer in Round 1 taught us many early, critical lessons about question and questionnaire design that still play a central role in our approaches today. The network switched to a fully harmonized "master" questionnaire in Round 2, with identical questions, responses, and question ordering, and has maintained this approach since.

Roughly 60–65% of the questions on the master questionnaire stay the same from one round to the next (though it is not necessarily the same 60% every time, and they are not necessarily in the same order). This enables tracking of key indicators such as lived poverty, evaluations of government performance, trust in government institutions, and support for and perceived supply of democracy. The remaining 35–40% of the questionnaire is comprised of "special modules," sets of questions on selected topics judged to be of particular interest and relevance at the time. Special modules in Rounds 8 and 9 have covered perceptions of China in Africa, taxation, environmental governance and climate change, gender equality and gender-based violence, child welfare, and several other issues.

We develop new questions through a rigorous, multistage process to ensure that they generate valid, comparable data across all countries. To begin, a questionnaire committee composed of both academics and survey professionals affiliated with Afrobarometer identifies *topics* based on emerging issues confronting African societies that are broadly applicable across all countries, that may engage new users, and that generate interest among stakeholders.

Specific questions are drafted by subcommittees with expertise in the proposed topics, drawing on external experts as needed. Our guiding principle is to develop questions that can be asked to incredibly diverse pools of ordinary citizens across countries and cultural contexts and in a multitude of languages. In this process, we pay particular attention to using terminology and question wording that is as simple, accessible, and brief as possible. We also consider the translatability of questions, phrases, or words into dozens of local languages. After thorough review by all committee members, the final draft questionnaire is shared with national partners, who have an opportunity to raise concerns regarding question relevance, understandability, and translatability.

New questions then undergo a series of pilot testing, both by phone and through face-to-face interviews. Once the final master questionnaire is completed (including questions and material on respondent selection, introductions, interviewer instructions, and post-interview feedback) in English, we update translations of the master questionnaire in other national languages: KiSwahili, French, Portuguese, and Arabic.

The next stage is for national partners to "indigenize" specific questions that require local referents (e.g. "President" versus "Prime Minister"), country-specific language (e.g. "parliament" versus "national assembly," the proper name of the tax authority, or the year of the most recent election), or country-specific lists of regions, ethnic groups, and political parties. In addition, any questions that are not relevant (e.g. questions about traditional leaders in Cape Verde, Mauritius, and Tunisia, questions about the ruling party in eSwatini, or questions about local government in Angola) are automatically assigned a "not asked in country" code.

Partners also add their "country-specific questions," or CSQs. The master questionnaire for each round has dedicated space for a number of country-specific items, which partners can use to explore current affairs or issues that are relevant in their own country (e.g. a constitutional reform process or the performance of a specific national anti-corruption body) or to explore existing topics in greater detail than the generic cross-national questionnaire can accommodate.

Occasionally more substantial adaptations have to be made during the indigenization process. For example, some questions refer to how the government has performed "in the past year," but if a new government has only been in office for six months at the time of fieldwork, the question text must be adapted. When adaptations are made in a given country, the survey management team and questionnaire committee will make a determination as to whether the question content is fundamentally comparable – meaning that the data can still be merged with and analyzed alongside data from other countries – or whether the difference is significant enough to warrant relabeling the question as a CSQ.

We also standardize some aspects of questionnaire management in order to minimize errors and inconsistencies and facilitate data management and data merging. The most significant of these practices has been the introduction of coding conventions. For example, Afrobarometer uses standardized codes for routine responses: "don't know" is always coded as 9, 99, 999, or 9999, and there are standardized codes for "other" responses, "not applicable," "no further response," "refused," and "not asked." This makes it easier to review questions and questionnaires and reduces coding errors. Another innovation is preassignment of a unique range of 40 country-specific codes for each country (e.g. Angola is assigned 1750–1790), which are used for coding localized lists – of regions, languages, ethnic groups, or political parties – as well as any post-coded category added on selected questions (e.g. on our "most important problems" question). We also use three-letter country codes in a number of places, including respondent numbers (e.g. LIB0555 in Liberia), interviewer identities (LIB08), and CSQs (Q80-LIB). In this way, codes and responses to these questions can be easily merged across all countries without overlap, which again reduces both management demands and opportunities for errors.

4.3.5 Translation

Afrobarometer was founded on the belief that even the least educated or literate among ordinary citizens can hold valid opinions, including complex ones, about the issues that they face and how their communities and countries are governed, and that capturing these opinions is an important part of our mission. But, we can only fully and accurately explore these opinions if we conduct interviews in languages in which respondents can fluently understand, consider, and respond to questions. And though many Africans are adept with languages and may speak several (Afrobarometer's Round 4 surveys [2008–2009] found the average respondent reported speaking 2.1 languages, see Logan 2018) – possibly including the dominant national language(s), many others do not. Afrobarometer policy is, to the greatest extent possible, to interview respondents "in the language of their choice." Our standard rule is that a translation should be produced for any language spoken by 5% or more of a country's population, although we may adjust once the final sample is drawn if none of the selected PSUs fall within a particular language zone. As a result, during a typical round of surveys, we translate the questionnaire into approximately 80 local languages.

It is of course essential that all of our translations into both national and local languages accurately and consistently capture the concepts and content of the original English questionnaire. Translation is thus one of the most critical harmonization steps in the entire survey process. Afrobarometer has developed a rigorous approach that involves both professional translators and trained fieldworkers in producing and fine-tuning translations into vernacular that is familiar to everyday speakers of the language. After double-blind forward and back translations, national teams hold "synchronization meetings" during which all translators meet to discuss any difficulties in translating specific questions or concepts and reconcile local translations to ensure that they all capture the same meaning. A final stage of review occurs as the fieldworkers use the questionnaires during training and pilot interviewers and provide final feedback on the translations for final fine-tuning. |

4.3.6 Data Management

Historically, Afrobarometer relied on national partner teams to do most data entry and cleaning. The network provided software, a template with standardized variable names, labels, and codes, as well as training and quality control through a network data manager. The migration to EDC in Round 7 has not only made data set production and finalization faster and more efficient as previously noted but also more centralized. Rather than waiting days or weeks for paper questionnaires to reach the capital, and data for all cases to be hand entered and checked before being sent to Afrobarometer data managers, data sets are constructed simultaneously with fieldwork. Many cleaning steps are also eliminated, so the time spent on data cleaning can be directed toward more substantive quality issues such as duplication checks, review of interview length and timing, response rates, and related issues, all of which provide feedback that can be used to continuously improves the network's protocols and data quality. As a result, the time required to finalize data sets after fieldwork is complete has been reduced from an average of more than two months to one month or less. The network data manager certifies final data sets and shares them with the national partner and other stakeholders. They are posted on the Afrobarometer website and shared with data archives one year after the completion of fieldwork.

In addition to the variables captured during the interview, final data sets also include several variables created *ex post* for the purposes of analysis. These include: (i) weighting variables for both country-level and multicountry analysis; (ii) several standard calculated indices used routinely by Afrobarometer, including our Lived Poverty Index, and indicators of Demand for Democracy and Supply of Democracy; (iii) condensed versions of age, education, religion, and poverty indicators for ease of reporting disaggregated findings.

Because of the standardization of the questionnaire and the country data set templates, and the use of both standardized and country-specific codes as described above, merging country data sets into a single multicountry data set for each round is a relatively straightforward process. CSQs are excluded from the merge, but all other variables are retained.

4.3.7 Documentation

In addition to a final questionnaire and data set, national partners are required to produce a set of deliverables for each survey aimed at thoroughly documenting the work. Required deliverables, some for internal consumption and some shared externally, include:

- A sampling report provided prior to fieldwork in collaboration with the network sampling specialist, and finalized after fieldwork is complete with the addition of information on any PSU substitutions made during fieldwork, including the reasons for each substitution (internal)
- A training report describing the program/content of the training (internal)
- A fieldwork report describing survey implementation and any special circumstances or situations encountered (internal)
- A technical information form (TIF) summarizing basic information on the survey including dates, sampling frame, weighting, implementing partners, response rates, and any special notes on events that may have occurred during fieldwork (external, incorporated into the Summary of Results and the codebook)

- A Summary of Results that reports topline results for all questions posed to respondents, disaggregated by gender and urban–rural location (external, shared via AB website)

In addition, the data management team produces a complete codebook for each survey, which also includes the TIF.

4.4 Harmonization and Country Selection

While they represent more than three-quarters of Africa's total population, the countries covered by Afrobarometer still do not represent the entire continent. Most of those that are excluded are among Africa's most closed and repressive societies. Afrobarometer's goal is to cover all 54 countries in Africa eventually. But before expanding, the eligibility of a candidate country is assessed on three key criteria, all of which connect directly to core issues of survey management and data harmonization.

First, we require a recent census frame, generally one completed within the past decade. A sampling frame that does not meet Afrobarometer criteria cannot produce survey results that are adequately comparable – i.e. equally representative – to those from other countries. To cite one example, completion of a new census in Angola in 2014 – the first in the country since 1984 – allowed Afrobarometer to include the country starting in Round 8 (2019–2021)

Second, a minimum level of political openness is essential. Specifically, we are concerned about the ability of respondents to speak their minds freely. Afrobarometer asks many potentially sensitive questions – about the performance of the government and the president, and about voting preferences and affiliation to political parties, among others. A standardized cross-national survey requires not only that we ask the same questions in all countries but that respondents are also similarly free to answer the questions honestly. If they are not, the collected data are not a reliable reflection of national views and are not comparable to that from more open countries.

Third, as mentioned, Afrobarometer works through local national partners. The availability of an organization with the skills and capacity to conduct a nationally representative survey tells us about the technical capacity of a candidate country. A partner's willingness to write a country report and release the results locally tells us a great deal about the political openness of that country.

When selecting new countries to which we can extend coverage, Afrobarometer relies primarily on information gathered during in-country feasibility assessments from local actors and experts, including policymakers and advocates, journalists, academics, and potential partner organizations. In addition, we scrutinize our data on questions about freedom of speech and political openness captured during interviews, as well as a series of questions answered by fieldworkers after the interview about interview conditions, including the overall attitude and demeanor of the respondent. Adhering to these three criteria for country selection ensures that we can collect data that is not only comparable but reliable.

4.5 Software Tools and Harmonization

Standardizing software has also been an important component in Afrobarometer's harmonization efforts. We have always required all national partners to prepare data sets using a common program.

From Round 1 to Round 6, Afrobarometer used SPSS software packages for preparing data entry scripts (SPSS Data Entry Builder), for capturing data (SPSS Data Entry Station), and for analyzing data (SPSS Base). These packages offered important advantages in user friendliness that accommodated less experienced members of the network. Although some partners would have preferred to use other software, this would require preparation of multiple templates, accommodating different rules for allowable codes, and other adjustments that work against harmonization.

The ability to harmonize software was a key reason that Afrobarometer waited until Round 7 to shift to EDC despite pressure to do so earlier: Prior to this, there was not a single, user-friendly, and widely available software platform that could be used across the entire continent. As noted, one key advantage of automated data capturing is minimizing individual localized errors, e.g. skipped questions or entry of incorrect codes. But one of the disadvantages is the potential to introduce global programming errors that can be relatively invisible but affect entire data sets. Protecting against programming errors requires multiple detailed reviews of programs and templates. Use of multiple software for data capture would have compounded this problem and presented major management and quality control challenges. Afrobarometer, therefore, delayed making the switch to EDC until computing software had improved sufficiently and could better meet our harmonization needs. We ultimately chose SurveyToGo software based especially on a combination of cost, ease of use, system support, and continental coverage. Although some national partners were initially reluctant to make the switch to EDC, in the end, the transition went remarkably smoothly, and many of our partners soon shifted all of their survey work to EDC.

Other important technological and software advances that have supported harmonization in Afrobarometer and increased data quality and accessibility have included:

- The ability to easily *geocode* data sets (for the sake of confidentiality, Afrobarometer geocodes at the level of the PSU rather than the household), which allows easier harmonization with macro-data and other data sources.
- The increasing ability to quickly and easily share large files such as sample frames and large merged data sets, and convert them to different formats.
- New data cleaning programs and quality evaluation syntaxes are increasingly accessible and effective in identifying potential problems with duplication or other fieldwork errors (see, e.g. Kuriakose and Robbins 2016).

4.6 Challenges to Harmonization

As we have demonstrated, Afrobarometer has expended extensive effort to produce results that are representative, valid, and reliable because they are generated by *both* an instrument and a process that are as harmonized as possible. Yet, we face continued challenges, both practical and philosophical. These challenges range from the more mundane problem of managing differences in the available census frames, projections, and maps encountered across different countries, to more profound concerns that the desire for (*ex ante*) harmonization may limit space for local knowledge and agency, or turn a network founded on democratic and inclusive principles into something of a dictatorship. Many of the more basic and obvious challenges arise from the vast differences in information and environment across the continent already discussed. Here, we will focus on some of the broader issues that continue to challenge harmonization, especially when we face difficult trade-offs between harmonization and other valued objectives.

4.6.1 Local Knowledge, Flexibility/Adaptability, and the "Dictatorship of Harmonization"

There has always been an understanding within Afrobarometer that implementers needed to share a common methodology that met rigorous standards. Yet, treating all partners as coinvestigators rather than simply as contractors requires us to also be flexible, adaptable, and to occasionally accommodate differences. Indeed, we have found this approach to be instrumental to the network's success. Implementing high-quality surveys is difficult and expensive, and while Afrobarometer core partners provide significant technical assistance to partners during fieldwork, they cannot be on the ground to police every step and ensure that partners adhere to every protocol. Rather, national partners who buy-in to our model and are committed to the network's success are much more likely to maintain strict adherence to standards than those who act as contractors who may see no harm in cutting the occasional corner. Collaborative partnerships are also a realistic way to establish the local legitimacy of the project. Local legitimacy is important not only where the formal approval of cabinet ministers or security officials, or the informal consent of local leaders, is necessary to conduct fieldwork, but also where the intended political impact of the project is premised on national leaders' acceptance of the survey results.

Approached incorrectly, harmonization can work against not just flexibility and adaptability, but partners' very sense of ownership. As laid out in previous sections, the intertwined imperatives of both quality control and harmonization can result in an increasingly expansive and strict set of survey protocols, as well as a fairly rigid demand that the AB master questionnaire be implemented "as is." As a result, the network has had to become less accommodating over time of partners who say, "Question XXX would work a bit better in our country if we could change this," or "The statistics bureau would like us to adjust the sampling protocol by doing YYY." We find ourselves saying "no," a lot.

But, we try to balance these strictures with opportunities for voice and input. Partners provide feedback during the debrief workshop held at the end of each round, can suggest topics during the questionnaire design phase, and have a significant opportunity to contribute during the planning workshop that launches each new cycle. Elected national partner representatives also sit on Afrobarometer's Senior Advisory Team (SAT), a key policy-making committee. Afrobarometer also tries to ensure that our partners enjoy nonmonetary benefits in addition to their contractual arrangements with the network, including by adding their own CSQs to the questionnaire, frequent training opportunities, and the chance to extend their profile and reach through Afrobarometer partnership in the network and dissemination of high-profile findings. But while we still see them as partners, we know they sometimes feel they are treated as contractors. It is a tension that cannot be entirely resolved.

4.6.2 The Quality-Cost Trade-off and Implications for Harmonization

Meeting the cost of doing high-quality face-to-face survey research in difficult environments is an ongoing struggle for Afrobarometer. Fundraising is a constant, and competition from less expensive – and less quality-controlled, less harmonized – surveys, is increasing. But in at least some respects, the imperatives of harmonization can help lower costs: In general, the more standardized our methods and processes are, the easier – and thus less expensive – they are to monitor and review.

But cost trade-offs can cut in both directions when it comes to harmonization. Across diverse survey environments, it is almost inevitable that for every harmonized/standardized protocol, there will be an exception where a more locally appropriate approach might be more cost-effective. For example, as noted, Afrobarometer generally requires field teams with six members to travel together around the country. However, in some countries, vehicles that can accommodate a team of this size are prohibitively expensive. Should we stand by our harmonized protocol, which will increase the cost of fieldwork, or make an exception to manage costs – an exception that could affect survey management, survey standardization, or survey quality – by sending a smaller team, perhaps without a supervisor on board? We face these kinds of trade-offs constantly.

4.6.3 Final Challenge: "Events"[5]

All of Afrobarometer's well-laid plans and protocols can be pushed aside in the seeming blink of an eye by unexpected circumstances. The most recent instance was the Covid-19 pandemic that erupted during 2020 and led us to suspend Round 8 fieldwork at the end of March 2020, halfway through our eighth round of surveys, with 18 of 34 countries completed. While fieldwork resumed six months later, and Round 8 surveys were eventually completed in July 2021, the comparability of the data across the full round – between the "before Covid" and "after Covid" countries, has been a core issue when it comes to analyzing the data for the full round. Citizens' experience of the pandemic – including government decisions to lockdown communities or close schools, their transparency in allocating resources and communicating with the public, and actual impacts in terms of economic and social hardship, illness, or death – may have deeply affected their evaluations of everything from presidential performance to the quality of health care systems. There was obviously no way that Afrobarometer could have prepared for this. But the Round 8 merged data set will need to be permanently marked and analyzed as a two-part, before and after, data set. This is true on the broad macro level, as well as on the more concrete micro level: for the countries that did fieldwork after the start of the pandemic, the Round 8 questionnaire was adapted to remove some questions in order to make space for a new special module on the impacts of the pandemic, vaccine hesitancy, and related issues.

While the Covid-19 pandemic provides a stark example, Afrobarometer has faced many similar challenges in the past, both at the country and regional levels. The 2008–2009 global economic crisis hit in the middle of our Round 4 surveys, while the 2014–2016 Ebola outbreak delayed fieldwork in several West African countries and likely shaped evaluations of healthcare systems throughout the region, both for better and for worse. Surveys have also been waylaid by the death of presidents shortly before and during fieldwork, and responses have been affected by the revelation of major corruption scandals while fieldwork was underway. There is not much that can be done in advance to prepare for these events except to realize that to some extent, disruptions of one sort or another are almost inevitable. But by keeping thorough records, and occasionally making small adaptations or additions to the questionnaire – something that is

---

5 When asked about the sort of things that could blow a government off course, British Prime Minister Harold MacMillan is reputed to have replied, "Events, my dear boy, events."

easier with the use of EDC methods – we adjust our data collection processes as we must. Occasionally, we make *ex post* adjustments as well, for example, by coding "before and after event X" variables into our data sets to allow for ready analysis of the implications of an event, as we have done in the case of the pandemic.

4.7 Recommendations

Afrobarometer's hands-on experience over the last 20 years, as outlined in the above discussion, has taught us several key lessons about survey research and data harmonization, lessons that are often about striking a balance – between stringency and flexibility, between constancy and change, between economy and quality, between boldness and overreach. The following recommendations summarize our advice to survey researchers pursuing similar goals:

(1) It starts with *country selection* based on real *country context*. Data from countries where respondents cannot speak freely, or where it is not possible to generate reliably representative samples, should not be incautiously compared to data from countries that do not experience these challenges. But there is no substitute for an on-the-ground assessment. Afrobarometer has worked successfully in some very authoritarian countries, but also had to set aside one data set (Ethiopia in 2013) that was ultimately judged "not comparable" – we probably should have waited to go there.

(2) *Harmonize not just for data comparability, but for data quality.* Standardization of everything from codes to training to data cleaning protocols does not just produce harmonized output – it is a critical tool in the quality control process.

(3) *Methods matter.* In complex, diverse survey environments, harmonizing data collection methods is equally important as harmonizing the questions asked and requires equal attention.

(4) *Ex ante harmonization* is almost always preferable to *ex post* when it is possible.

(5) *Standardize* everything possible, from the simplest coding practices to complex sampling processes. Multicountry surveys involve hundreds of often minute details, which introduce opportunities for differences and errors. Standardization reduces the degrees of variability and the need for review, so quality control efforts can focus elsewhere.

(6) *Spell it out/Write it down.* Despite how it sounds, "standardization" does not mean there will be one-size-fits-all solutions for each step of the survey process. Rather, it is essential to identify core principles and goals at the outset, and anticipate, to the extent possible, the different circumstances that will be encountered, and spell out – and write down – how the principles should be applied under different circumstances.

(7) *Be flexible and be prepared.* Standardization and harmonization are great, but they must be balanced against other critical goals, including cost, and especially respect for partners and their knowledge of their local context. Even after extensive efforts to develop standardized protocols, researchers will have to deal with the many exceptions, unique circumstances, and special conditions that may require adjustments or amendments. But it is better to have a well-grounded and extensive set of rules to govern implementation and then to carefully record and document any exceptions or adjustments that are required, rather than to leave things too open from the start.

(8) *Train and support.* High-quality methods, especially effectively harmonized high-quality methods, do not just come from a book, even one as great as the Afrobarometer Survey Manual. Especially for partners new to multicountry survey projects, investments in education, training, presence, and

support will pay off. Partner commitment and buy-in is a better quality control and harmonization tool than policing ever will be.

(9) *Harmonize software.* Harmonizing not just codes, question wording, and sampling protocols, but the software used for questionnaire design, data collection, data management, and analysis, can have profound advantages. Shared software simplifies up-front production *and* review (e.g. of templates), standardizes quality control processes, and reduces opportunities for cross-country differences and errors, thus leaving quality controllers to focus their time and attention on more important issues. The ideal timing for software transitions in multicountry survey research projects may be later than for single-country projects, to ensure that all countries can make the transition simultaneously.

(10) *Be cautious about but open to change.* Change is sometimes necessary, but the larger the project, the more important harmonization is, and the more difficult and potentially disruptive change is. We have tried to approach change with a mix of willingness alongside appropriate caution. The transition to EDC was a case in point. There was pressure to make the change earlier, but doing it too soon – before we could use harmonized software – could have been detrimental to data quality and to data harmonization.

(11) Finally, recognize that data harmonization has *both costs and benefits* – It may, at times, conflict with other goals, including capturing local nuance, promoting local ownership, and managing survey budgets. But, it can also have unexpected advantages. The cross-country comparability of Afrobarometer data has gained us traction with politicians and government officials who might have initially been reluctant to engage. For Afrobarometer, this has been one of the most critical "wins" of the harmonization process.

References

Bratton, M., Mattes, R., and Gyimah-Boadi, E. (2005). *Public Opinion, Democracy and Market Reform in Africa*. Cambridge University Press.

Bratton, M., Dulani, B., and Masunungure, E. (2016). Detecting manipulation in authoritarian elections: survey-based methods in Zimbabwe. *Electoral Studies* 42: 10–21.

Kuriakose, N. and Robbins, M. (2016). Don't get duped: fraud through duplication in public opinion surveys. *Statistical Journal of the IAOS* 32: 283–291.

Logan, C. (2018). 800 languages and counting: lessons from survey research across a linguistically diverse continent. In: *Tracing Language Movement in Africa* (ed. E.A. Albaugh and K.M. deLuna), 91–116. Oxford: Oxford University Press.

Mattes, R. (2012). Afrobarometer. *The Annals of Comparative Democratization*, 10/3 (October).

5

Harmonization in the National Longitudinal Surveys of Youth (NLSY)*

Elizabeth Cooksey, Rosella Gardecki, Carole Lunney, and Amanda Roose

Center for Human Resource Research, The Ohio State University, Columbus, OH, USA

5.1 Introduction

For over 50 years, the National Longitudinal Surveys (NLS) program has designed longitudinal surveys to capture the labor market experiences and other significant life events of several cohorts of men and women in the United States. This family of surveys has a strong focus on labor market activities but also provides a wealth of information on topics as varied as education and training, relationships, fertility, household composition, income and assets, government assistance program participation, and preparation for retirement, to name a few (Cooksey 2018).

Given that the NLS comprise a group of surveys that fall under the same umbrella of primary motivation and share core interests and emphases, it might be easy to imagine that harmonizing data concepts across the various studies would be relatively straightforward, especially when compared with cross-study harmonization, where data have been collected in various countries and often with different rationales and purposes. However, the long time period over which each cohort is followed, the magnitude of societal change that has occurred over the past five to six decades, and the transformations in survey methodology that have taken place, especially over the last 30 years or so, serve to chip away at this apparent ease. In this chapter, we highlight some of the unique challenges confronted when trying to harmonize data from long-running longitudinal surveys using the NLS as illustration and focusing on the NLSY in particular.

There are seven NLS cohorts in total and each cohort, with the exception of the NLSY79 children and young adults, was selected to be representative of all people born during a given period who were living in the United States just prior to the initial survey interview. The selection of each cohort was motivated by changing circumstances faced by different cohorts at different stages of transition into (or out of) the labor market. Of the four original cohorts, the **Older Men** cohort (ages 45–59 in 1966), for example, was designed to enable research on the employment patterns of men nearing the completion of their careers

---

*This is independent research that does not reflect the opinion of the Bureau of Labor Statistics.

Survey Data Harmonization in the Social Sciences, First Edition.
Edited by Irina Tomescu-Dubrow, Christof Wolf, Kazimierz M. Slomczynski, and J. Craig Jenkins.

Table 5.1 Description of original cohorts.

Cohort	Birth years	Year survey began	Year survey ended
Older men	1906–1921	1966	1990
Mature women	1922–1937	1967	2003
Young men	1941–1953	1967	1981
Young women	1943–1953	1967	2003

who were making decisions about when and how to withdraw from the labor force. The **Mature Women** cohort (ages 30–44 in 1967) was planned to understand employment patterns of women who were reentering the workforce and balancing the roles of homemaker, mother, and labor force participant. The **Young Women** (ages 14–24 in 1968) and **Young Men** (ages 14–24 in 1966) cohorts were intended to facilitate research on the employment patterns of women who were finishing school, making initial career decisions, and starting families, and of men who were completing school and entering the workforce or joining the military and were thus making initial career and job decisions that would impact their employment in the coming decades. Although these original cohorts are no longer active, their data remain publically available and continue to be a valuable resource for researchers (Cooksey 2018; www.nlsinfo.org).

When looked at longitudinally, each set of cohort surveys provides a rich source of information about important life course trajectories, and how changes in one area of life can lead to both short-term and longer-term changes in other domains. Together, the NLS surveys afford users unique opportunities to compare life course transitions across multiple generations of men and women in the United States (see Table 5.1).

Building on knowledge gained and lessons learned from these four original cohorts, two youth surveys (**NLSY79 and NLSY97**) were designed with samples drawn about 20 years apart. Both are large, nationally representative cohorts of male and female youth living in the United States who were 14–22 years old when first interviewed in 1979, or 12–17 years old when initially interviewed in 1997, respectively. These two ongoing studies were intended originally to trace the educational and labor market experiences of two generations of young U.S. residents as they transitioned from school to work. They were designed to be comparable to the Young Women and Young Men cohorts although clearly there is a big difference between being *designed to be comparable* and it actually being possible to *harmonize the data*.

The seventh NLS study is a little different as it is comprised of all children born to NLSY79 women with births occurring from 1970 to very young mothers, through 2014 to mothers who were in their late thirties and early forties when they gave birth. The more than 11,000 "Children" of the NLSY79 who have been followed since the study began in 1986, or from when they were born if during or after 1986, enter Young Adulthood the year they turn age 15. They are also still being interviewed. We concentrate on the two main ongoing NLSY studies in the rest of this chapter, however, as they serve to illustrate nicely not only some of the benefits but also some of the frustrations that researchers face when trying to harmonize data, even when collected within the same national boundaries and the same overall program.

5.2   Cross-Cohort Design

The **NLSY79** cohort is a nationally representative sample of young people, born between 1957 and 1964, and living in the United States at the end of 1978. The cohort originally included 12,686 respondents, but two oversamples were dropped early on (youth in the military and socioeconomically disadvantaged white youth) leaving 9964 respondents including oversamples of black and Hispanic respondents. In addition to the overarching goal of mapping the transition from school into the workforce, a secondary goal was to address research and policy questions related to recently legislated expanded employment and training programs for youth. As of early 2021, data are available from 28 survey rounds and fielding is ongoing for Round 29 (Rothstein et al. 2018).

The **NLSY97** cohort follows the lives of a nationally representative sample of 8984 adolescents in the United States born between 1980 and 1984. As of early 2021, this ongoing cohort has been surveyed 19 times. Like the NLSY79, this cohort also contains oversamples of black and Hispanic respondents. The NLSY97 cohort was designed to enable research on youths' transitions from school to the labor market and into adulthood for a generation born into a very different time period than youth in the NLSY79 cohort. The NLSY97 respondents are the oldest of the millennial generation and were first interviewed in the year when Titanic opened in movie theaters, Harry Potter was published, Princess Diana was laid to rest, Dolly the Sheep was cloned, and Pokemon was released. In contrast, the NLSY79 respondents are younger baby boomers and when first interviewed in 1979, the Deer Hunter won best picture at the Academy Awards, Margaret Thatcher became the first female British Prime Minister, Ronald Reagan announced his candidacy for US President, Pink Floyd released The Wall, the Sony Walkman was introduced, and Magic Johnson of Michigan State played against Larry Bird of Indiana State in the NCAA Basketball Championship. Not only did these two cohorts experience very different cultural environments, but they moved into early adulthood under very different educational, employment, and family-building circumstances.

In Section 5.1, we note that in order to achieve some degree of comparability between studies, and hence, enable cross-generational analyses, conscious decisions were made to model later studies on earlier ones – at least partially. For example, not only did the design of the NLSY97 take note of content in the NLSY79 but also both of these youth cohorts drew on the earlier Young Men and Young Women cohorts. Working tangentially against this ideal, however, is the need to recognize that each cohort experiences historical periods in ways that are unique to them given their life course stage: outside factors such as economic highs and lows, changing employment laws, increased educational opportunities, and shifting gender norms. All of these circumstances impact each generation's life choices and chances differently. We have to be especially aware of large "period" effects or societal shocks that can impact people's lives in very different ways depending on how old they are at the time.

Hope and Glory, a 1987 British comedy-drama-war film, exemplifies this latter point beautifully. Most films about wartime are written from the perspective of adults – of their worries and fears, of how their dreams are dashed and their lives put on hold or changed forever, of their experiences of death, injury, and illness. Written, produced, and directed by John Boorman, this film is based on his own experiences of growing up in the Blitz in London during WW2. *Hope and Glory* view these wartime experiences through the eyes of 10-year-old Billy, who remains in London with his mother and sister, rather than through those of his father who leaves to fight. The bomb that destroys Billy's school one day does not represent a terrifying event as it might to his mother, but rather the best thing that can happen to a young boy – it is the happiest of days because school is canceled!

Each of the NLSY surveys has been, and continues to be, carefully designed with cross-cohort research in mind. Both survey samples are based on birth year, are drawn from nationally representative area probability samples, and have similar needs to include certain topics, particularly those surrounding employment. During the design process, there is an emphasis on including questions and content that enhance opportunities for comparisons across time and across cohorts. As of late 2020, the NLS bibliography contained 111 research articles and chapters, 44 dissertations, and 74 oral conference presentations that use data from both the NLSY79 and NLSY97.

The overall strategy for facilitating cross-cohort projects across the various NLS cohorts focuses on design, the creation of conceptually similar variables, detailed documentation, and user outreach and support. Decisions rendered within each of these areas are made to help researchers who come from a wide variety of fields with different research interests to create their own harmonized datasets, while remaining cognizant of changed circumstances across cohorts, and taking into account methodological advances in survey research. For example, for each survey, we provide many summary variables that allow users to create comparable samples across the two cohorts (e.g. month and year of birth, interview month and year, and basic demographic characteristics).

As the NLSY79 and NLSY97 surveys are funded by the U.S. Bureau of Labor Statistics (BLS) with employment as a key emphasis, we focus our discussion on the approach, we use to create a consistent set of week-by-week work history variables known as the *work history arrays* for these two cohorts. Although employment is only one element that makes up the NLSY surveys, this discussion enables us to highlight differences in design both between the two cohorts as well as within each cohort across time that researchers wishing to undertake cross-cohort research projects must take into account when using the work history data as well as other information.

5.3 Applied Harmonization

The main directive of the NLSYs was to study the transition from school to work, although for the NLSY79 the current focus has now shifted to become the transition from work into retirement! From the first round, the NLSY79 and NLSY97 surveys have asked respondents about events that occurred during the reference period from the date of their most recent interview through the date of the current interview, even when a respondent has missed rounds. These surveys employ dependent interviewing in which respondents are asked about the outcomes of specific items or events reported during the prior interview (e.g. names of employers, schools, and children) using questions containing textual information from answers they gave previously. For example, a respondent who reports working at a particular job on the day of their last interview is reminded of that prior information when asked to provide any updated information in the current interview. This is true across numerous life events to create event timelines without gaps or overlap.

The data provide researchers with a number of opportunities to harmonize across cohorts and the work history arrays that are a set of employment-related event timelines created in-house to help researchers, are a prime example. These arrays consolidate information from within and across survey rounds to create a week-by-week accounting of three employment characteristics: (i) weekly labor force status; (ii) weekly hours worked; and (iii) additional jobs worked during each week. Each of these three work history arrays consists of a set of variables for each week starting with the week of 1 January 1978 for NLSY79 respondents, and the month after each respondent turns age 14 for NLSY97 cohort members. For each respondent, the arrays contain information from week 1 until the week of the respondent's most recent interview. For example, in the work status array, each week's variable contains either a unique job number for the main job held during that week if the respondent is working or an alternative labor market status. For example, the NLSY79 weekly status variables from January 1978 through the most recent interview are defined as:

0: No information reported for week
2: Not working (unemployed versus out of labor force not determined)
3: Associated with employment, gap dates missing, all time not accounted for
4: Unemployed
5: Out of labor force
7: Active military service
100–2815: Actual survey round/job number

The data used to generate these work history arrays are extracted from the detailed information on jobs held and periods not working since the date of the last interview, along with other employment information about each job, collected in each NLSY79 and NLSY97 survey round. Although the structure of the employment sections asked of each cohort is not identical, individual questions within the sections collect equivalent information. Through the employment questionnaire, respondents provide detailed job-specific information including start and stop dates, hours worked at that job, dates for any gaps when they were away from the job, and employment-related activities (e.g. searching for a job) during periods when they were not working. Because both surveys collect these elements for each job from the time of the last completed interview, there are no gaps in the information provided from round 1 to the most recent survey completed by each respondent. During fielding, our survey program calculates whether there are any gaps between reported jobs and prompts the interviewer to ask about employment-related activities during these periods.

From these responses, survey staff is able to construct a chronological week-by-week record for each respondent, even those who have missed one or more prior survey rounds, and create concepts that are directly comparable across cohorts. As noted above, three types of data are presented in separate, but related, arrays: a week-by-week labor force status already described, the corresponding hours worked at all jobs in those weeks, and information about multiple jobs held simultaneously. When a respondent worked more than one job during a given week, unique job IDs are recorded in a separate "dual jobs" array. Finally, the number of reported hours for each job worked in that week is summed to provide researchers with a weekly measure of time spent at work. The NLSY97 cohort was designed to follow the same general process from the collection of raw data elements to the creation of the variables that populate the arrays.

To illustrate the usefulness and time-saving nature of these arrays, consider the example of just one hypothetical respondent, we'll call Margaret. Suppose that Margaret worked for 40 hours a week at a job in a bakery between 1 January and 30 June, 2006. She started a second job on 1 May, where she worked 10 hours a week making cupcakes. She kept this second job all through the rest of the year, although at her interview on 4 September 2006, she reported increasing her time to 35 hours per week. Jennifer is a researcher and she wants to know how many hours per week her sample of respondents worked during 2006. Without constructed array variables this could be a very time-consuming task and only achievable at the month level, as days are not publicly available. She would have to look at information for start and stop months and years for potentially multiple jobs, along with hours worked for each job for each member of her sample. Margaret, for example, had her bakery job for the first 6 months of the year and her cupcake job for the last 8 months, with hours that varied between 10 and 50 hours per week throughout the year. The constructed arrays pull all this information into the same format and time frame enabling Jennifer to quickly calculate that Margaret worked 40 hours per week for 17 weeks, 50 hours per week for 9 weeks, 10 hours per week for 9 weeks, and 35 hours per week for 17 weeks. She held two different jobs in total during that time but only worked more than one job at a time for 9 out of the 52 weeks.

In this example, Jennifer wanted hours per week worked in a calendar year, but what if she had wanted to know how many hours per week her sample members worked in the first six months after their 25th birthday? Suppose, Margaret turned 25 in January, then the information that Jennifer now wants can again be quickly ascertained from the constructed work history arrays (starting from the first full week in February). The arrays provide Jennifer with the information that Margaret worked 40 hours per week for the first 12 weeks of this 26 week period, 50 hours per week for 9 weeks, and 10 hours per week for the last 5 weeks.

By consolidating employment-related information from not just one survey but from multiple rounds, these arrays, therefore, save researchers a lot of time and effort. If a respondent holds a long-term job, they will report that job across multiple rounds of data. Rather than a researcher having to pull that information together from multiple surveys across multiple years, our work history arrays report the number of hours worked per week during each round, along with any week-long gaps regardless of the round in which they occurred. Their creation also means that different researchers use the same cleaned and carefully checked information in their various analyses, and they enable timeline events to be combined. So a researcher interested in the effect of a training program on labor market supply can translate that training program's dates into the same event history format, and then compare labor force attachment through hours and weeks worked before and after the training.

The NLSY97 employment array structure mirrors that of the NLSY79 structure to allow for comparability across cohorts. While the survey sections asked to determine these arrays differ (see Challenges to Harmonization), the core concepts, such as whether a respondent is working in a particular week and the number of hours they work, are standardized. The process for creating these event history, arrays is quite similar in the NLSY79 and the NLSY97 and contains multiple steps of which the main ones can be summarized as follows:

- First, the dates for all jobs held are converted into week numbers, including within job gaps.
- Next, a *status* array is populated with the first job's ID for each week the respondent reported working, and an *hours* array is populated with information on hours worked at that job for each week worked.
- If more than one job is reported, information from the next job is then incorporated into the status arrays. For example, if a second job is held during a week where the respondent has reported a "within job" gap from the first job, then this second job information overwrites the first job's gap information in the main status array. This second job would also be entered into the main status array if the first job ended by that week. Otherwise, this job ID would be listed in the "dual" jobs array.
- Information from the second job's hours is incorporated into the *hours* array.

- The process is then repeated for up to 10 jobs for the NLSY79 and for all jobs for the NLSY97.
- If a respondent reports a gap of at least one week within a job and no other job is held during that week, additional information collected in the survey is used to determine if they were looking for work during those gaps so that either "out of the labor force" or "unemployed" can be assigned to the relevant weeks.
- Military service is included in the main status array only if no civilian jobs are worked in that week.
- Information collected during the survey is also used to determine the employment status between jobs.

Each stage includes additional steps to ensure data quality. For example, staff review interviewer comments post fielding if there is contradictory information, and update the data if the situation can be resolved. Other steps build on checks incorporated into the survey program. For example, respondents occasionally report that they do not know the day that a job began or ended, even with interviewer prompting. Since a day is required to calculate the gaps between jobs, the survey program imputes a day. These imputations are replaced with the code for "don't know" in the raw data but the array program incorporates the imputed information. Other checks are in place to ensure that dates are reported in the correct order (e.g. start date before stop date and gap dates fall within the start and stop dates of a job).

Additional rules ensure harmonization of core concepts across cohorts. For example, a job number appears in any week – defined as Sunday through Saturday – that the respondent reports working either a full or partial week. Both cohorts use a standard definition of active and passive search to assign unemployed versus out-of-the-labor force statuses. Although bounded interviewing combined with listing specific job names worked on the date of the last interview decreases seam effects, respondents do very occasionally report either a start or stop date prior to their last interview or deny ever working there. In these cases, the data reported through the date of the last interview remains unchanged. Standardizing these rules ensures that researchers can compare apples to apples.

All core data in these surveys have been collected using personal interviewers, regardless of whether the mode was a face-to-face paper and pencil interview (PAPI), a computer-assisted personal interview (CAPI), or a computer-assisted telephone interview (CATI). No core information has been self-administered over the web, although a Covid-19 supplement was fielded this way in the spring of 2021 to NLSY97 respondents – this was not a fielding period for them but the BLS believed information related to Covid-19 (primarily employment) was important to collect from both cohorts during the pandemic. Regardless of whether interviews are done by CAPI or CATI, the interviewer uses the same survey instrument that has pertinent information from prior surveys preloaded for each individual respondent. They record all survey answers on their laptops and data are then transferred securely to project servers. Surveys can be undertaken in either English or Spanish. The NLSY surveys, therefore, neither have to deal with the same concept being asked about in different parts of the survey nor with intra-wave harmonization as with CAPI/CATI interviewing and preloads of prior information, everything is checked against previous answers and consistency is enforced.

In many ways, the work history arrays are the cornerstone of these surveys. Researchers can use the employment data transformed into weekly arrays to easily match employer information by date with other events in the respondent's life. The weekly array format also allows researchers to construct variables according to their own timeframe of interest (e.g. monthly, quarterly, and yearly). This means that if a researcher wants to focus on the number of weeks or hours worked over a particular age range or set of months/years as in our earlier example, they are able to select just the relevant weeks from the arrays and construct their measures without having to piece information together across multiple rounds.

The work history arrays are particularly useful when researchers want to look at labor force participation with respect to a life event that may fall across a large range of timeframes for different respondents, such as degree completion or before and after the birth of a child. A week-to-date crosswalk is provided within the NLS documentation (described more fully below) to enable this form of harmonization. Presenting employment information as a continuous work history eliminates the need to look at start and stop dates across multiple rounds for a long-standing continuous or career job. For respondents returning to the survey after missing one or more rounds, information is collected about employment since the date of last interview and seamlessly used to update both the work history and hours arrays.

5.4 Challenges to Harmonization

Despite conscious efforts to enable harmonization of data across the various cohorts, challenges still face the researcher wishing to do so. When considering just the two ongoing youth cohorts, the NLSY79 has a wider age range of respondents than the NLSY97. Also important is that most NLSY97 respondents were initially interviewed when they were younger than their NLSY79 counterparts at their first interview, which affects the comparability of information available in the two cohorts. Other challenges can be illustrated using our example of work history arrays again.

Although the overall structure of the work history arrays is the same across the two cohorts, there are some differences between the NLSY79 and NLSY97 that have to be considered by users wishing to use the work history data from both cohorts. Some of these differences are historical artifacts related to changes in data collection, data storage, and dissemination. Others reflect design differences between the two cohorts.

As alluded to earlier, the NLSY79 began as a PAPI survey, and at a time when data storage space was limited and costly. During the PAPI years information was recorded in a single main questionnaire booklet with additional paper supplements to capture additional jobs. The main questionnaire collected information on labor force activity in the prior four weeks, along with information on the current or most recent job. Information was also recorded on up to nine additional jobs that had been held since the date of the previous interview. Due to data dissemination space constraints (these were the days when data were released on large reels of tape to be read on mainframe computers), employment information for only the five most recent job supplements was included on the public tape release. The work history arrays were accessible as a separate release that could be merged with the main data using the respondent ID. When the NLS moved to CAPI and there was more storage space available on CD-ROMs, the survey could both collect and disseminate more detail about jobs, and the releases were also combined. All data from all areas of interest, plus all created variables are released together for each survey round. However, to maintain historical consistency, details for only five jobs are presented on the main data set with the event history arrays containing weeks and hours data for up to 10 jobs.

In contrast, when the NLSY97 began, laptop computers were readily available enabling CAPI interviews from the start. CAPI interviews can be much more complex than when data are recorded on paper – they allow personalized interview pathways, and questions in a subsequent round that are determined by responses in a previous round. The NLSY97 survey collects full information on as many jobs as a respondent reports and all data are provided on the public data set.

Another key difference is how jobs are identified in the work history arrays of the NLSY79 and NLSY97. For the NLSY79, job numbers assigned to all jobs reported since the date of the previous interview are linked to the employer in a round-specific manner. As a consequence, in order to link a job in one round with the same job in a previous round, the researcher needs to use a variable identifying the previous job ID. This quirk stems from the PAPI years and can lead to overcounting unique employers when a respondent leaves an employer and then returns to that same employer in a future round. In 2013, a comprehensive "Employer History" roster was created to ease the burden of linking jobs through a separate variable for researchers; users still need to match IDs in the work history arrays across rounds but they can use the single ID available in the employer history roster rather than digging through each round's raw data. In contrast, the NLSY97 used CAPI from the start, and hence, was able to easily store and retrieve past rounds' information. In this cohort, the job ID for the NLSY97 is unique to the employer. If a respondent returns to a job after a number of rounds, the same Job ID is used, which gives a more accurate count of unique employers.

With changing times and research needs, survey content often requires modification and new questions to be introduced over time. This means that not all content is available in both cohorts or even within the same cohort across time. For example, both cohorts contain information about self-employment, but how this information is collected differs, which can impact a user's ability to harmonize across the two cohorts.

In the beginning years of the NLSY79, information about early self-employment jobs like babysitting or lawn mowing was collected using the same series of questions as are asked about regular jobs. This means that self-employed jobs are included in the NLSY79 work history arrays and in order to identify self-employed respondents, the researcher needs to examine the "class of worker" variables. Reflecting a new social reality, however, the NLSY97 made a distinction between self-employment and freelance jobs beginning in the early survey rounds. Respondents ages 12–14 did not go through the regular employment questions but instead reported on their freelance jobs in a separate section. Furthermore, respondents aged 16 or older who usually earned $200 or more per week at a freelance job were considered to be self-employed and were not asked the regular employment questions in the first three survey rounds, resulting in less information being available for these jobs at these ages in the NLSY97 than in the NLSY79. Starting in Round 4, respondents aged 18 and older skipped the freelance section completely and reported any self-employment, along with regular employee jobs. In Round 6, because all respondents were 18 or older, the freelance section was dropped completely. What all this information boils down to is that anyone who might want to compare self-employment activities *prior to age 18* across the two cohorts will have a lot of difficulty doing so, primarily because the two surveys collected self-employment data in different ways. From age 18 onward, the two sets of arrays present the same information. Because the youngest respondents in the NLSY79 were 14 at first interview whereas the youngest NLSY97 respondents were 12, however, freelance job information at these youngest ages is only available for some of the NLSY97 respondents – and of course this applies to other information collected from 12 and 13 year olds in 1997 for whom there are no counterparts in the NLSY79.

The fact that there is only partial overlap in the ages of the two cohorts in the early years of each study (i.e. only 14–16 year olds are in Round 1 of both the NLSY79 and the NLSY97), means that researchers will generally have more difficulties trying to harmonize variables across cohorts at the youngest ages but this eases as the NLSY97 respondents age. On the flip side, information on the teenage years of the NLSY79 older respondents had to be recovered retrospectively rather than contemporaneously.

For example, information on work history prior to the date of their first interview is collected from all NLSY79 youth who were 18 and older in Round 1. This again presents barriers to true harmonization.

Longitudinal studies outside of the United States often start at birth: the four British longitudinal cohort studies, the Dunedin Multidisciplinary Health and Development Study in New Zealand, and the Growing Up in New Zealand Study. Others begin when children are young such as Growing Up in Australia: The Longitudinal Study of Australian Children (LSAC), and Growing Up in Ireland, and both of these investigations have two cohorts, one of infants and one of older children. Each of these studies also has a much narrower range of ages than the NLSYs with age differences generally measured in months or even weeks rather than years (for example, in Ireland the two cohorts were ages 9 months and 9 years when the study began, LSAC commenced when children were 0–1 and 4–5 years old, and both the 1958 National Child Development Study and the 1970 British Cohort Study followed the lives of children born within a single week in Great Britain). The NLSY studies have, therefore, needed to pay particularly close attention to maintaining question wording over time to ensure that, for example, questions asked of NLSY79 respondents who were 25–26 year olds in 1985, and hence, who had been 19–20 year olds in 1979, were the same questions as those asked of 25–26 year olds 4 years later in 1989 who had only been 15–16 in 1979. Overall the NLS has been successful in this endeavor, but it has not always been possible to do so making even within-cohort harmonization tricky at times. However, when pitted against problems of harmonization that arise when trying to harmonize across different studies, from different countries, and with different purposes, these kinds of issues, although important to note, pale in significance.

5.5 Documentation and Quality Assurance

As mentioned in the introduction, NLSY97 and NLSY79 respondents experienced the transition into the labor market under very different educational, employment, and family building circumstances. The design of the NLSY97 relied on many of the key concepts collected in the NLSY79, however, both societal changes and advances in technology led to different types of questions asked. The processes used in the work history arrays smooth out these differences and, along with the documentation described below, provide a road map for researchers to create similarly harmonized variables in other domains.

Online user guides are provided for each NLS cohort at the NLS website, www.nlsinfo.org, and use a standardized section structure enabling users to navigate to similar content across the various cohorts. Each provides a detailed description of the sample including information about the sample design and screening process, interview methods, retention, and reasons for non-interview. A brief introduction to each cohort contains information on the age of cohort members at the time of first and most recent interview, which allows users interested in cross-cohort or cross-national research to find relevant initial information quickly and easily. A section on using and understanding the data provides material on survey instruments, types of variables, interviewer remarks, item nonresponse, interview validation, sample weights and design effects, a description of the data documentation, and how to use the NLS Investigator to access data. The Topical Guide segment contains a topic-by-topic description of the data, tables of topics across all survey rounds (known as "Asterisk Tables"), and a glossary of terms. For example, each cohort's topical guide includes a section titled "Employment," which is then broken down into subsections such as Industry, Occupation, Hours Worked, and Wages. Other main topical sections include Education, Family Background, Marital History and Fertility, Health, and Income and Assets.

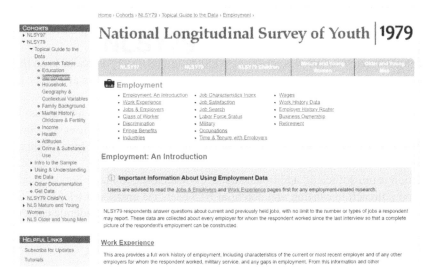

Figure 5.1   Screenshot of NLSY79 Employment Topical Guide section.

Topical guide sections list all created variables or event history variables at the top of the page, and provide a detailed narrative that includes information about any changes in the way questions were asked across survey rounds. Important information about using variables is highlighted, and links to other relevant sections are provided.

To aid researchers wishing to do cross-cohort research, tabs at the top of each topical guide for one cohort allow them to navigate to analogous content in other NLS cohorts. Questionnaires, tutorials, codebook supplements, errata, and appendices are included in a separate section. In Figure 5.1, we present a screenshot of the NLSY79 Employment section, with tabs to link to the employment sections for the other NLS cohorts, and links to other NLSY79 documentation items in the left hand navigation menu.

The NLSY79 user guide[1] has a separate Work History Data[2] subsection, which lists the created variables in the work history arrays, as well as the summary variables created from these arrays. Important information for using the work history data is highlighted, and the process by which the arrays are created is described. A separate Work History Data appendix[3] provides a detailed history of the programs used to create the work history variables across time, along with additional user notes. The NLSY97 user guide[4] includes information about the work history data in the introductory section to Employment,[5] and because NLSY97 data are collected in an event history format, how this data collection process works, is also explained here. Additional information about the work history variables, created variables,

---

1 https://www.nlsinfo.org/content/cohorts/NLSY79
2 https://www.nlsinfo.org/content/cohorts/nlsy79/topical-guide/employment/work-history-data
3 https://www.nlsinfo.org/content/cohorts/nlsy79/other-documentation/codebook-supplement/
nlsy79-appendix-18-work-history-data
4 https://www.nlsinfo.org/content/cohorts/nlsy97
5 https://www.nlsinfo.org/content/cohorts/nlsy97/topical-guide/employment

and important information concerning their use is contained in the Employers & Jobs[6] subsection. A separate Event History Creation and Documentation appendix[7] is also provided that presents additional detail about the event history creation and documentation.

A tutorial on "Constructing Comparable Samples across the NLSY79 and NLSY97"[8] is available on the documentation webpage. This page walks users through how to construct work status at age 20 in each cohort.

Finally, the US Census Bureau is contracted to code industry and occupational data to federal standards across the NLS cohorts. To maintain consistency across time and cohorts, the NLS uses the same code frames for longer periods. When the NLS adopts new standards to account for labor force changes in the economy, a crosswalk is provided by Census.

If the online documentation is not sufficient, additional support is available. NLS User Services staff provide direct assistance and outreach to existing and potential NLS users via email and telephone, as well as through workshops, posters, and exhibit booths held at multiple and varied academic conferences in the United States each year.

5.6   Software Tools

Although the programs used to create various NLS variables (such as the work status arrays described earlier) have been generated using different programming languages, this is invisible to data users. What is most important to the NLS program is that researchers should be able to understand, access, and download data as easily as possible to encourage their use. The vast majority of information from each of the NLS cohorts is available as separate databases and is free and easy to access through the NLS Investigator (http://www.nlsinfo.org/investigator) – an online search and extraction site that enables users to search and review NLS variables and create their own data sets. They are used by researchers from around the world and from a wide variety of disciplines including economics, sociology, criminology, public health, psychology, education, geography, nursing, family studies, and public policy.

As described below, eight search methods are available in NLS Investigator, which can be combined using "and"/"or" logic operators:

- *Area of interest.* helps narrow down variables to a particular topical area.
- *Word in title.* any word that appears in the variable title.
- *Question text.* any word (or portion of a word) that appears in the question text.
- *Question name.* a code that identifies each question – provides the location of the variable in the questionnaire or identifies it as a created variable. When possible, question names remain the same for the same question across survey years.

---

6 https://www.nlsinfo.org/content/cohorts/nlsy97/topical-guide/employment/employers-jobs
7 https://www.nlsinfo.org/content/cohorts/nlsy97/other-documentation/codebook-supplement/
appendix-6-event-history-creation-and
8 https://www.nlsinfo.org/content/getting-started/intro-to-the-nls/tutorials/tutorial-constructing-comparable-
samples-across

- *Reference Number.* abbreviated as RNUM, a letter and number combination uniquely assigned to each variable.
- *Survey year.* enables restriction of search to one or more survey interviews.
- *Codebook.* text appearing anywhere on the codebook page.
- *Variable type.* allows user to pick a class of variables such as created variables or roster variables (NLSY97 only).

Additionally, the NLS Investigator also organizes key variables into a number of topical areas such as education, employment, income, and health. This means that instead of searching among all variables using the traditional search indexes outlined above, the user can simply click on a general research area, and then choose among related subtopics. Each topic area then includes unfolding subtopics; for example, "Employment" unfolds into summary measures, employment history, employer-specific characteristics, and job search. In the NLSY79 example, a researcher would be able to click on "Employment" then "Employment History" and then "Weekly Arrays" to locate these arrays (see Figure 5.2). An identical index structure guides researchers to the NLSY97 work history arrays, which assist in cross-cohort comparisons.

Figure 5.2 Screenshot of NLS Investigator search function.

Researchers are, therefore, guided to comparable variables, both the harmonized created variables described above and other questions asked during the interview for which no created variables exist. The ability to browse through topical areas is especially helpful for researchers new to the NLS trying to find basic variables from the thousands of variables available to them, as well as making it easier for all researchers to understand the breadth of information available.

5.7 Recommendations and Some Concluding Thoughts

An underlying thread in this chapter is that the NLS has collected similar information from two groups of youth of close but not identical ages who lived nearly 20 years apart, and for whom a set of variables has been created that allow researchers to undertake both within- and cross-cohort analyses. We focused on the example of employment and variables that would allow researchers to calculate the effects of labor force transitions on various other events (education, marriage, parenthood, etc.) in the same way for both cohorts. On the other hand, life is not static and many changes occurred in the labor market in the 20 years between the start of the NLSY79 and the start of the NLSY97 – the decline of career jobs, union membership, and pension offerings, and the rise of alternative work arrangements and the gig economy are a few notable examples. To be responsive to societal change, survey questions have to change accordingly, which can cause havoc for harmonization! Furthermore, as illustrated by the movie *Hope and Glory*, an outside shock or a period effect will impact different cohorts in different ways, and survey design has to take this into account too: questions to ascertain the effects of a recession on people's lives will sometimes differ for 25 year olds and 45 year olds as, for example, 25 year olds are likely to be early in their careers, marriages and family building activities whereas 45 year olds are more likely to have mortgages and school-aged children. What the NLS has attempted to do is balance the need to be responsive to change, while still maintaining underlying core elements such as the basic structure of the employment arrays for consistency.

An interesting observation is that some of the global labor market changes occurred in response to changes in technology that also altered survey methodology including variable creation plus data collection, storage, and dissemination. As job profiles became more complicated, we also expanded our abilities to ask about them were able to store more data, program more complex algorithms, and disseminate variables more easily.

Although our focus on work arrays only covers one slice of a much broader set of employment-related factors, what might the future hold for data collection on employment? A cautionary note would be that alternative work arrangements are likely to play a larger part in people's working lives and hence create more of a challenge to future data collection efforts when trying to trace school-to-work transitions, or understand working lives across the life course. The gig economy is likely to become even more common in the future, and after Covid, it is not only likely that people will spend more time working from home, but employment patterns may well change in other unexpected respects. With traditional work situations changing in multiple ways, we believe it will be even more important in the future to be able to connect nonworking life event transitions to a variety of additional employment metrics such as work intensity and flexibility. Collecting information on "usual" or "typical" hours, or actual weeks worked at a job when sporadic gaps need to be taken into account is already time-consuming and expensive and likely to only become still more challenging.

In longitudinal research, a tension exists between maintaining consistency across time to enable harmonization and facilitate comparisons, and the desire to be responsive to changes in respondent's life circumstances and external contexts, or advances in measurement. As cohorts age, different questions may be needed to address changes in their life circumstances. Changes in policy or other external challenges may necessitate new survey content, and methodological developments can create unique opportunities for innovative design.

A similar tension occurs when designing surveys to maximize opportunities for cross-cohort comparison. Simply replicating existing questions for a new cohort may be insufficient to capture the same key constructs in a new generation of respondents. Are the questions asked of a prior cohort still relevant for the current life circumstances faced by a different generation? If you learn things that suggest you need to change the question to more accurately reflect the construct of interest, this could well be more important than keeping the same exact question wording. When designing the NLSY97, many of the questions and modules were the same or quite similar to those included in the NLSY79, but others were changed or added to address changes in circumstances faced by these more recent cohorts. Different generations not only face different opportunities, prospects, and challenges but also sometimes the same external factors impact multiple cohorts in real time when they are at different life stages making it impossible to make comparisons when the younger cohorts age up into the same life stages and ages as the older ones.

Anyone who has ever tried to design a survey where many participating voices are required to create a full and well-rounded chorus, but then seemingly infinite desires and multiple competing needs are expressed by those expert voices, knows that one hour of survey time begins with 5–10 times that length in initially proposed questions. In making decisions about what to keep, what to change, and what to add, it helps to remember that there is no single survey or set of surveys that can capture every story to be told across an individual's lifespan. By extension, no single data resource will ever be appropriate to answer every research question. With a user base as broad as the NLS possesses, it would be impossible to anticipate all the ways users might wish to harmonize across cohorts. It is also not just *what* is asked but *when* questions are asked that presents challenges for harmonization – decisions that can also be affected by funding availability and levels. What is important is to offer variables such as "age at interview" to aid researchers in making linkages across cohorts, provide consistent documentation and user support, and focus data harmonization efforts at the program level on the key elements, knowing that others may use the data for different purposes.

References

Cooksey, E.C. (2018). Using the NLSY to conduct life course analyses. In: *Handbook of Life Course Health Development* (ed. N. Halfon, C.B. Forrest, R.M. Lerner, and E. Faustman), 561–578. Springer.

Rothstein, D.S., Carr, D., and Cooksey, E. (2018). Cohort profile: the National Longitudinal Survey of Youth 1979 (NLSY79). *International Journal of Epidemiology*, dyy133. https://doi.org/10.1093/ije/dyy133.

6

Harmonization in the Comparative Study of Electoral Systems (CSES) Projects

Stephen Quinlan[1], Christian Schimpf[1], Katharina Blinzler[1], and Slaven Zivkovic[2]

[1]*Data and Research on Society, GESIS Leibniz Institute for the Social Sciences, Mannheim*
[2]*Johnson Shoyama Graduate School of Public Policy, University of Saskatchewan*
[3]*Survey Data Curation, GESIS Leibniz Institute for the Social Sciences, Cologne*
[4]*Department of Political Science, Johannes Gutenberg University of Mainz*

6.1 Introducing the CSES

The Comparative Study of Electoral Systems (CSES) is a collaborative research project among national election studies globally. Founded in 1994 (Thomassen et al. 1994) and devised by academic scholars, the CSES helps to understand how individual attitudes, institutional and electoral rules, and contextual conditions shape voting behavior. The core mission of CSES is to make electoral research global, multidimensional, and, given its durability, longitudinal. Participating election studies include a standard battery of survey questions (in CSES vernacular a Module) in their national postelection studies at least once each five years to a national sample of respondents representative of the eligible electorate. The Modules are designed to require a maximum of 15 minutes of interviewing time per respondent. CSES studies run in Module blocks of five years, with each Module having a dedicated theme. The resulting data are collated and augmented with contextual information about the electoral district, and other social, electoral, and economic conditions that characterize each participating polity and election. CSES has been a pioneer in combining macro data with individual cross-national survey data. Besides weights and study administration data, CSES data are cross-sectional and contains information from three distinct levels, namely:

1) microlevel variables generated by respondents' answers to the postelection survey fielded in each polity,
2) district-level variables containing election results from the constituencies that survey respondents are located in, and
3) macro-level variables imparting information about the polity's institutional character, and the contextual conditions at election time.

Table 6.1 provides an overview of the CSES Products. As of April 2022, CSES has fielded five completed modules with a sixth in the offing. It has been run in 57 polities, with at least 216 election studies

Survey Data Harmonization in the Social Sciences, First Edition.
Edited by Irina Tomescu-Dubrow, Christof Wolf, Kazimierz M. Slomczynski, and J. Craig Jenkins.
© 2024 John Wiley & Sons Inc. Published 2024 by John Wiley & Sons Inc.

Table 6.1 Summary of the CSES Data Products as of April 2022.

CSES data products		Time period	N Polities	N Election Studies	N Variables	N Cases
CSES modules	Module 1	1996–2002	33	39	304	62,409
	Module 2	2001–2006	38	41	434	64,256
	Module 3	2005–2011	41	50	448	80,163
	Module 4	2011–2016	39	45	484	75,558
	Module 5[a]	2015–2021	36	41	554	76,123
CSES integrated module Dataset (IMD)[b]		1996–2016	55	174	389	281,083

Note: Date correct to April 2022. In the calculation of N of variables, grouped variables, such as respondents' left–right placement of multiple parties, are counted individually here.
[a] CSES Module 5 is a preliminary release at the time of writing. More election studies are anticipated in the full release, scheduled for 2023–2024. N of polities, election studies, and cases refers to number correct as Advance Release 4 of Module 5.
[b] CSES IMD currently includes data from Modules 1–4. Future releases of the product will consist of subsequent Full Releases of Modules.

participating in CSES over the past quarter-century. In 2018, it launched the CSES Integrated Module Dataset (IMD), bringing together data common in at least three Modules into one longitudinal and harmonized product (for more, see Quinlan et al. 2018). CSES has served as the seed for the launching of several national election studies. Its data have been featured in over 1100 peer-reviewed research outputs, with over 15,000 users downloading CSES data products from over 140 countries since 2012. In sum, CSES is the go-to dataset for scholars wishing to study electoral behavior cross-nationally and longitudinally.

CSES comprises the collective efforts of the scholars who administer the CSES questionnaire – in CSES parlance, the Collaborators. Collaborators raise their own funding and collect the data while adhering to the CSES guidelines in administering the CSES Module questionnaire. The CSES is governed by the Planning Committee (PC), elected every five years by Collaborators upon suggestions from this group and the CSES user community. The PC is tasked with designing a thematic CSES Module after submissions from the user community and is also charged with overseeing the CSES Secretariat's work. The CSES Secretariat is the project's administrative and data hub. Since 2011, the Secretariat has been a partnership of the GESIS – Leibniz Institute for the Social Sciences in Germany and the University of Michigan, USA, with financial support provided by the American National Science Foundation, GESIS, and the University of Michigan.[1] CSES prides itself on the principles of being a public good, and the project's philosophy incorporates a democratized bottom-up approach. Data are distributed at no cost with no embargoes.

This chapters' objective is to explore how CSES approaches data harmonization. We tackle this thematically, using the industry-standard differentiation between ex-ante and *ex-post* harmonization (Granda et al. 2010; Wolf et al. 2016). While CSES overwhelmingly fits the ex-ante classification, the recent creation of the CSES IMD and the need to conquer new frontiers in comparative research has

---

1 For more details, see CSES website at https://cses.org.

resulted in the project's scope evolving to incorporate *ex-post* dimensions. Our chapter explores themes common to other projects: the creation of the Module questionnaire, the project's guiding methodological principles, and demographic variable harmonization. More innovatively, we deep dive into harmonization issues more custom to the CSES, including macro data harmonization, the coding of parties/coalitions, and the interplay between ex-ante and *ex-post* harmonization CSES encountered in the development of IMD. Before advancing further, we briefly outline five key principles, the CSES adheres for data harmonization and technical infrastructure used to achieve harmonization objectives.

6.2 Harmonization Principles and Technical Infrastructure

CSES has at its core five principles for harmonization, namely:

1) *Strong relationship with Collaborators.* Communicate to Collaborators CSES methodological principles and provide clear instructions on administering Module questionnaire to ensure maximum harmonization. Encourage Collaborators to liaise with CSES Secretariat and PC in advance about potential non-conformities.
2) *Rigorous documentation.* Applies internally to processes used by the CSES Secretariat (e.g. log files to keep track of harmonization decisions with deposited datasets and working with Collaborators to understand nonconformities). There is an external component, too – harmonization decisions and challenges are highlighted for users of the CSES in its Codebook. CSES Codebooks are among the most comprehensive in volume vis-à-vis other comparative social science datasets. The CSES approach is "documentation is our friend" – the more detail provided to users, the better.
3) *Replication.* In line with the social science gold standard of replication (King 1995), all harmonization decisions made by the CSES Secretariat have to be reproducible internally through standard programming scripts with extensive layman instructions on the coding decision processes for each variable. CSES Secretariat members must track all data changes in this script, ensuring replicability of harmonization decisions across personnel and time, and these are detailed externally in the CSES Codebook.
4) *Extensive data checks before data release.* Uniform programming scripts facilitate standardized data-checking procedures to ensure harmonization is reliable and valid.
5) *User empowerment and transparency.* Any deviances hampering comparability are highlighted, allowing users to decide whether this impacts their research question. CSES operates on the maxim number of users who are best equipped to determine what is suitable for their analysis.

To achieve our harmonization goals, CSES principally relies on Microsoft Office and STATA, both software tools chosen because of their broad availability, popularity, and cross-platform compatibility.

6.3 Ex-ante Input Harmonization

The CSES is unusual compared to other comparative projects as it has two ex-ante input components – the advancement of the Module questionnaire (microlevel data) and the development of an extensive suite of macro data. We deal with each in turn.

6.3.1 Module Questionnaire

Previously, Howell (2010) has described the CSES questionnaire design process. Rather than recapitulating this contribution, we stress relevant features concerning harmonization. While each CSES Module was initially considered independent of the Module that preceded it, there has been a recent appreciation of longitudinal comparative analysis. In response, the PC now defines "a core" set of questions that are asked in each Module, thus facilitating the possibility of time series analysis. Whereas Theme Module questions aim to capture *zeitgeist* issues during which the module is in the field, core variables constitute what could be classified as the essential variables in electoral research – turnout, vote, party identification, satisfaction with democracy, partisanship, and political efficacy.

A subcommittee of the PC is tasked with devising a Module theme that is intellectually rigorous and relevant and is also applicable to a variety of states. Questions may be sourced from other studies to ensure the validation of question measures, although, in certain circumstances, questions have been devised anew. Still, only after consideration of the pretest results are these questions included. Once the PC reaches a consensus regarding a draft questionnaire, they devise a stimulus paper detailing the substance behind the proposed theme. Several pretests are fielded, ideally in a random sample postelection study in at least three different polities to maximize regional and linguistic variation (Howell 2010), optimally encompassing variation in region and language. The pretests are conducted by volunteer Collaborators, with the analysis of the tests carried out by the CSES Secretariat and the pretest Collaborators. These pretest endeavors are the tool for CSES to evaluate the Module in various contexts. Their results and the proposed questionnaire are showcased to the CSES Plenary, the project's governing assembly, which includes all national Collaborators of the project, for discussion. Upon deliberation of the stimulus paper, the questionnaire, and the pretests, the Plenary adopts a final version of the questionnaire, usually through a consensus vote.

While deviations from the master questionnaire should be kept to a minimum to enhance comparability and proposals to diverge should be discussed with the CSES Secretariat and PC in advance, in reality, there are some nonconformities for legitimate reasons. It falls to the CSES Secretariat to address these issues, adapting the country-level coding decision to CSES standards to ensure maximum harmonization. Two of the most common deviations concerning the master questionnaire relate to when a two-pronged filter question is combined into one item or an additional answer category is offered to respondents. For example, the retrospective sociotropic economic measure in CSES Module 4 was devised by the CSES as two separate questions – the first asking *"Would you say that over the past twelve months, the state of the economy in [COUNTRY] has gotten better, stayed about the same, or gotten worse?"* If the respondent answered gotten better (or gotten worse), a follow-up question probed the extent to which the respondent considered the situation improved or worsened. However, some studies administered the item without the filter, offering just one question with five categories to respondents.

Internally, the CSES Secretariat employs two key devices in addressing these nonconformities. The first is the use of *Collaborator Questions*. These are exchanges between Collaborators and the CSES Secretariat after conducting a preliminary cleaning of the data. Collaborator Questions enable the CSES Secretariat to probe Collaborators on several issues, including nonconformities to the master questionnaire and the rationale behind these deviations.[2] These questions and answers are recorded in the *CSES*

---

2 On average, the CSES Secretariat poses between 20 and 25 questions to Collaborators covering both micro, district, and macro data components.

log file – the second device. It is an internal form of documentation for the CSES Secretariat detailing all issues arising in harmonizing national-level data and back-and-forth discussions between the CSES Secretariat and the Collaborators. It allows a long-standing record of the data cleaning process to be maintained, enhancing replication and data quality.

Externally, the CSES relies on the documentation of its harmonized coding decisions. For example, take the sociotropic economic voting deviation we highlighted above. This was dealt with by breaking up single questions into two parts to accord with CSES standards. For variations akin to the satisfaction with democracy measure, CSES adds an additional value category to the satisfaction with democracy question labeled "see Election Study Notes (ESNs)" – this alerts users that this is a deviant category specific to a study and users need to consult the documentation.

Figure 6.1a,b illustrate this point and introduce the CSES *ESNs*.[3] An ESN is additional information in the CSES Codebook attached to a variable or a specific case. They detail information concerning deviations and details of national coding schemes for demographic variables (see Section 6.4). Knowledge contained in ESNs allows the user to determine the impact of the deviation on their analysis.

Harmonization endeavors do not stop at administration of the master questionnaire, however. All surveys appearing in CSES must meet a minimum set of methodological standards to ensure data quality and enhance harmonization, some of which we elaborate on below.

Principal among these is all election studies' samples must be random at all selection stages and have sufficient geographic coverage (Howell 2010).[4] A minimum of 1000 age-eligible respondents per study is required, and the CSES Module must be principally based on a postelection component, specifically the turnout and vote choice questions, to ensure reported behavior rather than reported intent is measured. There is scope for multi-wave studies to ask about demographics and other items before the election, but this decision needs the PC's agreement.

Nonetheless, given the CSES survey is most often an add-on to a national election study, CSES embraces a degree of methodological pluralism applying to certain methodological aspects. Mode for one. While CSES strongly recommends face-to-face interviewing, studies using mail-back, telephone, or multimode are acceptable if they meet representativeness standards described above.[5] The project encourages Collaborators to work together on translations as happens with the German, Swiss, and Austrian CSES Collaborators. While the CSES favors the survey is administered as one block of uninterrupted questions and fielded within three months of the election date, exceptions can be made with the PC's agreement. To ensure all this information regarding the module's administering can be gleamed, national Collaborators are required to complete the CSES Design Report. It contains a variety of information to assess the quality of data collection, including material on the sample design and sampling procedure, sample selection, and the response rate. Thus it is the

---

3 These notes can also come as "polity notes" if the issue applies to multiple studies from the same polity, a common occurrence with CSES IMD.

4 Naturally, when conducting a survey, some groups are over/underrepresented in the sample. The CSES Secretariat performs data quality checks to ensure the data are representative of the intended population of interest. To adjust for over and/or underrepresentation of particular groups, CSES products include up to three sets of weights for each study where applicable – design, demographic, and political, which are created by the Collaborators.

5 Most CSES studies (~70%) up to an including Module 4 have fielded the Module face-to-face.

```
--------------------------------------------------------------------------
D3003_1     >>> Q03. STATE OF ECONOMY
--------------------------------------------------------------------------

        Q03. Would you say that over the past 12 months, the state of
             the economy in COUNTRY has gotten better, stayed about the
             same, or gotten worse?
        ..................................................................

            1. GOTTEN BETTER
            3. STAYED THE SAME              -> GO TO Q4
            5. GOTTEN WORSE                 -> GO TO Q3b

            7. VOLUNTEERED: REFUSED         -> GO TO Q4
            8. VOLUNTEERED: DON'T KNOW      -> GO TO Q4

            9. MISSING
        | ELECTION STUDY NOTES - GREAT BRITAIN (2015): D3003_1
        |
        | The state of the economy (D3003_1-D3003_3) was asked as a
        | single question: How do you think the general economic situation
        | in this country has changed over the last 12 months?
        | Answers to this question were recoded as follows:
        |
        | CSES Code       Election Study Code/Category
        |       01.       a little better
        |                 a lot better
        |       03.       the same
        |       05.       a little worse
        |                 a lot worse
```

Figure 6.1 Example of Election Study Note concept in CSES Codebook. Source: CSES Module 4, variables D3003 (State of the Economy), and D3017 (Satisfaction with Democracy).

primary tool for judging whether the data collection meets the methodological standards for inclusion in the study. This document also provides vital information to the CSES Secretariat in its harmonization efforts.

To draw users' attention to the variety of administrative approaches used to field the study cross-nationally, CSES embraces two methods. The first is data-driven and involves the inclusion of variables in the dataset classifying the interview mode of respondents, the fieldwork dates, the interview timing vis-à-vis the number of days since the election date, the interviewer's gender, and the language of interview. It allows users to control for methodological variety in their analyses. The second approach is documentation. The Design Report and the original language questionnaires for each study are made public for transparency. The CSES Secretariat devises a written overview of each study's design in CSES (see Figure 6.2). *ESNs* are also used to highlight issues in greater detail to users in each particular variable.

6.3.2 Macro Data

CSES has been the pioneer of having a macro data component since its inception in 1996. Macro data are a means of capturing the contextual environment survey respondents operate in. Macro data have two different variants: system-level data and aggregated data (Fortin-Rittberger et al. 2016). The former delineates systems or institutions' properties and are not aggregated up from lower-level units but are sourced from constitutions, electoral commission documents, legal rules, or government information pages. Examples include the features of a country's electoral system, the type of governing structure, or governments' composition. Aggregated macro data blends data about lower-level units into a higher-level unit that summarizes the features of the lower-level units. Economic indicators such as gross domestic product (GDP) or unemployment data for a country fall into this classification. CSES collects both these macro data types, with the responsibility for the curation and harmonization of this data falling to the CSES Secretariat.

```
-------------------------------------------------------------------------
>>> OVERVIEW OF STUDY DESIGN AND WEIGHTS - GERMANY (2013)
-------------------------------------------------------------------------

<<>> STUDY DESIGN

The CSES module is part of the German postelection cross-sectional study,
for which the fieldwork was conducted between 23 September and 23 December,
2013, by MARPLAN Media- und Sozialforschungsgesellschaft mbH. Data
collection started one day after the federal election on 22 September and
lasted for 92 days in total. The questionnaire was administered in German
and in person (CAPI).

The sample is designed to be representative of all German citizens resident
in the Federal Republic of Germany, who were living in private households,
aged 18 or older, and who were eligible to vote. Originally the sample also
included an additional number of 19 persons of age 16-17, but these were
excluded from the CSES as they are not eligible to vote.

The sampling method used was the Address-Random Method (ADM) design, with
all private German households as the base population. 306 sampling points
(voting districts) were randomly selected - 211 in West Germany and 95 in
East Germany. In West Germany, 1,400 interviews or 6.6 interviews per
sampling point on average were aimed for and in East Germany 700, or 7.4
interviews per sampling point on average, resulting in an oversampling of
East Germany. In every sampling point, 60 households' addresses were
randomly selected using the random  route sampling. From this list of
confirmed addresses, 25 households per sampling point were selected
randomly. The third step was the selection of the target person via the
Kish-Selection-Grid. A minimum of four attempts were made to contact a
potential respondent, at different times of the day and days of the week,
before declaring it a noncontact. There are 1889 completed interviews in
the sample.

<<>> POLITY WEIGHTS

The German Election Study provides two weights: a SAMPLE WEIGHT and a
DEMOGRAPHIC WEIGHT. The sample weight controls for the oversampling of East
Germans and household size. The demographic weight controls for education,
age, gender and size of the communities. User are advised that the weights
were calculated by the German Election Study while still including 19
respondents who were ineligible to vote in the 2013 election.
```

Figure 6.2 Example of Study Overview of an Election Study detailed in a CSES Codebook. Source: CSES Module 4, Germany 2013 study.

The CSES Secretariat employs a three-pronged approach to ensuring the best harmonization standards are realized. The first is the project employs macro experts on its staff. These individuals have strong expertise (usually to postdoctoral level) of the political dynamics and electoral systems of many of the polities in CSES.

Second, system-level data in part are sourced from the CSES *Macro Report*, which each Collaborator is responsible for completing. The *Macro Report* allows CSES to collect certain macro information consistently from national-level experts. Moreover, as polities participating in CSES are not always available from other comparative datasets, it ensures macro data are available for all elections and polities. Initially, most of the macro data emerged from this report. But as CSES has broadened its suite of macro variables and other macro sources have become available, the Macro Report is now only partly the source, now principally focusing on information unavailable from other sources.

Third, CSES relies on a suite of macro data sourced from reliable comparative sources like the World Bank, the United Nations, or projects like Freedom House and Polity IV. Since CSES Module 3, there was a move by the project to include macro data beyond the electoral and political system and to incorporate macroeconomic and social conditions. Hence, economic indicators like GDP, unemployment, democracy (e.g. Freedom House and Polity IV), fractionalization (Alesina et al. 2003), development (UNHD index) indices, and social measures (e.g. immigration statistics) are collected.

Macro data harmonization faces challenges. Concerning the Macro Report, one of the most common difficulties is ensuring consistency across time, especially if Collaborators have changed between Modules. In response, the CSES Secretariat independently verifies most of the information against other official sources where possible. It also cross-checks the scores assigned to variables in a particular polity with previous classifications in CSES Modules. These serve as critical quality controls (Howell 2010). Where inconsistencies are found, and where several methods of cataloging are possible, the CSES Secretariat interacts with the Collaborator to discuss how to classify the variable. In arriving at an outcome, the guiding principle is to ensure that the value is an appropriate fit in a comparative analysis. *ESNs* are commonly used to alert users to issues and to provide more contextual information to allow them to understand the coding decision.

On the aggregate macro data, there are two main challenges to consistency. The first is that there is often a time lag between when this data become available and when the CSES Secretariat may be gathering it. When this happens, the CSES will code it as missing with an ESN in the Codebook detailing it will be available in a forthcoming release. Another challenge to this kind of data is that it is often retroactively updated as more information becomes available. While such changes generally only result in a small revision to the data, they still pose a challenge. Take GDP for example – this data are often updated as more economic information becomes available in a specific polity. In the Codebook, CSES alerts users that data estimates for variables like GDP are those available when the data are processed and that the CSES does not retroactively update these estimates as to do so might impede replication. Internally, the CSES Secretariat notes the particular date on which the aggregate macro data are collected to ensure replicability. Another challenge that CSES can face is other comparative datasets like the World Bank do not have data for particular polities included in CSES. Take Taiwan, which has appeared in all CSES Modules since its inception, its status is contested, and its presence in comparative datasets is not always guaranteed. The CSES response looks to other sources for macro items in Taiwanese data when it is not available in the source dataset. If located, this data are included with an ESN in the Codebook detailing the deviation and providing the original source. The rationale behind this is to ensure that data were available and follow the CSES principle to give the user the power to determine the source deviations' impact on their analysis.

6.4 Ex-ante Output Harmonization

The CSES takes an ex-ante output harmonization approach for three types of variables: demographics, vote choice variables in Modules, and derivative variables – a variable created by the CSES Secretariat with reference to an existing variable in the dataset. We deal with each in turn.

6.4.1 Demographic Variables in CSES Modules

CSES approaches demographic variables in Modules similar to the International Social Survey Programme (ISSP) (see Chapter 5, this volume). Collaborators are provided with "ideal" conceptualizations and coding frames for the intended target variables but may deviate due to local conditions. As the CSES is, in many instances, an add-on questionnaire to more extensive election studies, some of which are older than CSES with a longstanding history of asking questions in a particular way. Suppose CSES was to impose a specific form of asking for an item. In that case, it would not only interrupt the time series but may diminish a study's willingness to participate in the CSES.

Upon receipt of data from Collaborators, the CSES Secretariat verifies coding for each demographic variable. Most variables arrive in a state where harmonization is effortless, with a relatively easy mapping to the standards set out by the CSES PC. However, we highlight two variables where we continually encounter challenges – namely education and income, often noted for their complexity (see Canberra Group 2011; Ortmanns and Schneider 2016). Education is difficult as the varying national education systems can be complex to harmonize, even though since Module 4, CSES has adopted the international standard – ISCED. CSES offers income data to users in quintiles. Nevertheless, this can be troublesome in mirroring to precise quintiles. In response, CSES Module 5 provides the income quintiles and a variable with the original income classification asked by each election study. This gives users power to decide whether to use provided quintiles, the original income variable (unharmonized), or to devise a comparative classification on the basis of what CSES offers.

The CSES Secretariat applies a multistep procedure in navigating these challenges, performing several data quality assessments, which Howell (2010) describes more. Once preliminary processing of the data is complete, the CSES Secretariat uses *Collaborator Questions* to obtain clarifications and explanations as to why the demographic was measured in a particular way and to explain nonconformities or oddities. CSES also performs extensive checks in the so-called "cross-national" phase when all processed individual-level studies (both micro and macro components) are merged into one single file.

One of the most important checks is the theoretical check, assessing whether frequently used variables in the dataset are theoretically sound after harmonization. For example, the CSES Secretariat investigates if the correlations between income and education and between education and interest in politics are in the expected direction at both the Module and polity levels. While identified deviations might point to processing errors in the data, sometimes anomalies can be perfectly reasonable and explained by country-specific conditions, so the project documents this to users in the CSES Codebook.[6] These checks are relatively original to the project and have been devised in-house by the CSES Secretariat (and shared with other projects like the ISSP). These checks are especially relevant for demographic variables, as these typically undergo most of the changes during data processing.

6.4.2 Harmonizing Party Data in Modules

One of the biggest challenges for CSES harmonization efforts concerns parties/coalitions and vote choice data, a data component many other cross-national social science projects do not incorporate. The CSES assigns parties two types of codes: numerical and alphabetical. Numerical codes are assigned to every party/coalition in the system mentioned in any of the relevant variables (for example, vote choice and/or party identification). Alphabetical codes, assigned to the most popular parties in the system as determined by the election results (up to nine parties), are referred to as the relational component of the dataset. By relational data, we mean data which connect a single respondent to multiple entities of the same variable. We illustrate this in Figure 6.3 with an example of a respondent asked to rank a suite of parties on the left–right ideological scale. As there are nine parties in this hypothetical example, the respondent will give nine separate ratings – one for each party. In CSES, there are several relational dimensions with respondents asked to ideologically place parties and rate their likeability of leaders and parties. The relational data are also connected to the macro component of the study, with election results and expert ratings by Collaborators provided for several parties.

Initially, harmonization of parties/coalitions was not a goal of the CSES. Instead, the numerical coding of parties/coalitions for Modules 1–4 was unique to each election study and Module and the numerical values assigned to parties/coalitions had no particular meaning. Consequently, users had to consult

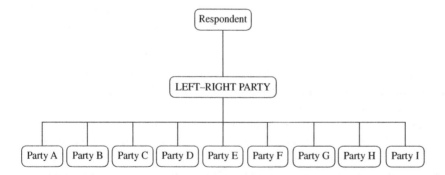

Figure 6.3 CSES relational data structure for variables assigned alphabetical codes A-I. The example is based on a hypothetical respondent asked to rate multiple parties on the left–right ideological scale. In the hypothetical example, there is the maximum nine-parties to rate. Source: Quinlan et al. (2018)/CSES.

---

6 We conduct similar tests for non-demographic variables including running models for turnout and vote choice.

CSES Codebook to decipher what party is assigned, which particular numerical code. In the later stages of CSES Module 4, the project adopted a new convention in the vein of providing a more harmonized approach to numerical coding. Numerical assignment depended on the parties/coalitions vote share in the election. For example, the numerical code 1 was assigned to the party/coalition winning the most votes in the main election, code 2 to the party/coalition winning the second most votes in the election, etc. However, with IMD's development (see Section 6.4), CSES adopted a new coding scheme for numerical party codes for CSES Module 5. Each party is now assigned a unique six-digit numerical code. The first three digits represent the UN ISO country code, the fourth digit distinguishes if there are one or more studies from the same country in the module dataset, and the last two digits are party codes. The latter follows the previous convention – lower numbers signify parties with more votes in the main election.

While the conventions applied to coding of parties and coalitions have evolved across Modules, some difficulties of harmonizing vote choice and party data have remained constant. A key challenge is how to operationalize coalitions and their constituent members. Consider an electoral alliance, the Alliance of Hungarian Solidarity, formed in the 2018 elections between the Fidesz Party and the Christian Democratic Party (KDNP). Vote choice data are for the alliance, but the relational data (e.g. the likeability of the party) are provided for Fidesz only. In other instances, the relational data are provided for all individual members of the alliance. This challenge becomes more complicated when relating this to how a respondent voted in the previous election if the said alliance did not contest the previous contest. CSES handles these scenarios by providing as much data as possible to the user (i.e. data related to all members of the alliance). Most important is to give detailed information in the CSES Codebook (see Figure 6.4 for an example) as to how alliances are classified for each election study.

6.4.3 Derivative Variables

In recent products (e.g. Modules 4–5, IMD), the CSES Secretariat has created several derivative variables. A derivative variable is one created with reference to an existing variable in the dataset. For example, creating a variable classifying vote for an incumbent government versus vote for opposition parties, based on the original vote choice variable. The CSES inspiration for the types of derivative variables comes from user feedback. Not only does inclusion of these variables save users valuable time as they often involve significant intellectual and coding resources, but their creation also provides a harmonized and replicable go-to source variable(s). In the Modules, the most prominent derivative variable captures whether respondents vote for incumbent governments or not, based on the premise elections are principally referendums on the incumbent administration (e.g. Tufte 1975). Other examples of derivative variables in CSES products include those generational cohorts, vote switchers, turnout switchers, or voters for left and right parties.

Coding of derivative variables poses significant harmonization challenges. We highlight two with a variable capturing whether a respondent reported voting for a leftist/rightist party or candidate. The first challenge is which standard to follow. For some parties, a left/right ideological classification is straightforward. But for others, categorization differences exist as we know parties can often shift positions over time (e.g. Adams et al. 2004; Adams and Somer-Topcu 2009). Moreover, there is debate as to whether expert judgments (e.g. Castles and Mair 1984; Benoit and Laver 2006) or content analysis of manifestos (Klingemann et al. 2006; Volkens et al. 2016) provide the most accurate indicators. The CSES took a

```
--------------------------------------------------------------------
>>> PARTIES AND LEADERS: HUNGARY (2018)
--------------------------------------------------------------------

348001.  PARTY A  Fidesz - Magyar Polgari Szovetseg - Keresztenydemokrata
                  Neppart
                  Fidesz - KNDP
         LEADER A  Victor Orban


348002.  PARTY B  Jobbik Magyaporszagert Mozgalom (Jobbik)
                  Jobbik - Movement For a Better Hungary
         LEADER B  Vona Gabor


348003.  PARTY C  Magyar Szocialista Part -
                  Parbeszed Magyarorszagert Part (MSZP) - SEE ELECTION
                                                     STUDY NOTES
                  Hungarian Socialist Party - Dialogue for Hungary
         LEADER C  Molnar Gyla
         LEADER I  Karacsony Gergely - SEE ELECTION STUDY NOTES


348004.  PARTY D  Lehet Mas A Politika (LMP)
                  Politics Can Be Different
         LEADER D  Szel Bernadett


348005.  PARTY E  Demokratikus Koalicio (DK)
                  Democratic Coalition
         LEADER E  Gyurcsany Ferenc


348006.  PARTY F  Momentum Mozgalom - SEE ELECTION STUDY NOTES
                  Momentum Movement


348007.  PARTY G  Magyar Ketfraku Kutya Part - SEE ELECTION STUDY NOTES
                  Hungarian Two-tailed Dog Party


348008.  PARTY H  Egyutt-A Korszakvaltok Partja - SEE ELECTION STUDY NOTES
                  Together


348009.           Munkaspart
                  Hungarian Workers Party


348010.           Osszefogas - SEE ELECTION STUDY NOTES
                  Unity


348011.           Zoldek
                  Hungarian Social Green Party


         | ELECTION STUDY NOTES - HUNGARY (2018): PARTIES AND LEADERS
         |
         | The alphabetic order of parties is based on their share of votes
         | in the second segment - the proportional national party list
         | tier.
         | Following CSES conventions, in countries where voters have two
         | votes (i.e.: a constituency and a list vote) simultaneously,
         | parties are ordered by the national share of the party list vote
         | (tier 2).
         |
         | Code 348003. refers to the electoral coalition between Hungarian
         | Socialist Party (MSZP) and Dialogue for Hungary. Variables
         | denoted with the alphabetical letter C, however, only refer to
         | one of its members, the MSZP, which was the dominant member in
         | the coalition.
         |
         | Karacsony Gergely was the leader of party "Dialogue for Hungary,"
         | which was part of the coalition and contested the election with
         | 348003. Hungarian Socialist Party (PARTY C).
         |
```

Figure 6.4 Example of how electoral alliances and their coding are detailed in a CSES Codebook. Source: CSES Module 5, Hungary 2018 study.

conciliatory approach and created two different derivative measures: the first reflects the expert left–right party placement by CSES Collaborators on a 0–10 scale. The second variable is based on the left–right scoring of parties in the MARPOR (Volkens et al. 2016) project based on a content analysis of the election manifestos. This approach provides users with the freedom to choose among various well-established options and selecting the best suited to their analysis. It also easily opens up the possibility for appropriate robustness checks. CSES products also include the original scores for these variables, further empowering users to diverge from the provided derivative harmonized measures.

The second difficulty concerned the left–right categories themselves: can all parties be aligned to these two categories, or should we have a center category too? Here, we decided to adopt an original approach in tailoring individual solutions for the different variables. Left–right party placement based on CSES has three categories (left, center, and right), whereas the variable based on MARPOR is dichotomous (left–right). The idea was to alert users this decision was not trivial, that there are other possible recodes, and to encourage users to closely consider the variable they choose after consulting with the CSES Codebook.

6.5 Exploring Interplay Between Ex-ante and *Ex-post* Harmonization

The literature on data harmonization clearly separates ex-ante from *ex-post* harmonization. However, we identified a strong interplay between the two. This reciprocity can occur when harmonizing datasets which initially had ex-ante intentions and no coordinated longitudinal design in mind, but subsequently decided to introduce a longitudinal aspect. This occurred with the CSES with the decision in 2015 to explore the creation of a cumulative integrated file incorporating all Modules fielded. While harmonizing overtime has thus far been conceived of as *ex-post* harmonization (Wolf et al. 2016, p. 507), this overlooks the ex-ante aspect underlying the project. To illustrate this interplay, we focus on the CSES IMD, the integrated CSES data product, highlighting some difficulties encountered when harmonizing demographic, parties/coalitions, and vote choice variables.[7]

6.5.1 Demographic Variables in CSES IMD

Remember, the CSES demographic variables are *ex-post* harmonized by design. This created some difficulties when it came to harmonizing for the CSES IMD. First, scales across the CSES Modules for identical constructs varied in some cases. Take education – in Modules 1–3, it was measured using a nonstandard scheme, but for Modules 4 and 5, CSES has adopted the international standard ISCED measure. Consequently, harmonizing education for CSES IMD required the CSES Secretariat to design a new coding scheme, integrating the different conceptualizations. While the new scale adhered to the principle of coding educational attainment, harmonization came at the expense of aggregating preexisting categories to the lowest common denominator. Three considerations reassured the CSES Secretariat in this endeavor:

- An internal content analysis of CSES publications identifies the most common ways users incorporate education into their analyses.
- The assessment that aggregated codes for education was sufficiently differentiated to be useful for turnout and vote choice models, the critical objective of CSES.

---

7 One key principle at play in the development of IMD was to keep dataset changes between Modules and IMD to a minimum.

- The opportunity for researchers to go back to Module Datasets should they require more fine-grained or diverging education measures.

To achieve the latter, the CSES Secretariat focused on providing users with the most transparent and comprehensive documentation.

Another difficulty was inter-coder reliability. Even with consistent coding schemes for target variables, country-specific scales were not always coded consistently by different coders. Take the variable respondents' employment status. The target measure aimed to distinguish respondents within and outside of the labor force separately, with those in employment further differentiated by whether they worked full time or part-time. However, the CSES Modules reserved two codes for country-specific categories. Reviewing these classifications revealed most referred to either respondent on a job leave (maternity leave, etc.) or to respondents engaged in military/civil service. Additionally, many studies did not differentiate between full time and part-time employment but classified all employed respondents into one general category. The IMD incorporates both of these insights by extending the preexisting coding scheme to separate codes for respondents on job leave, military/civil service, and employed respondents for which no information on working hours is available. These additional classifications deepened harmonization.

A different harmonization challenge arose concerning variables for which CSES Modules did not have a standardized code scheme. Take respondent ethnicity as an example. CSES Modules detail the national coding's of ethnicity, with no attempt at harmonization. So code 1 in the 2013 Australian study included in CSES Module 4 refers to Aboriginals or Torres Strait Islanders while code 1 in the 2011 and 2015 Canadian studies, also in CSES Module 4, refers to Canadians. IMD remedies this with a harmonized comparative scale. Due to sensitives concerning the construct, harmonization efforts started with a systematic review of coding frames employed by other comparative projects. The most promising standard identified was the "European Standard Classification of Cultural and Ethnic Groups" (ESCEG), developed by Heath et al. (2019) for the European Social Survey (ESS). Although the ESS focuses on European countries, ESCEG also covers ethnicities originating from outside Europe. An initial round of coding confirmed most ethnicities distinguished by CSES Collaborators fitted categories foreseen by ESCEG. Before extending ESCEG to incorporate ethnic groups not prevalent in the European context, e.g. a differentiated list of Chinese and Philippine ethnicities, the CSES Secretariat consulted the classification authors. Based on their valuable feedback, the CSES IMD now includes a harmonized scale for ethnicity, facilitating adopting a common standard across projects.

6.5.2 Harmonizing Party Data in CSES IMD

The first challenge was parties within the same countries had not been assigned identical codes across time. These codes had first been assigned without ordering and later, based on the election outcome in a given election. Thus, the same party within a polity could have been assigned as many different codes as there were election studies from that polity and in which the party occurred. For example, across the four different CSES Modules included in the first version of the CSES IMD, the German party *Christian Democratic Union* (CDU) had been assigned the numerical codes one (Module 1), two (Module 2), one (Module 3), and two (Module 4). Retaining the different codes would violate the basic principle of harmonization to minimize redundant standards. Hence, the CSES Secretariat decided each party would receive a unique numerical code. It was further decided to use a seven-digit identifier to cap the number of numerical codes and to allow users to identify parties in the data quickly. The first three digits represent the three-digit UN Polity Identifier Code. The last four digits contain numerical codes, ranging from zero to 9999, which uniquely identify a party within a polity. Table 6.2, provides examples that illustrate this solution.

Once again, while the data had been collected, the development of a numerical party coding scheme that accounts for the element of time is primarily part of *ex-post* harmonization. At the same time, it also matches the ex-ante output harmonization feature of developing a common coding scheme, knowing the codes would vary across the CSES Modules. Like in the case of demographics, one might be hard-pressed to discern the two harmonization forms when it comes to harmonization efforts across time, which highlights the interplay between the two forms.

In addition to dealing with varying numerical codes within polities, harmonizing party codes also entailed addressing parties' changes across time. These changes arose from different sources such as the participation of a party in various electoral alliances or coalitions, the emergence of a new party, the merger between two or more parties, and the splitting of parties. As a general rule, CSES decided that coalitions, new parties, and new parties resulting from mergers or party splits would be assigned a unique numerical party code. This approach fits the underlying principle of code harmonization. It treats parties remaining identical across time as one entity while treating emerging or vanishing entities (i.e. new parties or parties becoming part of mergers) different from each other.

CSES also devised different tools to collect the relevant information to assign and code the parties, to check the coding in the dataset, and to document the parties in accessible ways for the user community. The first tool is an excel spreadsheet containing every party and the numerical codes assigned to these parties in the CSES Modules and additional information such as party abbreviations and information about mergers and coalitions. The data were taken from the documentation of existing Modules and complemented by publicly available information from sources such as ParlGov (Döring et al. 2022). The spreadsheet, used for internal purposes, contains all the relevant information to assign each party the harmonized numerical code and provided the necessary internal documentation to track coding decisions. The CSES Secretariat further developed an internally used software script (STATA DoFile) to

Table 6.2 Harmonizing party codes across time in CSES IMD.

Polity	IMD code	Party name	Acronym
Germany	2760001	Christian Democratic Union/Christian Social Union	Union
Germany	2760002	Christian Democratic Party	CDU
Germany	2760003	Christian Social Union in Bavaria	CSU
Germany	Etc.	Etc.	Etc.
Australia	0360001	Liberal Party	LP
Australia	0360002	National Party of Australia	NPA
Australia	0360003	Australian Labor Party	ALP
Australia	Etc.	Etc.	Etc.

Note: First three digits = UN Country Identifier; Last four digits = Party identifier. Source: Quinlan et al. (2018)/CSES.

check whether the harmonization of the numerical party codes from the CSES Modules was successful. Finally, the CSES Secretariat documented all of the numerical party codes in the IMD Codebook, complemented by value labels in the dataset to provide the user with the best possible information.

6.6 Taking Stock and New Frontiers in Harmonization

Harmonization for CSES rests on five touchstone principles: strong relationship with Collaborators; rigorous documentation; replication, replication; extensive Data Checks; and user empowerment and transparency. The project's harmonization objectives are realized by predesigned statistical program scripts and programs like MS Office, complemented by well-trained staff with appropriate expertise. This mixture of ingredients would be the key takeaways in terms of coming up with an ideal recipe for successful harmonization in a decentralized, user empowered, and comparative longitudinal social science project.

But CSES is also thinking ahead to the next frontiers in harmonization endeavors – what we call user harmonization through connection of various datasets. In response, CSES is at the forefront of ensuring its products can be linked with other comparative datasets, acknowledging the growing demand in social sciences (e.g. Döring and Regel 2019) to combine multiple data sources. We call this data bridging and achieve this by providing bridging identifiers in CSES data products enabling easy linkage between other datasets at multiple levels. All CSES datasets include polity identifiers in line with international standards (for example – three-digit UN ISO polity code and a variable classifying polity name in English in string format), allowing users to easily link data from the OECD, the World Bank, or the Quality of Government database with CSES.

Modules 4 and 5 and the CSES IMD allow for data bridging at a party level with the inclusion of party identifiers from the Parliaments and Governments database (ParlGov) and the Comparative Manifesto Project (MARPOR) within the data products. CSES IMD has gone further enabling users to easily bridge data from a multitude of sources at the party and country levels, including the Chapel Hill Expert Surveys (CHES, party level), the Varieties of Democracy (V-Dem, polity level), and Party Facts, an online database linking data on political parties across 48 different datasets.

The CSES website includes a specific page dedicated to these data bridging efforts (CSES 2020). Starting with a concise description of how to data bridge, the website provides users with detailed information on how to bridge CSES data to 15 other major projects and databases. To lower the entry barriers for data bridging further, CSES provides well-commented example syntaxes for STATA users, demonstrating hands-on how CSES data can be merged to other sources by polity or party identifiers. Creating these syntaxes was motivated by the experience that the structure in which relational party identifiers are presented in CSES often differs from other projects, calling for some guidance provided by the CSES Secretariat. Simultaneously, these syntaxes benefit the discipline more broadly by providing practical advice on how to engage in data bridging generally.

References

Adams, J. and Somer-Topcu, Z. (2009). Do parties adjust their policies in response to rival parties' policy shifts? Spatial theory and the dynamics of party competition in twenty-five postwar democracies. *British Journal of Political Science* 39 (4): 825–846.

Adams, J., Clarke, M., Ezrow, L., and Glasgow, G. (2004). Understanding change and stability in party ideologies: do parties respond to public opinion or to past election results. *British Journal of Political Science* 34: 589–610.

Alesina, A., Devleeshauwer, A., Easterly, W. et al. (2003). Fractionalization. *Journal of Economic Growth* 8: 155–194.

Benoit, K. and Laver, M. (2006). *Party Policy in Modern Democracies*, 1e. London, UK: Routledge.

Canberra Group (2011). *Handbook on Household Income Statistics*, 2e. Geneva, CHE: United Nations.

Castles, F. and Mair, P. (1984). Left-right political scales: some 'expert' judgements. *European Journal of Political Research* 12 (1): 73–88.

CSES (2020). The Comparative Study of Electoral Systems Data Bridging. https://cses.org/data-download/data-bridging (accessed 21 December 2021).

Döring, H. and Regel, S. (2019). Party facts: a database of political parties worldwide. *Party Politics* 25 (2): 97–109.

Döring, Holger, Constantin Huber and Philip Manow. 2022. Parliaments and governments database (ParlGov): information on parties, elections and cabinets in established democracies. Development version.

Fortin-Rittberger, J., Howell, D., Quinlan, S., and Todosejevic, B. (2016). Supplementing cross-national survey data with contextual data. In: *The Sage Handbook of Survey Methodology* (ed. W. Christof, J. Dominique, W.S. Tom, and F. Yang-chih), 668–677. Sage.

Granda, P., Wolf, C., and Reto, H. (2010). Harmonizing survey data. In: *Survey Methods in Multicultural, Multinational, and Multiregional Contexts* (ed. J. Harkness, M. Braun, and B. Edwards), 315–334. Hoboken, NJ: Wiley.

Heath, A., Schneider, S.L., and Salini, L. (2019). European Standard Classification of Cultural and Ethnic Groups (ESCEG) ESS9 ESS Round 9 Documentation Report Appendix A6 Ed. 1.0 - Classifications and Coding Standards, ESS9-2018. European Social Survey: Data Archive NSD - Norwegian Centre for Research Data, Norway.

Howell, D. (2010). Enhancing quality and comparability in the comparative study of electoral systems (CSES). In: *Survey Methods in Multicultural, Multinational, and Multiregional Contexts* (ed. J. Harkness, M. Braun, and B. Edwards), 525–534. Hoboken, NJ: Wiley.

King, G. (1995). Replication, replication. *PS: Political Science & Politics* 28 (3): 444–452.

Klingemann, H.D., Volkens, A., Bara, J., and Budge, I. (2006). *Mapping Policy Preference II: Estimates for Parties, Electors and Governments in Eastern Europe, the European Union and the OECD 1990–2003*. Oxford: Oxford University Press.

Ortmanns, V. and Schneider, S.L. (2016). Harmonization still failing? Inconsistency of education variables in Cross-National Public Opinion Surveys. *International Journal of Public Opinion Research* 28 (4): 562–582.

Quinlan, S., Schimpf, C., Blinzler, K., et al. (2018). The Comparative Study of Electoral Systems (CSES) Integrated Module Dataset (IMD). https://cses.org/wp-content/uploads/2019/03/CSES_IMD_StimulusPaper.pdf (accessed 21 September 2020).

Thomassen, J., Rosenstone, S., Klingemann, H.D., and Curtice, J. (1994). The Comparative Study of Electoral Systems: A Conference Sponsored by The International Committee for Research into Elections and Representative Democracy (ICORE). Berlin. https://cses.org/wp-content/uploads/2019/03/CSES-_-Reports-and-Papers-_-Comparative-Study-of-Electoral-Systems-Conference.pdf.

Tufte, E. (1975). Determinants of the outcomes of midterm congressional elections. *American Political Science Review* 69 (3): 812–826.

Volkens, A., Lehman, P., and Matthieß, T., et al. (2016). *The Manifesto Data Collection. Manifesto Project (MRG/CMP/MARPOR)*. Berlin: Wissenschaftszentrum Berlin für Sozialforschung (WZB).

Wolf, C., Schneider, S.L., Bewhr, D., and Joye, D. (2016). Harmonizing survey questions between cultures and over time. In: *The SAGE Handbook of Survey Methodology* (ed. C. Wolf, D. Joye, T.W. Smith, and Y.C. Fu), 502–524. London, UK: Sage.

7

Harmonization in the East Asian Social Survey

Noriko Iwai[1], Tetsuo Mo[2], Jibum Kim[3], Chyi-In Wu[4], and Weidong Wang[5]

[1]*Faculty of Business Administration, Japanese General Social Survey Research Center, Osaka University of Commerce, Japan*
[2]*Japanese General Social Survey Research Center, Osaka University of Commerce, Japan*
[3]*Department of Sociology, Survey Research Center, Sungkyunkwan University, South Korea*
[4]*Institute of Sociology, Academia Sinica, Taiwan*
[5]*Department of Sociology, National Survey Research Center, Renmin University of China, China*

7.1 Introduction

The East Asian Social Survey (EASS) project is organized as a cross-national survey research effort, which designs a common module and creates an integrated data set from four East Asian societies, China, Japan, South Korea, and Taiwan. The issue of harmonization in designing a common module, data collection, and creation is pivotal for the success of the project.

In an arena of cross-national surveys in which European and American researchers tend to lead the discussion, the EASS project establishes common questions based on issues and concerns unique to East Asian societies, produces academic survey data sets, and disseminates them for cross-national analyses. The EASS Project was launched after the "Japanese General Social Survey (JGSS) International Symposium" in 2003 where the principal investigators of the Korean General Social Survey (KGSS), the Taiwan Social Change Survey (TSCS), and the JGSS gathered. These three teams had already started or just started repeated national surveys, modeled after the U.S. General Social Survey (GSS) (Smith et al. 2006); TSCS started in 1984, JGSS in 2000, and KGSS in 2003, respectively.

At the time the EASS project was launched, TSCS and KGSS had just participated in the International Social Survey Program (ISSP). ISSP has been conducted by the Japan Broadcasting Company in Japan since 1993. China joined the ISSP in 2008. While sharing the philosophy of the ISSP, three teams wanted to start the project, which focuses particularly on questions and issues that are commonly relevant to East Asian societies, bringing them together into culturally and theoretically meaningful topics for comparative research (EASSDA 2019), just as the European Social Survey started to pursue in their region. Three teams decided to launch the EASS project and asked CGSS, which started the Chinese General Social Survey (CGSS) in 2003 to join the project.

Survey Data Harmonization in the Social Sciences, First Edition.
Edited by Irina Tomescu-Dubrow, Christof Wolf, Kazimierz M. Slomczynski, and J. Craig Jenkins.
© 2024 John Wiley & Sons Inc. Published 2024 by John Wiley & Sons Inc.

The EASS project is to make a module of about 60 questions and incorporate it into the already recurring social surveys, the JGSS, CGSS, KGSS, and TSCS. Surveying a common module in these four societies made it possible to conduct international comparisons without creating a new independent cross-national survey. The project team planned to incorporate these modules every two years in the four national surveys starting in 2006. The theme for the first EASS (EASS 2006) was "Family in East Asia." The themes for the second (EASS 2008), the third (EASS 2010), the fourth (EASS 2012), and the fifth survey (EASS 2015) were "Culture and Globalization in East Asia," "Health and Society in East Asia," "Network Social Capital in East Asia," and "Work Life in East Asia," respectively. Since 2016, the EASS has taken up previous themes again with some modifications: "Family" in 2016, "Culture and Globalization" in 2018, and "Health and Society" in 2020. EASS data have been deposited to the East Asian Social Survey Data Archive (EASSDA) at Sungkyunkwan University (South Korea) and Inter-university Consortium for Political and Social Research (ICPSR) at the University of Michigan.

In this chapter, we discuss the process of input and output harmonization in the EASS project. Input harmonization can be seen in every step of survey preparation, such as deciding module themes, subtopics, questions, answer categories, scales, and standard background variables (SBVs) and their translations. Output harmonization of EASS is implemented through ex-ante strategies. Each team is responsible for coding data based on a format for the EASS modules and SBVs from their original data, in which there may be slight variations across the EASS teams. A team that is responsible for integrating data files from the four teams then checks through each file for accuracy and details on target respondents and makes revisions as needed. Issues on harmonizing data will be discussed with concrete examples.

7.2 Characteristics of the EASS and its Harmonization Process

7.2.1 Outline of the East Asian Social Survey

Table 7.1 summarizes the EASS and the four constituent teams. EASS generally holds meetings twice a year in spring and in autumn. The meeting is hosted by four teams in rotation. The host team pays for travel expenses for two persons from each team.

The four teams of EASS agree to field approximately 15-minute long interviews for the module surveys, and the module generally consists of 60 questionnaire items (not including the SBVs). The questionnaires are designed for the interviews of either interviewer-administration or self-administration. TSCS, KGSS, and CGSS adopt the face-to-face interview method, while JGSS combines the face-to-face interview method for demographic questions and the self-administered method for questions on attitude and behavior. JGSS adopts the mixed method because completing all the questions in an interview would take more than 40 minutes, which increases the difficulty of obtaining respondents' cooperation in Japan.

As mentioned previously, TSCS, KGSS, and CGSS participate in the ISSP. EASS referred to ISSP heavily for deciding SBVs. CGSS adopts a split-ballot method, so that the ISSP module and the EASS module are included in the different questionnaires most of the time; TSCS includes two modules in the same questionnaire when the theme for EASS and for ISSP are the same or fairly close. On the other hand, KGSS always includes the ISSP module and the EASS module in the same questionnaire. This situation sometimes affected the EASS harmonization process, which will be discussed later.

In principle, the sample of EASS should be a national representative probability sample of the adult population aged 18 or over in each country/region, and a minimum sample size is 1000 cases, preferably

Table 7.1 Summary of East Asian Social Survey and their Source Surveys.

		Japan	Korea	Taiwan	China
Source Survey		Japanese General Social Surveys (JGSS)	Korean General Social Survey (KGSS)	Taiwan Social Change Survey (TSCS)	Chinese General Social Survey (CGSS)
		JGSS Research Center, Osaka University of Commerce	Survey Research Center, Sungkyunkwan University	Institute of Sociology, Academia Sinica	National Survey Research Center at Renmin University of China
	Principal Investigator	IWAI, Noriko	KIM, Jibum	WU, Chyi-In	WANG, Weidong
	Sample population	20–89 Years of Age	18 or over	18 or over	18 or over[a]
	Sampling method	Two-stage stratified random sampling	Four-stage stratified random sampling[b]	Three-stage stratified random sampling	Four-stage stratified random sampling
	Fieldwork method	Interview and self-administered[c]	Interview	Interview	Interview
	First survey and interval	1–3 years since 2000	Annually 2003–2014; biennially after 2016	Almost every year since 1984	Annually since 2003 except for 2007, 2009, 2014, and 2016
	Inclusion of ISSP	Not included (implemented by NHK)	Included since 2003 except 2015, 2017, and 2019	Included since 2002[d]	Included since 2009 except 2010, 2013, 2014, 2016, 2018, and 2019[d]
EASS 2006	Fieldwork dates	Oct–Dec., 2006	Jun–Aug, 2006	Jul–Aug, 2006	Sep–Nov, 2006
	Sample size	3998	2500	5032	7872
	Valid responses	2130	1605	2012	3208
	Response rate[e]	59.8%	65.7%	42.0%	38.5%
EASS 2008	Fieldwork dates	Oct–Dec, 2008	Jun–Aug, 2008	Jul–Sep, 2008	Sep–Dec, 2008
	Sample size	4003	2500	4601	6300
	Valid responses	2160	1508	2067	3010
	Response rate[e]	60.6%	61.0%	44.9%	47.8%
EASS 2010	Fieldwork dates	Feb–Apr, 2010	Jun–Aug, 2010	Jul–Nov, 2011(18–72) Feb–Apr, 2012(over 72)	Jul–Dec, 2010
	Sample size	4500	2500	4424	5370
	Valid responses	2496	1576	2199	3866
	Response rate[e]	62.1%	63.0%	49.7%	72.0%

(Continued)

Table 7.1 (Continued)

		Japan	Korea	Taiwan	China
EASS 2012	Fieldwork dates	Feb–Apr, 2012	Jun–Aug, 2012	Jul–Oct, 2012	Jun–Dec, 2012
	Sample size	4500	2500	4104	8200
	Valid responses	2335	1396	2134	5819
	Response rate[*]	58.8%	55.8%	52.0%	71.0%
EASS 2015	Fieldwork dates	Feb–May, 2015	–	Aug–Nov, 2015	Jun–Dec, 2015
	Sample size	4500		4090	2500
	Valid responses	2079		2031	1743
	Response rate[e]	52.6%		49.7%	69.7%
EASS 2016	Fieldwork dates	Jan–Mar, 2017; Feb–Apr, 2018	Jun–Oct, 2016	Aug–Nov, 2016	Jun–Dec, 2017
	Sample size	1500; 4000	2240	4076	6000
	Valid responses	744; 1916	1051	2024	4132
	Response rate[e]	55.6%; 54.3%	46.9%	49.7%	68.9%
EASS 2018	Fieldwork dates	Nov–Dec, 2017; Nov–Dec, 2018	Jun–Oct, 2018	Jul, 2018 to Jan, 2019	Jun–Dec, 2018
	Sample size	1500; 1200	2400	4093	6000
	Valid responses	860; 678	1030	1960	4499
	Response rate[e]	64.0%; 62.7%	43.0%	47.9%	76.0%

JGSS: https://jgss.daishodai.ac.jp; TSCS: https://www2.ios.sinica.edu.tw/sc/; KGSS: http://kgss.skku.edu/; CGSS: http://cgss.ruc.edu.cn. EASSDA: https://www.eassda.org/; JGSSDDS: https://jgss.daishodai.ac.jp/english/jgssdds/jgssdds_top.html.

[a] 18–69 in 2006.

[b] Three-stage stratified random sampling up to 2012.

[c] Self-administered only for EASS 2018.

[d] Basically EASS and ISSP modules are asked with different samples.

[e] Figures are based on calculations by each team. Please refer to Codebook of EASS Surveys edited by JGSS Research Center for details.

to have 1400 cases. The sample for JGSS does not include 18 and 19 years old because those under the age of 20 were not on the voter registration list, which JGSS had been using for the extract of sample population until 2006. Although the minimum voting age was lowered to 18 in 2016, JGSS continued to use the same age groups of the sample population to maintain the continuity.

There are no rules for the fieldwork period. As shown in Table 7.1, KGSS, TSCS, and CGSS start their survey in summer. The fieldwork period for JGSS is either fall or spring depending on research funding. In order to secure the sample size, JGSS has conducted a survey with the same questionnaire for the second straight year for EASS 2016 and 2018.

7.2.2 Harmonization Process of the EASS

In the following sections 7.2.2.3 to 7.2.2.5, the process of the harmonization of the EASS, unexpected events and errors, as well as how these were handled, will be described.

7.2.2.1 Establishing the Module Theme

The EASS module theme has been changed in every round. It is usually decided in an autumn meeting half a year before conducting the survey. Each team proposed themes and subtopics with theoretical rationale and related materials. Each team tends to pick up contentious social issues in each society. The four societies have shared Confucian values, but different political, social, and industrial paths led their people to take different perceptions and approaches to the same social issues. Principal investigators of four teams sometimes discussed intensely to harmonize the theme. Considering common phenomena of sharply declining fertility rates, rapid population aging, and shrinking household sizes in four societies, and reflecting the fact that principal investigators of TSCS and JGSS specialized in family, "Family in East Asia" was chosen as the theme for the first EASS module (EASS 2006; Iwai and Yasuda 2011). "Culture and Globalization in East Asia" was chosen for the second round (EASS 2008) in keeping with the increased cultural exchanges among the four societies through, for example, Korean TV drama, Japanese animation, and Chinese movies around 2005 (Iwai and Ueda 2012). After deciding the theme, each team invited several experts in the field to help design the module questions.

EASS takes care not to overlap its theme with that of the ISSP in the same year, but sometimes it is unavoidable. In 2007, the EASS selected "Health and Society in East Asia" as the 2010 module theme. The following year, in 2008, "Health" was also selected as the theme for ISSP 2011. As stated before, TSCS usually splits the questions for the EASS and the ISSP into separate surveys, but with the "Health module," the TSCS decided to include them both in the 2011 questionnaire. The KGSS and CGSS teams included the EASS 2010 Health module in their surveys conducted in 2010 and included the ISSP 2011 Heath module in those conducted in 2011. As a result, greater harmonization efforts were made in the teams for choosing subtopics as well as the alteration of some questions and indicators (Hanibuchi 2009). While the ISSP Health module focused on one's perceptions on the health care system and services in one's country, EASS focused on more detailed physical and mental health status using the SF-12 health-related quality of life scale and asked respondents to specify their disease or symptom name if they suffer. The EASS health module also includes more questions about alternative medicine treatment, social support, and physical and socioeconomic environments. In addition, it deals with epidemiology (concern about the new strain of influenza outbreak in 2009 and vaccination against influenza) and addictions.

Overlapping of the ISSP and EASS themes became more serious for the 2015 module. The theme for the ISSP was "Work Orientation" and that for the EASS was "Work Life in East Asia." This time, TSCS and KGSS, which usually incorporate two modules in different years, had to include them in the same questionnaire due to a shortage of research funds. Furthermore, CGSS decided to include both modules in the same questionnaire, so that EASS had to wait for the ISSP module to be confirmed to avoid duplications of questions. It was really challenging to harmonize the EASS 2015 questionnaire (Iwai and Uenohara 2015). JGSS asked permission from NHK to include some of the ISSP work orientation questions used in the past rounds into the JGSS survey. As a result, the number of comparable variables excluding background variables was only 16. In addition, KGSS could not conduct the survey itself because of financial shortage.

7.2.2.2 Selecting Subtopics and Questions

After the theme has been decided, each team prepares proposals for subtopics. In case of JGSS, new members are recruited through an open call for research proposals in addition to experts on the theme. Invited experts, as well as principal investigators of each team, conduct lively discussions on the selection of the subtopics at the two consecutive drafting meetings (usually in spring and autumn).

In designing the module, some teams prefer to set the theoretical and conceptual frameworks of the whole module completely first and then to discuss subtopics. Other teams prefer not to spend much time on theoretical discussion but to discuss subtopics and specific questions once a rough framework is established. Some teams prefer a diversified strategy of incorporating a few questions across many topics. Other teams prefer a centralized strategy of asking about a small number of topics in detail. JGSS prefers a diversified strategy since the JGSS Research Center is designated as Joint Usage and Research Center by the Minister of Education, Culture, Sports, Science and Technology, and is required to provide chances for researchers to join the JGSS project through an open call for questions or analyzing data, which could be utilized from many different perspectives. Since EASS consists of only four teams, EASS did not set a particular rule to proceed discussion. Depending on the theme, the EASS as a group either spent time on making a rigid theoretical framework such as in the case of the EASS 2012 Social Network and Social Capital module, or move to discuss subtopics and specific questions early such as in the case of EASS 2006 Family module, EASS 2008 Culture and Globalization module, and EASS 2010 Health module.

EASS questions are included subtopic by subtopic in each team's questionnaire, and sequences are kept the same as the source questionnaire. Each team sometimes adds country-specific response categories, items, or questions. All teams carefully check that these additions would not influence the results of EASS questions.

For the 2006 Family module, KGSS and TSCS were extremely interested in intergenerational support, while JGSS insisted on a diversified strategy of incorporating a few questions across many topics. An agreement was not reached via discussion, so the questions were narrowed down one by one through a voting process. The final draft delved deep into intergenerational support, but it also included questions on topics such as spouse selection, gender preference for children, and the division of roles within married couples. Considering that researchers on the four teams are involved in a variety of research topics, including a somewhat diverse set of topics is necessary.

With the 2008 Culture and Globalization module, the teams started to include some questions in the module for which each team can set their own answer options. In a question asking how much a respondent likes each musical genre, for example, the answer categories were "Classical," "Rock," "Jazz," "Pop," and "Traditional Music (of that society)." For the last of these, JGSS listed "Enka," and TSCS and CGSS listed "Traditional Opera."

Starting with the 2010 Health module, it became customary for teams with room on their questionnaires to include optional questions in cases where agreements could not be reached. For the health module, there were several questions included by two or three teams, such as a question about chronic illnesses a respondent suffers from in multiple-choice method. "Hypertension," "diabetes," "heart disease," and "respiratory problem" were choices common to all teams, while additional choices were optionally included based on their prevalence in each society.

7.2.2.3 Harmonization of Standard Background Variables

The EASS sets SBVs separately from the module questions. In deciding the SBVs, the EASS referred to the ISSP heavily. Since three out of four EASS teams participate in the ISSP, most of the ISSP SBVs were adopted in the EASS. Depending on a module, the EASS makes some adjustments. The SBVs for the latest harmonized EASS data (EASS 2018 Culture and Globalization II) are shown at https://jgss.daishodai .ac.jp/surveys/sur_quest/EASS2018_SBV.pdf. The JGSS publishes an EASS Codebook for each module, which contains both the original and English versions of the questionnaires used by the four teams. The SBVs, which are not in ISSP 2017 but in EASS 2018 are as follows: one's strength of religious belief, general happiness, self-rated household income level, occupation (ISCO/ILO-08) of the last job, self-rated health, household composition, total number of children, spouse's age and earnings, and father's and mother's completed education level.

Ideally, the four teams would use identical questions and answer choices for these variables. However, as the four teams include the EASS module into their own repeated cross-sectional surveys, some of the SBV questions of four teams differ in wordings and answer choices for the sake of the continuity of questions in respective surveys. Therefore, the EASS takes both input harmonization and ex-ante output harmonization procedures for the SBVs.

Input harmonized SBV are as follows: respondent's sex, marital status, strength of religious belief, general happiness, occupation (ISCO/ILO-08), self-rated household income level, self-rated health, number of household members, total number of children, and place of living. Age, hours worked weekly, number of employees, and type of organization (public versus private) are asked for respondent and spouse. A respondent's place of living was not asked consistently in the first module, but the four teams started to use an identical format in 2008, such as a self-assessment of the community type (a big city, the suburbs or outskirts of a big city, a town or a small city, a country village, and a farm or home in the country).

Ex-ante output harmonized SBV are as follows: highest completed education level, years in school, work status, employment relationship, type of organization (for-profit versus nonprofit), main employment status for respondent and spouse, and household members composition. Religious affiliation, place of living, highest completed education level, and years in school are asked for father and mother. These variables are recoded from the original codes used in each team's survey to the codes fixed by the EASS teams. For example, the highest education level is originally measured with 13 categories in Japan and China, 7 categories in Korea, and 21 categories in Taiwan. These are changed into a harmonized variable with seven categories (no formal qualification to graduate school). A target variable for religious affiliation is also agreed upon ex-ante. In JGSS, it is asked with an open-ended question, and approximately 60 categories are created after coding. While in Korea, religious affiliation is measured with 5 categories, it is measured with over 10 categories in Taiwan and China. In CGSS, it is asked as a multiple-choice question in 2012, 2015, and 2016. These data are recoded into 11 categories including no religion.

Regarding questions about work and occupation, there are more variations in wordings and answer choices across four teams in their original questions, requiring more complex procedures for harmonization. Three teams except TSCS combine data for two questions, work status and reasons for not working, and to create a variable of main employment status. A similar procedure is used for creating a variable on the employment relationships. In the 2006 survey, CGSS used two different questionnaires for urban and rural respondents and asked them differently about employment status. This required several more steps to recode the data and to harmonize the variable with data from other teams.

Questions about income differ in form and units across the four teams. KGSS and CGSS ask open-ended questions to gather respondents' personal, spouse, and household income, while JGSS and TSCS ask closed-ended questions. While KGSS and TSCS ask about monthly income, JGSS and CGSS ask about yearly income. For Japanese employees, biannual bonuses account for a considerable proportion of the annual income. Variables on income are not harmonized but kept as country-specific.

7.2.2.4 Harmonization of Answer Choices and Scales

One of the biggest issues in the process of making the EASS module was how to harmonize the answer choices for questions that ask about the respondent's opinion. Compared to questions on objective properties or behavior, those on subjective awareness are more easily influenced by the number of answer choices. The Japanese tend not to express their opinions clearly, and their answers tend to concentrate on a midpoint, such as "neither agree nor disagree" (Hayashi and Hayashi 1995). It is often pointed out that the concentration on a midpoint response category would make the interpretation of results difficult, and therefore, a midpoint category is undesirable (Otani et al. 1999). JGSS confirmed this in two preliminary surveys before starting the main survey (JGSS-2000; Iwai 2003; Sugita and Iwai 2003). Therefore, surveys conducted in Japan typically utilize a four-point scale with no midpoint. However, researchers from other societies are not very receptive to using it in international comparative surveys.

Shishido (2011a) compared the types of answer choices for questions on subjective awareness by the ISSP, the World Values Survey (WVS), the East Asia Value Survey (EAVS), the East Asia Barometer Survey (EABS), the Asia Barometer (AB), and JGSS. He revealed the following three common characteristics: (i) a frequent use of verbal scales with labels attached to each category, (ii) a high proportion of bipolar scales, and (iii) an infrequent use of scales with more than five points. On the other hand, the use of a midpoint answer choice varies according to the survey. A five-point scale with a midpoint choice is widely used on the ISSP, while two-point and four-point scales with no midpoint are often used on the WVS, which targets countries from different cultural spheres. Surveys such as the EAVS restricted to Asia contain a relatively large share of scales with no midpoint.

JGSS and other teams (TSCS and KGSS; CGSS joined after this discussion) had a different opinion on agree/disagree questions, which often appear in the EASS module. JGSS insisted on a four-point scale of "agree, somewhat agree, somewhat disagree, and disagree," while KGSS and TSCS sought to use a five-point scale of "strongly agree, agree, neither agree nor disagree, disagree, and strongly disagree," which is commonly seen on the ISSP. Three teams conducted pretests using a draft version of the module. TSCS used a split-ballot method to test a four-point and a five-point scale. KGSS tested a four-point scale with some questions. JGSS split the sample into four to test the following four scales: a four-point, a five-point, a seven-point starting with "agree," and a seven-point starting with "strongly agree." The four teams, including CGSS examined the distributions of responses and concluded to adopt a seven-point scale ("strongly agree, agree, somewhat agree, neither agree nor disagree, somewhat disagree, disagree, and strongly disagree"), which alleviates the tendency for responses to be centralized around the midpoint (Iwai and Yasuda 2011: p. 5, pp. 96–98). The idea that using bilaterally symmetrical answer choices with a midpoint will ensure compatibility with Western data and make it easier for respondents to rank their own opinions (Smith 1997) was considered.

7.2.2.5 Translation of Questions and Answer Choices

As with the other international comparative surveys, the EASS pays close attention to the translation of the questions and answer choices. While TSCS and CGSS can communicate in Chinese, English is the EASS four teams' common language. Once the English version of the module is finalized, each team checks whether it can be appropriately translated into the questionnaires for their respective languages, including both key concepts in the question text and answer choices.

In launching the JGSS project in 1998, the JGSS team compared the Japanese questionnaires of the ISSP as well as other international comparative surveys, which had been conducted in Japan with their source questionnaires (mostly in English). JGSS noticed that some of the Japanese translations of answer choices for the ISSP did not correctly communicate a nuance of their original English wordings; "strongly agree" is translated as "*so omou* (I think so)" and "agree" is translated as "*dochirakato ieba so omou* (I tend to agree)." These improper translations bring about the situation that distributions of responses for the ISSP in Japan are heavily concentrated on "strongly agree." JGSS conducted two preliminary surveys to test different wordings and check the distributions of responses to decide on answer choices for the main survey (Iwai 2003; Sugita and Iwai 2003).

Shishido (2011b) compared the response distributions in six societies including Japan, for the question "A working mother can establish just as warm and secure a relationship with her children as a mother who does not work" that appears on both the ISSP and the WVS. The response scale used in the ISSP is a five-point scale, while the WVS uses a four-point scale (Figure 7.1). Although the question is identical, the distributions vary considerably for Japan. Of the six countries and regions, Japan shows the highest proportion of respondents who choose "strongly agree" in the ISSP but the lowest percentage for the same response category in the WVS. As noted above, "strongly agree" is translated improperly in the ISSP, while it is literally translated in the WVS. Differences in translation lead to dramatically different results. The EASS teams are careful to arrive at the same nuance of questions and answer choices by sometimes writing them in Chinese characters that four teams can understand. Since 2010, there has been constantly a Chinese native and a Korean native research fellow in JGSS team who are very fluent in Japanese. They check the compatibility of the EASS module in every revision.

7.3 Documentation and Quality Assurance

7.3.1 Five Steps to Harmonize the EASS Integrated Data

As the EASS teams do not collect data solely for the international comparisons, each team first prepares the data for domestic usage and then creates the data file for the EASS integrated data. Each team has a person who is responsible for preparing data for the EASS. This individual, via email or phone, communicates with the JGSS chief data producer, who supervises integration and cleaning of the EASS data. JGSS has come to lead this process and compile a codebook for the following two reasons. It must publish

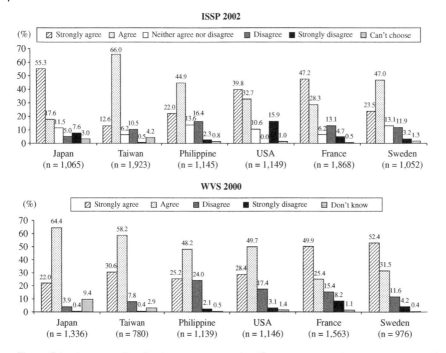

Figure 7.1 Response Distributions to the question "Do you agree or disagree that 'a working mother can establish just as warm and secure a relationship with her children as a mother who does not work'" for ISSP-2002 and WVS-2000 among six countries. Source: Shishido (2011b, Figure 2, p. 100)/With permission of Nakanishiya.

the Data Summary Table and Codebook according to the plan submitted to research funding organizations. The team continually has research fellows who are fluent in the EASS four languages and skillful in data cleaning. The data harmonization of the EASS is achieved through the following steps.

Step 1: Format for the integrated data is created

Once the four teams begin conducting the surveys that contain the EASS module, JGSS promptly creates the SPSS data entry matrix (including target variable names, labels, value labels, missing values, and numbers of digits) and shares it with the other teams. This process serves not only to develop the framework for actual data entry but also to reconfirm that all four teams are on the same page about their understanding of the variables and codes. EASS modules contain some optional questions that are only asked by two or three teams; variable names and value labels for these questions are also determined in advance. JGSS also creates a comparison table for the question numbers on the EASS module and those on the original questionnaires of each team and asks other teams to fill it in. The teams are also asked to list information about the target respondents for each question on the comparison table.

Step2: EASS data are created by each team

Once the four teams have created and cleaned their respective data, they create the data for the EASS integrated data, referring to the above SPSS data entry matrix and comparison table. Most of the variables in the EASS modules can be directly converted from the full domestic data sets, but SBVs such as those related to occupations require recoding. The source variables in the domestic data and the target variables in the EASS data sometimes feature reverse direction of the scale (for example, in the case of relative household income), and thus recoding must be applied following the EASS SPSS matrix described above.

Step 3: Creating the program for cleaning the integrated data

The JGSS team creates a program for cleaning the EASS data with the SPSS syntax. The detailed procedure will be presented in Section 7.5 (software tools).

Step 4: Data from the four teams are cleaned separately

The cleaning program is applied to the JGSS data first to ensure that it have the appropriate features. Since the data file has already been cleaned before being converted for the EASS, there are almost no problems except for potential conversion errors. The data sent from the other three teams are checked prior to cleaning with the program. If the data are improperly formatted or unclear, the team which sent it is contacted and asked to correct the data. Then the cleaning program is applied to the data one by one. Detected errors are recorded in a Word file or Excel file and sent to the corresponding team along with suggestions for correcting,

Step 5: The integrated data are created and cleaned

Once the EASS data for each team have been cleaned, the data are integrated and checked for the last time. In this stage, the frequency distributions for all four data sets are computed and compared for each variable. Then, the target respondents and variables are checked for mistakes in reference to the comparison table created in Step 1. If a problem is discovered, each team is contacted and asked to correct the issue.

7.3.2   Documentation of the EASS Integrated Data

At the same time the integrated data are created, JGSS asks the other teams to send materials related to their surveys. JGSS drafts a codebook and prepares for releasing the data. The EASS data along with the codebook and related materials are then deposited to the EASSDA and ICPSR. It will also be released via JGSS Data Download System (JGSSDDS; https://jgss.daishodai.ac.jp/english/jgssdds/jgssdds_top.html)

The deposited materials include:

- *Data Set File*. SPSS format and Excel CSV format
- *Questionnaires*. EASS module questions, source questionnaires, and their English translation
- *Study Description Form* (*SDF*). a brief description of data, including study title, fieldwork dates, principal investigators, fieldwork institution, population, sampling methods, fieldwork methods, initial sample size, number of respondents, response rates, language, weighting procedure, and national population characteristics
- *Study Monitoring Questionnaire* (*SMQ*). an extensive description of data, including questionnaire translation, pretest, sampling, fieldwork, response rate, and weighting
- *Codebook*. in addition to the above three materials, list of the EASS project members, variables list, data documentation and cross-national frequencies, and comparison table of variables across four teams

In the SDF section, the following information about known characteristics of the national population from the relevant high-quality data sources, such as census or other government surveys are provided: gender, age groups, years of schooling (15 years or older), employment status (15 years or older), national statistics on distribution of individual work income, and national statistics on distribution of household income. This information can be used for examining the validity of the data.

The codebook, questionnaires, SDF, and SMQ for each survey are uploaded to websites of the EASSDA (https://www.eassda.org) and the JGSS (https://jgss.daishodai.ac.jp/english/surveys/sur_top.html).

7.4 Challenges to Harmonization

Despite careful preparations, the EASS has faced difficult situations, one of which was a difficulty in harmonizing translations of the answer choices for a question that is taken from a copyrighted medical survey. Another one is a difficulty in synchronizing the data collection phase, which is affected strongly by the funding situation of each team.

7.4.1 How to Translate "Fair" and Restriction by Copyright

As mentioned before, the EASS teams decided to adopt a 12-Item Short-Form Health Survey (SF-12), which is a patient-reported health survey and a reduced version of SF-36 for the EASS 2010 Health Module. The SF-36 is an internationally tested and validated scale and can comprehensively measure health status, including both physical and mental health. It has been translated from the original English version into many other languages and validated.

Among the EASS four teams, JGSS incorporated the copyrighted Japanese version of the SF-12 (Fukuhara and Suzukamo 2004) into the questionnaire. In China, at the time of the EASS 2010 Health Module development, there was no Chinese version that had been validated, so CGSS purchased a copyright for the original English version and translated it by themselves. While there was already a validated version of SF-12 in Taiwan, the TSCS decided not to include the SF-12 into their questionnaire. Paying for questions of a measurement scale might be common in medical science, but it is not yet common in social sciences.

The SF-12 consists of 12 questions, of which literal meanings of 11 questions and their answer choices are straightforward. However, the answer choices for the subjective health question, "In general, would you say your health is: excellent, very good, good, fair, or poor," could lead to inadequate translation. This question is also included in ISSP 2011 Health Module, as well as in ISSP 2007, 2015, and 2018. TSCS included this question into TSCS 2011 as one of ISSP and EASS Health Module questions.

Table 7.2 shows the translated answer options for the question in Japanese, Chinese, Korean, and Taiwanese. Parentheses under the translations in each language include the literal translation of the word back to English. This table illustrates that the wording of the translations is not identical across the four languages. Translations in Taiwanese are accurate, while those in Japanese, Chinese, and Korean are not (Iwai 2017; Iwai and Yoshino 2020). The original English scale is unbalanced, skewed toward good health, and "poor" is the only one on the negative side. The TSCS correctly translated the scale with these features. Both CGSS and KGSS translated the answer choices as if they are for a balanced bipolar scale: *very healthy, somewhat healthy, general, somewhat unhealthy,* and *very unhealthy* by CGSS; and *very*

Table 7.2 Original and translated answer responses for the subjective health question in EASS 2010 Health Module (SF-12), ISSP, and EASS other modules.

SF-12/ISSP Original Answer Choice	Translated Answer Choice (literal meaning of the choice)					
	EASS 2010				ISSP in Japan by NHK	
	JGSS[a]	CGSS	KGSS	TSCS	2007/2011	2012/2015/2017/2018
1 Excellent	最高によい (Excellent)	很健康 (Very healthy)	매우 좋다 (Very good)	非常好 (Excellent)	かなりよい (Very good)	最高によい (Excellent)
2 Very good	とても良い (Very good)	比較健康 (Somewhat healthy)	다소 좋다 (Somewhat good)	很好 (Very good)	よい (Good)	とてもよい (Very good)
3 Good	良い (Good)	一般 (General)	좋지도 나쁘지도 않다 (Neither good nor bad)	好 (Good)	普通 (Fair)	よい (Good)
4 Fair	あまり良くない (Somewhat poor)	比較不健康 (Somewhat unhealthy)	다소 나쁘다 (Somewhat poor)	普通 (Fair)	あまりよくない (Somewhat poor)	普通 (Fair)
5 Poor	良くない (Poor)	很不健康 (Very unhealthy)	매우 나쁘다 (Very poor)	不好 (Poor)	かなりよくない (Very poor)	よくない (Poor)
Answer choice	Translated Answer Choice (literal meaning of the choice) in EASS Other Modules					
	JGSS	CGSS	KGSS	TSCS		
1 Very good	非常に良い (Very good)	很健康 (Very healthy)	매우 좋다 (Excellent)	非常好 (Very good)		
2		比較健康 (Somewhat healthy)	다소 좋다 (somewhat good)			
3		一般 (General)	좋지도 나쁘지도 않다 (Neither good nor bad)			
4		比較不健康 (Somewhat unhealthy)	다소 나쁘다 (Somewhat poor)			
5 Very bad	非常に悪い (Very bad)	很不健康 (Very unhealthy)	매우 나쁘다 (Very poor)	非常不好 (Very bad)		

[a] The Japanese translation is set by Quality Metric Incorporated and Shunichi Fukuhara, which cannot be changed.

good, somewhat good, neither good nor bad, somewhat poor, and *very poor* by KGSS. Japanese answer choices are set by Quality Metric Incorporated and Shunichi Fukuhara, who have the copyright of the SF-12. In checking the translation, JGSS noticed that the word "fair" is not translated properly. "Fair" has a connotation of a positive meaning, not a negative one so it should not be translated as "somewhat poor." JGSS asked the organization to allow us to change the translation, but it was not approved.

Separately from this subjective health question, the EASS teams have included another subjective health question, "how would you rate your health," with a bipolar balanced choice scale starting from very good to very bad in EASS 2006, 2012, and 2016. As shown in Table 7.2, answer choices for this question are translated almost equally in nuances across the four teams.

Figure 7.2 shows the distributions of the subjective health question for Japanese, Chinese, Korean, and Taiwanese respondents in EASS 2010 on the left side and those in EASS 2012 on the right side. Response distributions for Chinese and Korean respondents in the two EASS modules are fairly similar as the response scales were the same. On the other hand, response distributions for Japanese and Taiwanese are different. As for Japanese respondents' concentration at the middle value is substantial in EASS 2010 compared to the response distributions in EASS 2012. Since Japanese respondents have a strong tendency to avoid extreme responses (Behr and Shishido 2016), having two categories with highly positive health status "excellent" and "very good" resulted in less variance. Response distributions for Taiwanese

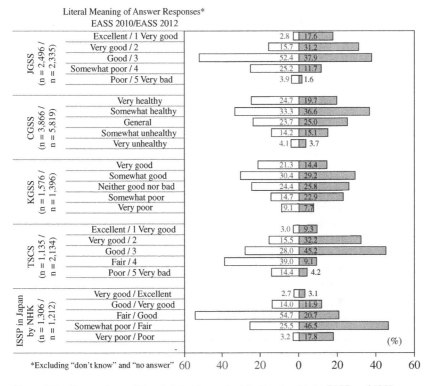

Figure 7.2 Comparison of the distributions of subjective health in EASS and ISSP.

respondents in EASS 2010 are also considerably different from those in EASS 2012. The highest point in the distributions in EASS 2010 is at "Fair" whereas in EASS 2012 the responses peaked at the midpoint.

Looking at the response distributions for EASS 2010, Taiwanese appear to be the least healthy, with the highest proportion of respondents choosing the fourth category "Fair" (39%). However, based on the response distributions for EASS 2012, they are not the least healthy; the proportions of the fourth category in Taiwan are less than those in Korea and China. The differences in the distributions are likely to be attributed to the different connotations in translations of answer choices. While Taiwanese translation of 'Fair' implies "so-so" in English, Japanese, Chinese, and Korean translations imply "somewhat poor" or "somewhat unhealthy."

As previously mentioned, the subjective health question has been included in ISSP several times. The NHK Broadcasting Culture Research Institute, which conducts ISSP in Japan, has also difficulty translating its answer choices (Murata 2020). Initially, the NHK translated the question as not exactly but close to a bipolar balanced scale: *very good, good, fair, somewhat poor,* and *very poor* (Table 7.2). For ISSP 2012 the translation was revised to more closely match the original English meaning: *excellent, very good, good, fair,* and *poor.* Response distributions change drastically between ISSP 2011 and 2012 (Figure 7.2). Murata, a member of the NHK team, points out that translation changes can lead to confusion for data users. Analyses performed abroad (in English) might note that "good" responses declined from 55% to 21% in one year and that the "fair" response became the most common in Japan. However, this drastic decline must be caused by changes in the translation of answer choices. Only those analysts who are careful enough to check the original and the translated version of the questions and their responses could notice the background of this drastic decline.

While JGSS had recognized the translation problem of the JGSS questionnaire, JGSS did not check other teams' translations seriously until October 2016 when JGSS happened to check the translation of the "Grit" scale in four societies for preparing questions for EASS 2018 Globalization Module. In August 2019, the four EASS teams decided to assign three different variables for the subjective health question in the EASS 2010 (one variable for Korea and China, one for Japan, and one for Taiwan), so that data users will notice the differences in translations. In preparing the EASS 2020 Health Module, the four teams decided that each team would adjust the translations to the original English. JGSS, which cannot change the wordings of the Japanese SF-12, will add one more subjective health question with properly translated answer responses.

7.4.2 Difficulty in Synchronizing the Data Collection Phase

In recent years, the critical issue for harmonizing the EASS is obtaining research funds to conduct national surveys, including the EASS module on time. Although there is no rule for the timing of the survey implementation, the synchronization of the data collection phase is essential to ensure data compatibility. The four EASS teams are supported by the university or the institute (TSCS) to which they belong and have applied and received a research fund from a national foundation. While TSCS and CGSS can conduct national surveys constantly with more secured research funds, KGSS and JGSS, which have managed to keep the project by scraping up several research funds had to give up or waive a field survey. KGSS was unable to field a survey in 2014, which was supposed to include the EASS 2014 Work Life Module (Table 7.1). With the EASS 2016 Family Module, JGSS could not provide enough cases only by JGSS-2017, so JGSS conducted another survey in 2018 to ensure the size of the valid responses. Regarding the EASS 2018 Globalization Module, JGSS modified the data collection mode, shifting from a combination of a personal interview and a self-administered questionnaire to a self-administered questionnaire. Since the EASS module questions, apart from SBVs, have always been included in a self-administered part in JGSS, this method change does not affect data quality.

7.5 Software Tools

For creating data files, JGSS and KGSS use SPSS, while CGSS and TSCS use STATA. As data files can be transferred from SPSS to STATA and vice versa, there is no problem with harmonizing data files. JGSS creates an EASS data entry matrix with SPSS and shares it with other teams.

As mentioned in Section 7.3.1, JGSS prepares a comparison table in Excel, which enlists EASS variable names, target populations, and question numbers in source surveys, is a useful reference for data users. Each team is responsible for filling in the information, and JGSS double-checks the information. The comparison table can be found in the codebook on the website (https://jgss.daishodai.ac.jp/english/research/codebook/EASS2016_Codebook.pdf, pp.101–113).

JGSS develops a SPSS syntax as well to check data from each team before merging them. The program checks whether (i) all variables are within their defined ranges, (ii) the target respondents for each question have responded, and (iii) there are any contradictory responses for variables that relate to each other. Regarding the third point, some of the detected cases cannot be definitively stated as errors and are not changed after detailed exploration. Frequent errors found by this program are cases related to the treatment of missing values, and mixing codes for not applicable with no answer. With these errors, JGSS makes suggestions for revisions along with their case numbers and data on related variables in Word and Excel files and sends them to respective teams.

7.6 Recommendations

For those who want to organize a cross-national comparative survey project in East Asia, our recommendations are as follows.

(1) Harmonizing process, especially in deciding module themes and questions, requires in-depth discussion. People in East Asia who share a Confucian sense of seniority to some extent sometimes avoid intense disputes. Establishing a reliable relationship that enables truthful communication through research seminars, and social gatherings after seminars are desirable. Before starting the EASS project, each team invited other teams to their research institutions and held research seminars to introduce other teams to their own research team members. A principal investigator and the whole team should be motivated to share the project aim and participate in the discussion.

(2) Researchers should be cautious in translating questions and answer choices. Most of the time, English is the common language for a survey project in East Asia. There should be persons who are fluent in other teams' languages. As mentioned, EASS sometimes uses Chinese characters to understand the meaning of wordings in other teams' source questionnaires. It is helpful, but one should be aware that the same Chinese character can have different meanings in the four societies.

(3) Researchers should be aware that there are variations in response tendencies among East Asian societies.

For those who plan to use the EASS data, it is highly recommended to check the comparison table that enlists EASS variable names, target populations, and question numbers in source surveys before starting the data analyses. Data users also should check the original and the translated version of the questions and their responses. Since the EASS data have been corrected several times, especially occupation in EASS 2006 and subjective health in EASS 2010, data users should be careful to obtain the latest version of the data and read the note attached to the data file.

Acknowledgment

The EASS is based on the CGSS, JGSS, KGSS, and TSCS, and distributed by the EASSDA. JGSS is supported by the Minister of Education, Culture, Sports, Science and Technology (JPMXP0620335833) and the Japan Society for the Promotion of Science (KAKENHI Grant Number JP17H01007, and Program for Constructing Data Infrastructure for the Humanities and Social Sciences Grant Number JPJS00218077184). The KGSS is supported by the National Research Foundation of Korea Grant funded by the Korean Government (NRF- 2020S1A3A2A03096777). TSCS has been sponsored by the Ministry of Science and Technology (formerly known as the National Science Council) in Taiwan since 1985.

References

Behr, D. and Shishido, K. (2016). The translation of measurement instruments for cross-cultural surveys. In: *The SAGE Handbook of Survey Methodology* (ed. C. Wolf, D. Joye, T.W. Smith, and Y.C. Fu), 269–287. London: Sage.

EASSDA (2019). EASSDA Newsletter, July 2019, http://www.eassda.org.

Fukuhara, S. and Suzukamo, Y. (2004). *SF-8TM Nihongo Manual* [SF-8TM A Manual in Japanese]. iHope International.

Hanibuchi, T. (2009). EASS 2010 Health Module no Sakusei: JGSS ni yoru Pretest no Kekka o chushin ni [Development of EASS 2010 Health Module: Results of JGSS Pretest]. *JGSS Research Series No.6 (JGSS Monographs No.9)*, pp. 211–242.

Hayashi, C. and Hayashi, F. (1995). Kokuminsei no Kokusaihikaku (Comparative study of national character). *Proceedings of Statistical Mathematics* 43 (1): 27–80.

Iwai, N. (2003). JGSS project (2) Chosa-hoho to Chosa-komoku (JGSS project (2) research designs and questionnaires). *Statistics* 54 (11): 48–55.

Iwai, N. (2017). Effects of Differences in Response Scale in Cross-national Surveys. Paper presented at the 1st RC33 Regional Conference on Social Methodology, Taipei, September 11–14, 2017.

Iwai, N. and Ueda, M. (2012). *Culture and Values in East Asia: A Comparison among Japan, South Korea, China, and Taiwan Based on East Asian Social Survey 2008*, trans. Sasaki T. and Shinohara, S. Nakanishiya Shuppan.

Iwai, N. and Uenohara, H. (2015). JGSS-2015 oyobi EASS 2014 Work Life Module no Sakusei [The Development of JGSS-2015 and EASS 2014 Work Life Module]. *JGSS Research Series No.12 (JGSS Monographs No.15)*, pp. 63–84.

Iwai, N. and Yasuda, T. (2011). *Family Values in East Asia: A Comparison Among Japan, South Korea, China, and Taiwan Based on East Asian Social Survey 2006*, trans. Sasaki, T., Shinohara, S., and Hayashi, H. Nakanishiya Shuppan.

Iwai, N. and Yoshino, S. (2020). Pitfalls in the construction of response scales in cross-cultural surveys: an example from east Asian Social Survey. In: *Advanced Studies in Behaviormetrics and Data Science* (ed. T. Imaizumi, A. Nakayama, and S. Yokoyama). Singapore: Springer Nature https://link.springer.com/book/10.1007/978-981-15-2700-5#toc.

Murata, H. (2020). Sekai o Yomitoku Kokusai Hikaku Chosa ISSP: Sono Igi to Kadai (the International Social Survey Programme (ISSP) analyzing the world with its cross-national surveys: the significance and the challenges). *The NHK Monthly Report on Broadcast Research* 11: 36–48.

Otani, S., Kinoshita, E., Goto, N. et al. (1999). *Shakai-Chosa eno Approach: Ronri to Hoho [Approach to Social Surveys: Logic and Method]*. Tokyo: Minerva Publishing.

Shishido, K. (2011a). The Japanese preference for "neither agree nor disagree". In: *Family Values in East Asia: A Comparison among Japan, South Korea, China, and Taiwan Based on East Asian Social Survey 2006*, 96–98. Kyoto: Nakanishiya.

Shishido, K. (2011b). The elaboration of translation in cross-national surveys. In: *Family Values in East Asia: A Comparison among Japan, South Korea, China, and Taiwan Based on East Asian Social Survey 2006*, 99–101. Kyoto: Nakanishiya.

Smith, T.W. (1997). Improving Cross-National Survey Research by Measuring the Intensity of Response Categories. GSS Cross-National Report, 17. Chicago, IL: National Opinion Research Center, University of Chicago.

Smith, T.W., Kim, J., Koch, A., and Park, A. (2006). Social-science research and the general social surveys. *Comparative Sociology* 5 (1): 33–43.

Sugita, H. and Iwai, N. (2003). JGSS project (3) Sokutei Shakudo to Sentakushi (JGSS project (3) measurement and wording of questions). *Statistics* 12: 49–56.

8

Ex-ante Harmonization of Official Statistics in Africa (SHaSA)

Dossina Yeo

Abbreviations

ACBF	African Capacity Building Foundation
AfDB	African Development Bank
ACS	African Charter on Statistics
AFRISTAT	Observatoire économique et statistique d'Afrique
AFRITAC	African Regional Technical Assistance Center
AfSS	African Statistics System
AIH	African Information Highway
APAI-CRVS	Accelerated Improvement of Civil Registration and Vital Statistics Civil Registration and Vital Statistics
APRM	African Peer Review Mechanism
AQAF	African quality assurance framework
ASY	African Statistics Yearbook
ATSY	African Trade Statistics Yearbook
AUC	African Union Commission
AUDA-NEPAD	African Union Development Agency/New Partnership For Africa's Development
COMESA	Common Market for Eastern and Southern Africa
CSOs	Civil Society Organizations
CVRS	Civil registration and vital statistics
ECA	Economic Commission for Africa
ECCAS	Economic Community of Central African States
ECOWAS	Economic Community of West African States
e-GDDS	Enhanced General Data Dissemination System
ERETES	Système d'aide à l'élaboration des comptes nationaux
EU	European Union
EUROSTAT	Statistical office of the European Union

Survey Data Harmonization in the Social Sciences, First Edition.
Edited by Irina Tomescu-Dubrow, Christof Wolf, Kazimierz M. Slomczynski, and J. Craig Jenkins.
© 2024 John Wiley & Sons Inc. Published 2024 by John Wiley & Sons Inc.

EUROTRACE	Software for the collection, compilation, and dissemination of trade data at regional and national levels
FAO	Food and Agriculture Organization of the United Nations
GPS	Governance, Peace and Security
HCPI	harmonized Consumer Price Index
ICP	International Comparison Program
IGAD	Intergovernmental Authority on Development
ILO	International labor Organization
IMTS	International Merchandize Trade Statistics
IRD-DIAL	French National Institute for Sustainable Development-Developpement, Institutions, Mondialisation
NSDS	National Strategies for the Development of Statistics
NSOs	National Statistics Offices
NSS	National Statistics System
PARIS21	Partnership in Statistics for Development in the twenty-first Century
PHC	Population and Housing Census
PPP	purchasing power parity
RECs	Regional Economies Communities
RSDS	Regional Strategies for the Development of Statistics
SADC	Southern African Development Community
SDDS	Special data dissemination standard
SDMX	Statistical Data and Metadata exchange
SHaSA	Strategy for the harmonization of Statistics in Africa
SNA	System of National Account
STATAFRIC	African Union Institute for Statistics
STCs	Statistical Training Centers/Schools
STGs	Specialized Technical Groups
TIMXLS	Interface that uses IMTS data to calculate unit value trade indices
UNDP	United Nations Development Program
UNDP	United Nations Development Program
UNSD	United Nations Statistics Division
WAEMU	West African Economic and Monetary Union
WHO	World Health Organization
UNECA	United Nations Economic Commission for Africa
UNEP	United Nations Environment Program
UNFPA	United Nations Population Fund
UNICEF	United Nations Children's Fund

8.1   Introduction

Similar to other regional economic unions, the African Union is striving to promote sustainable development and regional integration in Africa. Policies and strategies at regional and continental levels have been initiated to converge toward an *"integrated, prosperous and peaceful Africa"* with a single currency.[1] The successful implementation, monitoring, and evaluation of these policy frameworks depend heavily on the availability of comparable data across all African countries: common policies, such as macroeconomic convergence, common market, and economic and monetary union, as well as performances made by countries toward the attainment of their commitments at regional, continental, and international levels such as Africa's Agenda 2063: The Africa We Want[2] (AU 2015) or United Nations Transforming our world: 2030 Agenda for Sustainable Development[3] (UN 2015), cannot be effectively tracked or measured without harmonized data.

However, despite all recent efforts, African statistics are still compiled using different methodologies and norms across countries. Statistics from African countries are primarily considered unfit because they are often outdated, disseminated with significant delays, do not cover all areas of socioeconomic life, and do not meet user's needs. As a result, most reports on Africa continue to rely on international data and make no reference to data produced by countries. Estimated data in international databases are always different from data in African countries' publications and do not often reflect the socioeconomic and cultural realities of the continent.

To address these challenges and enable African countries to produce quality data based on international norms and standards, the Strategy for the harmonization of Statistics in Africa (SHaSA)[4] was adopted in 2010 and revised in 2018 to take into consideration Agenda 2030 and Agenda 2063, emerging statistical areas and new types of data. The main purpose of SHaSA is to generate timely harmonized statistical information covering all dimensions of political, socioeconomic, environmental, and cultural development and integration of Africa. It has four strategic objectives, including to: *(i) produce quality statistics for Africa; (ii) coordinate the production of quality statistics for Africa; (iii) build sustainable institutional capacity in the African statistical system; and (iv) promote a culture of quality policy and decision-making.* Implementation, monitoring, performance evaluation, and reporting mechanisms that include all different stakeholders at national, regional, and continental levels with clear responsibilities and based on the principle of subsidiarity have been put in place to ensure the attainment of these objectives (AUC et al. 2018).

SHaSA is unique in the sense that it is a collective initiative of all data producers[5] on the continent and backed by policymakers to provide quality data to all users including individuals, civil society organizations (CSOs), academics, private sector, and decision-makers. It offers a clear approach to

---

1 See The Vision of the African Union, https://au.int/en/about/vision.

2 Africa's Agenda 2063 is Africa's blueprint and master plan for transforming Africa into the global powerhouse of the future, adopted in January of 2015 by the 24th African Union (AU) Assembly of Heads of State and Government.

3 The United Nations Transforming our world: 2030 Agenda for Sustainable Development was adopted by the General Assembly on 25 September 2015. Global Goals are a collection of 17 interlinked global goals designed to be a "blueprint to achieve a better and more sustainable future for all."

4 https://au.int/en/ea/statistics/shasa

5 Data producers: Government statistical agencies, Academicians/Researchers, CSOs, NGOs, RECs continental and international organizations, etc.

make a paradigm shift to ensure the continuous availability of comparable and reliable statistics that genuinely reflect the reality of countries, regions, and the whole continent in all domains of statistics. It is worth mentioning that Africa is the only region in the world that has such a comprehensive strategy for statistical harmonization endorsed by Heads of State and Government of the African Union through the Executive Council of the African Union in July 2010 (Dec. No EX.CL/Dec.565 (XVII)). The adoption of SHaSA marked a decisive turning point for the continent and showed the rising awareness and growing understanding of the vital role of statistics as an objective tool for the informed decision-making process, policy analysis, investment and business management, and measurement of the progress of societies.

The purpose of this chapter is to discuss the value-addition that the SHaSAe[6] has brought statistical harmonization in Africa, by examining harmonization approaches used and the achievements and challenges encountered in the production of comparable data in various statistical domains on the continent.

8.2 Applied Harmonization Methods

SHaSA follows the need for a holistic strategic approach for the production and use of quality statistical information in developing countries, and links to initiatives of the 2000s. To monitor and evaluate progress toward the attainment of sustainable development, the Millennium Development Goals (MDGs) were adopted in 2000, followed in 2004, by the Marrakech Action Plan for Statistics: "Better Statistics for Better Results." "Statistics for Transparency, Accountability, and Results: A Busan Action Plan for Statistics" renewed the action plan for statistics in 2011. Since then, PARIS21 has been engaged in expanding global statistics comparability and statistical measures with regional organizations, to foster the development of national and regional statistical strategies in developing regions such as Africa, the Association of Southeast Asian Nations, the Caribbean Community, and the Pacific Community.[7] The AUC, together with PARIS21, the United Nations Economic Commission for Africa (UNECA), and the African Development Bank (AfDB) have been assisting African countries in the development and the implementation of national strategies for the development of statistics (NSDS).

NSDS provides a framework to coordinate and harmonize statistics processes and procedures, including broader coverage of the sources and production of statistics beyond National Statistics Offices (NSOs) to include other producers of official statistics. At the regional level, several Regional Economies Communities (RECs) are developing and implementing Regional Strategies for Development of Statistics (RSDS) as frameworks for the coordination and harmonization of the statistics at their levels, to respond to specific objectives of regional integration including monitoring macroeconomic convergence and multilateral oversight. RSDS encompass national and regional priorities and are aligned with the NSDS of the member countries. In the same vein, SHaSA, which is the continental strategy, mainstreams the

---

6 https://au.int/en/ea/statistics/shasa
7 https://paris21.org/rsds

priorities of NSDS and RSDS. All three-level strategies are integrated and are implemented in a coherent and concomitant manner.

These three strategy frameworks came with coordination and follow-up mechanisms at all levels (country, RECs, continental, and international), to ensure the development and application of harmonization methods. The mechanisms include clear roles and responsibilities defined for each stakeholder with respect to the principle of subsidiarity, by capitalizing on the gains and complementarity according to their comparative advantages. This approach ensures better synergy, consensus, and collective efforts in the implementation of the strategies with the limited available resources on the continent.

In addition to the political decision bodies that provide the political support required, 18 Specialized Technical Groups (STGs) have been established, comprising statisticians from National Statistics Systems (NSSs), Pan-African institutions, RECs, specialized agencies, regional and international agencies, and other experts in the areas (see Table 8.1: List and composition of the Specialized Technical Groups). The primary roles of STGs are to develop robust methodological guidelines in every domain of statistics. These methodological guidelines are twofold: manuals for the production of data (ex-ante harmonization) and manuals for reprocessing, adjustment, and re-estimation of existing data to ensure comparability (ex-post harmonization). The guidelines include quality assessment frameworks and best practices for production, verification of the accuracy of data already produced, and their validation process. They also contain concepts, standards, classifications, etc., which are based on internationally recognized norms and standards for the production of comparable data across the world. Based on the methodological guidelines, common data collection templates, tools, frames, or formats are prepared for conducting statistical operations such as censuses, surveys, administrative records, population, and business registration systems.

With the support of development partners, the AUC and its partner organizations have made tremendous progress over the past years in the harmonization of statistics in key areas including governance, peace and security (GPS), labor, migration, external trade, national accounts, harmonized consumer price index (HCPI), civil registration and vital statistics (CRVS), and international comparison program (ICP) just to mention a few. In these sectors, the continent has come out with common concepts, methods, and methodological manuals for the collection of statistics and reprocessing/adjusting the existing statistics for comparability purposes. The proposed norms and methodologies are new for emerging statistics, such as GPS statistics or adaptations of international norms and methods to African realities. They have been used by African countries to conduct surveys and exploit administrative sources, where necessary.

Besides the efforts made by the producers of official statistics, there are many other data producers operating on the continent who are not statistics authorities.[8] They include researchers/academic institutes, CSOs, and non-governmental organizations. Examples of such efforts are the Corruption Perception Index (CPI) data from Transparency International and the Afrobarometer. The following subsections discuss only the different efforts deployed by statistics authorities on the continent in producing comparable data using ex-ante harmonization and ex-post harmonization.

---

8 See the African Charter on Statistics. "Statistics Authorities," national statistics institutes and/or other statistics organizations in charge of official statistics production and dissemination at national, regional, and continental levels.

Table 8.1 List and composition of the Specialized Technical Groups.

N°	Specialized Technical Groups	Secretariat	Composition (other members)
01	STG-GPS (Governance, Peace and Security)	AUC/AfDB	ECA, ACBF, RECs, NEPAD, APRM, Member States
02	STG-ES. External Sector (External Trade and Balance of Payments)	AUC/AACB	ECA, AfDB, ACBF, RECs, AFRITAC, Member States
03	STG-MF (Money and Finance)	AACB/AUC	AUC, ECA, AfDB, ACBF, RECs, AFRITAC, Member States
04	STG-NA&P (AGNA) (National Accounts and Price Statistics)	ECA/AfDB/AUC	RECs, AFRISTAT, AFRITAC, Member States
05	STG-II&T. (Infrastructure, Industries & Tourism)	AfDB/NEPAD	AUC, ECA, ACBF, RECs, Member States
06	STG-PFPS&I. (Public Finance, Private Sector, and Investment)	AfDB	AUC, ECA, ACBF, RECs, AFRISTAT, AFRITAC, Member States
07	STG-STE. (Science, Technology & Education)	AUC/ACBF/NEPAD	AfDB, ECA, RECs, Member States
08	STG-So. Demography, Migration, Health, Human Development, Social Protection & Gender	ECA/AUC	AfDB, ACBF, RECs, AFRISTAT, Member States
09	STG-Env. (Agriculture, Environment, Natural Resources, & Climate Change)	AfDB/AUC	ECA, ACBF, AFRISTAT, RECs, NEPAD, FAO, UNEP, Member States
10	STG-CB (AGROST). Statistical Training and Capacity Building.	ECA/ACBF/AUC	AfDB, RECs, AFRISTAT, STCs, ACBF, AFRITAC, Member States
11	STG-Labor and Informal Sector Statistics	AUC/AfDB	ECA, RECs, ILO, Member States
12	STG-Classification	ECA/AFRISTAT	AUC, AfDB, RECs, Member States
13	STG-Statistics on Civil Registration	ECA/AUC	AfDB, AFRISTAT, RECs, UNICEF, UNFPA, WHO, Member States
14	STG-Sustainable Development	AUC/AfDB/ECA	AFRISTAT, NEPAD, RECs, UNDP, Member States
15	STG-ICT for Statistical Production	AfDB/AUC	ECA, AFRISTAT, RECs, Member States, STCs
16	STG-Mobilization of Political Will	AUC/AfDB	ECA, AFRISTAT, RECs, Member States
17	STG-Emerging Statistical Issues	AfDB/ECA/ACBF	AUC, AFRISTAT, RECs, UNDP, STCs, Member States
18	STG-National Strategies for the Development of Statistics	AUC/AfDB/ECA	ACBF, AFRISTAT, RECs, PARIS21 STCs, Member States

Source: Adapted from AUC et al. (2018).

8.2.1 Examples of Ex-ante Harmonization Methods: The Cases of GPS Data and CRVS

8.2.1.1 Governance, Peace and Security (GPS) Statistics Initiative

Following the adoption of the GPS methodology in December 2012 by the Committee of Directors General NSOs in Côte d'Ivoire, two survey instrument modules and two administrative sources of data instruments were developed by STG-GPS. Twelve pilot countries used the survey component to conduct a GPS survey by adding the modules to already scheduled official household surveys, or use the administrative component to generate comparable official data on GPS with the support of AUC, United Nations Development Program (UNDP), and French National Institute for Sustainable Development-Developpement, Institutions, Mondialisation (IRD-DIAL).

Subsequent to the success of the first round of the GPS initiative, the UN Statistical Commission established the Praia Group on governance statistics, to address the conceptualization, methodology, and instruments for producing GPS statistics at the global level. GPS instruments were revised and used for the second round of the survey. The modules were integrated into the harmonized 1–2 survey on labor force and informal sector in the eight countries of the Economic Community of West African States (WAEMU) (Benin, Burkina Faso, Côte d'Ivoire, Guinea-Bissau, Niger, Mali, Senegal, and Togo). A new phase of the GPS-Action Plan: methodology, survey program, capacity building, and outputs and institutionalization was launched in 2019. Graph 8.1 below presents an example of data collected for Cabo Verde using the GPS instruments in 2013 and 2016. These data have helped to assess the Aspiration 3 of AU Agenda 2063: an Africa of good governance, democratic, respect for human rights, justice, and rule of law.

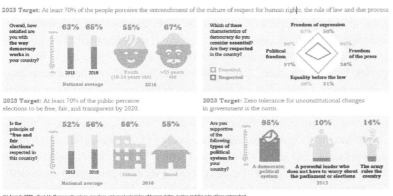

Graph 8.1 Governance, Peace and Security Statistics in Cabo Verde (2013 and 2016). Source: Adapted from Instituto Nacional de Estatistica (2015); https://au.int/en/agenda2063/aspirations, last accessed November 10, 2022.

8.2.1.2 Development of Civil Registration and Vital Statistics (CRVS)

With the support of the regional CRVS Core Group, African countries have put in place civil status registration systems that they are gradually strengthening to better meet international norms and recommendations by taking into account the specific characteristics and sociocultural realities of the continent. The implementation of the African Program for the Accelerated Improvement of Civil Registration and Vital Statistics (APAI-CRVS) has made it possible to attain remarkable results, notably through political will and commitments at the highest level of the state: the modernization and digitalization of CRVS, systematic and coordinated approaches at the national and regional levels; strengthening the capacities of civil registry officers; and the promoting the transfer of knowledge and sharing of experiences among countries. A CRVS Digitization Guidebook is developed and used by all African countries. The World Health Organization (WHO) and other regional partners have helped Asia and the Pacific region and the Eastern Mediterranean region to develop regional strategy for the improvement of CRVS systems.[9]

Today, many African countries have a biometric civil registration system and are generating harmonized vital statistics from their civil registration following APAI-CRVS harmonized norms and standards. Countries are moving toward holistic, innovative, and integrated CRVS and digital identity management systems to close the identity gap in Africa, where more than 500 million people have no legal identity.

8.2.2 Examples of Ex-post Harmonization: The Cases of Labor Statistics, ATSY, ASY and KeyStats, and ICP-Africa Program

Guideline for producing labor market indicators from existing data sources in Africa. In the area of labor statistics, the 2012 labor market information systems harmonization and coordination framework led to a methodological guideline to assist African countries in producing labor market indicators from existing data sources in Africa.[10] The guideline proposes a harmonized approach to producing and analyzing a minimum of employment and labor indicators based on already collected data and in line with the new international standards adopted in 2013 by the 19th International Conference of Labor Statisticians (ICLS). The document analyzes the availability of data; presents a list of 20 priority indicators; defines the main concepts and data sources; and describes the way to calculate the indicators. The proposed list has six categories: headcounts, rates, labor underutilization, informational economy, poverty, and gender. The first phase started with pilot countries in the five regions: Northern Africa: Mauritania, Tunisia, and Algeria; West Africa: Benin, Burkina Faso, Cote d'Ivoire, and Nigeria; Central Africa: Cameroon, Chad, and Gabon; Eastern Africa: Ethiopia, Rwanda, and Tanzania; and Southern Africa: Angola, Mozambique, Namibia, and Zimbabwe. After two weeks of implementation, 15 pilot countries met for an evaluation workshop held in Abidjan, Cote d'Ivoire, in June 2017. Based on the results of the pilot countries, a new version of the guideline was prepared for data processing and calculation of harmonized indicators. After the evaluation of the pilot operation and the finalization of the approach, the new guideline was shared with all countries for their use in order to produce harmonized time-series data on labor market indicators.

---

9 See https://www.who.int/data/data-collection-tools/civil-registration-and-vital-statistics-(crvs).

10 https://au.int/sites/default/files/documents/32846-doc-guidelines_for_producing_labor_market_indicators.pdf

The production of the African Trade Statistics Yearbook (ATSY). Following the launch of the African Continental Free Trade Area in 2012, the need for comparable data on external trade produced by African countries has increased tremendously. The AUC, with the support of the EU, has been producing the ATSY, a time series of trade data. To standardize the processing and uploading of harmonized data in the Continental Trade Database (AFRICATRADE 2020), a "trade data transmission rules and formats" document was prepared, to be used by countries. Trade raw data and metadata are collected from countries, and estimates are made. STG-External Sector meets annually and discusses the draft publication before its release.

Methodological notes for production of KeyStats and the African Statistics Yearbook (ASY). KeyStats and ASY are annual publications, which measure progress and assess the regional integration process on the continent in selected thematic areas, such as the free movement of persons, external trade, infrastructure development, production, and macroeconomic and social indicators. Data from the ASY are used to produce the Joint European Union and the African Union-A statistical portrait.[11] KeyStats provides socioeconomic data on each country and aggregated data on each REC. The methodology presents the definition, data sources, remarks/comments, and mode of calculation of each indicator at national, regional, and continental levels, as well as verification methods. Data are collected by a designated responsible unit/division in the NSO in each country and RECs. Validation meetings are held with representatives from African countries, RECs, Pan-African organizations and other stakeholders to discuss, verify, check, estimate missing data, and validate the data by ensuring their comparability among countries before publication.

International Comparison Program for Africa (ICP-Africa). One of the largest initiatives in the world, ICP aims at generating estimates of purchasing power parity (PPP), which facilitates comparisons across countries in the level of prices and economic aggregates in real terms. Since the launch of ICP-Africa in 2002, the number of participating countries stood at 48–50 for the 2005 and 2011 rounds, respectively, on the continent. While at the global level ICP is coordinated by the World Bank, the ICP-Africa program is led by the AfDB.

PPPs are calculated based on the price of a typical basket of goods and services that make up final consumption expenditure and gross capital formation in each participating country. They are a measure of what a country's local currency can buy in another country. Market exchange rate-based conversions reflect both price and volume differences in expenditures and are thus inappropriate for volume comparisons. PPP-based conversions of expenditures eliminate the effect of price level differences between countries and reflect only differences in the volume of economies.

The 2017 ICP report was based exclusively on the prices and national accounts expenditures provided by the participating countries. Purchasing power standards and real expenditures were compiled in accordance with the established ICP methods and procedures. PPPs and estimates of PPP-based GDP and its major expenditure components in aggregate and per capita terms were published for the 176 economies, including 50 African economies (World Bank 2020).

---

11 https://ec.europa.eu/eurostat/documents/3217494/7142142/KS-FQ-15-001-EN-N.pdf/a829bd49-eaec-4656-ac04-9028c9 3d6258?t=1453225026000

8.3 Quality Assurance Framework

The production of quality and comparable data requires the application of quality assurance frameworks. Data are harmonized and reliable if their collection and processing methods remain stable over time and across countries. Over the past decades, the absence of a quality assurance framework in many countries has created mistrust, lack of integrity, and suspicion about data. Many people doubt data produced by African countries. African data are recognized as being outdated, incomparable, etc.; and are not disseminated promptly and covering all areas of social and economic life. It is for these reasons that international data sources are often used in preparing reports on Africa and make no reference to data produced by African countries. Furthermore, data in many international databases differ significantly from data in African country publications.

To achieve and maintain public trust in African statistics requires that those statistics are produced in an objective, transparent, and professionally independent manner. Recently, in the framework of SHaSA, the AUC has developed and is implementing a generic African quality assurance framework (AQAF) to safeguard data integrity and public trust. The AQAF is based on the principles of the African Charter on Statistics (ACS); which captures the principles and best practices of the UN Fundamental Principles of Official Statistics.[12] The template and guidelines developed by the United Nations Statistics Division (UNSD) for the development of the National Quality Assurance Framework were used. AQAF is structured around the following elements: the strategic context for the framework; alignment of the framework to the ACS; and a mapping of the quality standards/indicators. Accordingly, this framework meets the desired international quality requirements. Each STGs of SHaSA is expected to adopt the AQAF to meet the requirements for the production of harmonized data in their respective domains.

Strategic context for the quality framework. SHaSA provides the content for the framework. The AQAF describes who should do what and at which level. It envelops all the stakeholders of the African Statistics System (AfSS)[13] pyramid (primary, secondary, and tertiary levels) (AU 2009) and across the four strategic objectives of SHaSA. The different levels of AfSS require different types of quality standards, although this differentiation is not mutually exclusive. Nevertheless, each level of the pyramid is relatively dominated by certain types of standards, as indicated below.

- **Quality standards at the primary level of data production**: The majority of data are produced by the NSSs. It is where the quality standards/indicators are mainly applied to assure the quality of the primary production of statistics. The quality standards/indicators cover the content of censuses, surveys, and registers.
- **Quality standards at the secondary level of data production**: Quality standards/indicators focus on the comprehensiveness of data collection at a regional level. Harmonized data are required to merge data from each country and to have aggregated indicators. RECs are expected to check, verify, and indicate data gaps in a systematic way before aggregating them.

---

12 See unstats.un.org.
13 See African Charter on Statistics. https://au.int/en/treaties/african-charter-statistics.

- **Quality standards at the tertiary level of data production**: Here, the AU Institute for Statistics (AFRISTAT) or any Pan-African organization will assemble data from the primary production level (NSSs) and, to a lesser extent, from RECs, depending on the indicator. The quality assurance indicators assess the quality of the scope, processes, and methodologies of data harmonization in each domain of statistics and data aggregation at the continental level.

AQAF is an implementation instrument for the ACS and assures the quality of the entities of the data harmonization strategy identified in SHaSA. The six principles of ACS provide for quality dimensions, as indicated below:

- Professional independence
- Quality
- Mandate for data collection and resources
- Dissemination
- Protection of individual data, information sources, and respondents
- Coordination and Cooperation

Each principle of ACS is divided into sub-principles. Under each sub-principle, elements to be measured are defined, including quality indicators for each element to provide that the quality of the element has or has not been assured; and also methodology on how to assess the quality. AQAF is designed to address three components that constitute a data quality system. These components are: (i) institutional environment; (ii) statistical process; and (iii) statistical output. The components of the data quality system of the AfSS and their association with ACS principles and sub-principles are mapped in Table 8.2.

As stated, AQAF is a generic guideline for statistical quality assurance that all African countries can adapt and use to ensure that the statistical production process and dissemination practices are of quality standards that meet international expectations. Its use has yet to be generalized. The next steps would consist of its widespread dissemination and building the capacity of the experts from countries on its use to produce and disseminate data.

In addition to AQAF, African countries are encouraged to adhere to Special Data Dissemination Standard Plus (SDDS Plus) and Enhanced General Data Dissemination System (e-GDDS), which provide appropriate tools for judging the statistical capacities of African countries. The purpose of the e-GDDS is to: (i) encourage member countries to improve data quality; (ii) provide a framework for evaluating needs for data improvement and setting priorities in this respect; and (iii) guide member countries in the dissemination to the public of comprehensive, timely, accessible, and reliable economic, financial, and sociodemographic statistics.

All countries that participate in the GDDS automatically participate in the e-GDDS. The IMF Board approved the use of AfDB's African Information Highway (AIH) portal system for implementing the e-GDDS, using the SDMX protocols in AIH to facilitate data collection and sharing across Africa and with development partners. AIH's main objective is significantly to increase public access to official and other statistics across Africa, while at the same time supporting African countries to improve data quality, management, and dissemination. In Africa as of October 2020, there are 39 e-GDDS participants, 4 SDDS subscribers, but no SDDS plus adherents. African countries should seek to develop their statistical systems to upgrade from e-GDDS to SDDS.

Table 8.2 ACS Principles grouped by data quality system components.

Components	ACS principles	ACS sub-principles
Institutional Environment	Professional independence	• Scientific independence • Impartiality • Responsibility • Transparency
	Mandate for data collection and resources	• Mandate • Resource adequacy • Cost-effectiveness
	Protection of individual data, information sources and respondents	• Confidentiality • Giving assurances to data providers • Objective • Rationality
	Coordination and Cooperation	• Coordination • Cooperation
Statistical Process	Quality	• Relevance • Sustainability • Data sources • Accuracy and reliability • Continuity • Coherence and comparability • Timeliness • Topicality • Specificities • Awareness building • Statistical process
Statistical Output	Dissemination	• Accessibility • Dialogue with users • Clarity and understanding • Simultaneity • Correction

Source: Adapted from AUC (2018).

8.4 Challenges to Statistical Harmonization in Africa

Despite the concrete successes mentioned above in the harmonization and dissemination of data in many domains, there are still numerous constraints to the availability of comparable data on the continent. African statistics are still poor in terms of quality and quantity as they do not cover all domains. They also vary from one source to another. If the well-thought and well-designed initiatives mentioned in previous sections were fully implemented, African statistics would not suffer from the lack of trust from the public. The challenges to harmonization can be structured around the implementation of NSDS and ex-ante and ex-post harmonization processes.

8.4.1 Challenges to the Implementation of NSDS

NSDSs, are expected to facilitate the coordination and the harmonization of statistics processes and procedures for all types of statistics beyond the NSO, including other data producers. They provide for the regular organization of Population and Housing Censuses (PHCs), household surveys, the organization of periodic economic censuses and surveys, the development of CRVS, and the exploitation of administrative records. The quality standards at the primary level are applied to ensure the quality of data generated through censuses, surveys, and registers. Nevertheless, most NSDS are not effectively implemented, and quality procedures are not always applied to statistics operations because of several reasons, including:

(i) *Lack of ownership and poor coordination.* At the initial stage throughout all different phases of NSDS, all implementing agencies are not fully involved and are called to implement a strategy that they are not aware of, or they poorly understand the purpose of. Sometimes stakeholders also lack the required expertise to perform the tasks they are expected to do. Furthermore, most statistical laws are obsolete. They do not include all domains of statistics and all data producers. This creates poor coordination among data producers due to unclear roles and responsibilities for each stakeholder, which compromise the implementation of NSDS. Another problem is that NSDS and regional and continental statistical initiatives are not aligned. This creates an additional burden for countries as they are supposed to implement their strategies and international ones without adequate accompanying measures, particularly financial resources, and expertise.

(ii) *Poor political support and insufficient allocation of resources.* In many cases, the implementation of NSDS and all statistics operations lack sufficient political support leading to the inadequate allocation of resources for the full implementation of all data harmonization goals. The lack of statistics culture or statistical awareness among all spheres of African society reduces the support for investing in improvements in harmonization of high-quality and comparable statistics. The successful initiatives observed on the continent are donor-driven and funded. A good example is the PHCs. Except for two countries, all African countries conducted their PHC during the 2005–2014 cycle and close to 50 countries plan to conduct their PHC during the 2020 census cycle (2015–2024 decade). PHCs are supported by partners, who provide technical and financial assistance. In contrast, many countries are unable to carry out agricultural censuses regularly and very few are conducting economic censuses and surveys. Furthermore, African countries still need financial support and technical expertise in many domains related to civil registration, keeping administrative records, business registers, etc. Data are public goods, and citizens have the right to get access to them. When data are produced, they must be analyzed and disseminated in an effective manner using adequate formats, dissemination, and communication tools such as user-friendly databases, websites, social media, and other tools to ensure data reaches all users. This has not been the case in many countries.

(iii) *Limited understanding of international concepts and standards.* There is little understanding of several statistical concepts and methodologies in many domains because some of them are not relevant or fit to an African context. African experts do not always participate in international conferences where statistical concepts and methodologies are discussed and defined. This makes it difficult for countries to apply international standards in the production of data.

(iv) *Internal data harmonization within the NSS and use of administrative records.* If there is no coherence and internal harmonization throughout the entire NSS, data are not disaggregated by district and it is impossible to have coherence in data produced at all levels. It is easy to see data discrepancies from government agencies on the same indicator. Furthermore, many African countries put more emphasis on the population censuses and surveys; which are very costly instead of using administrative data, civil registration, or establishing business and population registers, etc. These constraints the availability of sufficient timely and updated data.

8.4.2 Challenges with Ex-ante Harmonization: Examples of GPS and ICP Initiatives

The challenges with ex-ante harmonization will be analyzed by comparing the GPS initiative with ICP initiative. The GPS initiative came at a favorable moment with the launch of the African Peer Review Mechanism (APRM) for African countries to self-assess themselves on a set of standards, norms, and codes of best practices on various aspects of democratic, economic, and corporate governance. The initiative was entrenched in Agenda 2063, Aspiration 3: an Africa of good governance, democracy, respect for human rights, and justice and the rule of law; Aspiration 4: a peaceful and secure African; and the UN Sustainable Development Goal 16: peace, justice, and strong institutions. It generated a lot of interest among policymakers, CSOs, and international partners and got political support, legitimacy, and credibility right from the beginning. The production of GPS data by NSOs, as compared to other organizations, such CSOs, was well received because of the sustainability, technical expertise, scientific independence, and professionalism approaches that NSOs bring in. Using regular socioeconomic surveys as vehicles for the GPS modules helps to ensure GPS data are collected regularly and the establishment of time series on GPS statistics. Furthermore, GPS modules were prepared in a participatory manner involving all stakeholders (statisticians, technical experts, policymakers, etc.) and got approved. With the support of the AUC, UNDP, and other partners, 12 countries were able to collect GPS data through household surveys and administrative records.

However, since the launch of the initiative, GPS data are not produced regularly in the pilot countries, and other countries have not initiated the exercise. The institutionalization of the production of GPS statistics by creating a specialized statistics unit in NSOs and the mainstreaming of GPS modules in household surveys to be conducted regularly to avoid additional financial resources has not been an easy task. GPS data are neither widely integrated into statistics production systems nor produced regularly by all African countries. It is obvious to say that the GPS initiative has not been entirely a success. The main reasons behind this situation are mostly the poor funding of NSS, political problems, and lack of interest of policymakers in statistics.

As compared to the GPS initiative, the ICP-Africa program is driven by the AfDB funding. With this funding, almost all countries are participating in the program. However, some countries viewed the activities related to ICP as separate work and an extra effort. ICP activities can be easily mainstreamed into regular consumer price index operations to ensure the success of both operations and cost-effectiveness since there is no need to employ extra data collectors and build new systems. Quality assurance and validation processes are employed to guarantee that data collected by each participating country in Africa are comparable by checking and correcting all errors.

8.4.3 Challenges with Ex-post Harmonization: Examples of KeyStats and ATSY

The production of KeyStats and ATSY obeys the same process. These publications are based on existing data from countries. The methodological guidelines of KeyStats present the definitions of concepts, data sources, reprocessing of data, and modes of the calculation of aggregated data at regional and continental levels. However, the insufficient quantity and quality of data produced by countries using these guidelines and the existence of data gaps make their production a complex and challenging task. The irregularity of the organization of population and agricultural censuses and households and economics surveys and the nondevelopment and nonexploitation of administrative records make it difficult for countries to have updated and timely data on all indicators of the publications.

Furthermore, sometimes micro-data of censuses and surveys exist, but they are not correctly processed, analyzed, and disseminated on websites, and there is also no database where the data can be extracted from. The problem of resources and technical expertise are the leading causes of the unavailability of data. It is not because countries refuse to share the data, but rather that data are unavailable, or of low quality. This makes it challenging to reprocess and estimate missing data sent by countries. Furthermore, as the required data for KeyStats cover other sectors beyond NSOs, therefore, focal persons or units in charge of this work in the NSOs have to collect data from various producers of national statistical systems. Oftentimes, they are confronted with un-harmonized data within the NSS, and they do not always possess the technical expertise to quality check and make the necessary adjustment. Data on the same indicator can also be different from one producer to another because of the utilization of different methodologies, classifications, and definitions of concepts.

In the same vein, the data transmission file for ATSY consists of the detailed annual results of statistics on trade in goods. The main source of trade data remains customs. To maximize the customs revenue, customs authorities put more emphasis on the recording of the customs values of the products to the detriment of the weights or the supplementary units of the products. In addition, African countries do not always use the current versions of the Harmonized System nomenclature (HS 2017) for coding goods. Customs use several versions of the harmonized system at the same time, this makes the task very difficult for the production of trade statistics. Many agricultural and livestock products escape from customs controls; there is a need for surveys on informal cross border trade to capture such data.

8.5 Common Software Tools Used

The harmonization process is not only standards, methods, procedures, and classifications; it also implies the use of the same statistical software. This facilitates efficient processing, integration, analysis, and calculation of comparable data over time and across space. It also presents an opportunity for making data integration possible while reducing costs, storage of data, and presentation of the outputs. In the framework of SHaSA, under Objective 3.3: put in place an effective technological environment, several actions have been taken to promote the development and the use of common software in various domains of statistics to facilitate the accurate and timely collection, processing, analysis, and dissemination of data.

In recent years, RECs and regional organizations such as AFRISTAT have been promoting the use of common software in their member countries. ECOWAS, like other RECs, is supporting in its region the use of ERETES for national accounts and EUROTRACE for collection, compiling, and dissemination of trade data at the national and regional levels used for IMTS in conjunction with ASYCUDA. ASYCUDA is software for registering customs data. The software package Chapo is used in WAEMU countries for the calculation of consumer price index. AfDB has installed the Africa Information Highway and Open Data Platforms in 54 African countries and 15 African subregional and regional organizations to improve data dissemination and public access to a raft of diverse statistics on Africa.

Based on the success of the use of EUROTRACE and ERETES in several regions, the AUC is working with the support of its partners to promote their use in the continent. EUROTRACE is a generic and open system able to be adapted to national and regional requirements and most types of statistics. EUROTRACE allows to: (i) import and manage the data necessary for the development of the external trade statistics (in particular customs data); (ii) treat these data, in particular, by carrying out quality controls and settings to the standards; and (iii) work out and calculate a certain number of aggregates of the foreign trade and export them for dissemination and publication.

A EUROTRACE User Group was established as a platform for exchanging experiences, together with ideas on the mid- and long-term development of the software. Currently, EUROTRACE is used in many countries in COMESA, ECOWAS, IGAD, ECCAS, and SADC regions. Efforts are made to extend the use of EUROTRACE to all African countries for harmonization purposes. **ERETES** is used for the production of harmonized and integrated national accounts data in compliance with the international standards of SNA 2008. It has three components, including: a software tool; a methodology to build national accounts; and an assistance and user community platform to share concerns and find solutions for national accounts compilation. The AUC is cooperating with Eurostat and the French Cooperation for broad use of ERETES by all African countries. Through the STG on National Accounts, training workshops are conducted to build technical capacity of national accounts compilers covering: IT tools, ERETES software, supply and use tables, quarterly GDP, the System of Environmental-Economic Accounting, etc. With ERETES, the users should apply rigorously the methodological requirements of 2008 SNA, which is not the case with EUROTRACE. EUROTRACE does not impose any specific methodology on the users.

Recently, the AUC has developed, in the framework of the EU-Pan African Statistics Program, an improved version of **TIMXLS** to provide the users with an interface to manage steps leading to the production of trade indices easily. TIMXLS uses IMTS data to calculate unit value trade indices. It is customizable and allows the production of user-defined external trade indices (parameterization of the data filtering, flexible base year definition, and flexible output nomenclature). The use of **SDMX** is also currently promoted to foster standards for the exchange of statistical information. NSOs, central banks, and international organizations are using SDMX at various stages of their statistical business processes.

8.6 Conclusion and Recommendations

The critical importance of quality data is well-recognized. Data have the power to change the entire society for the better. They are a vital instrument not only for the development of nations but also for all segments of society. Citizens need reliable data available on time to hold the state to account and ensure good governance and transparency in the management of state affairs. Similarly, quality statistics facilitate the attainment of better development results as they drive evidence-based decision-making. Many structural socioeconomic problems, epidemic diseases, hunger, multidimensional poverty, impoverished and marginalized populations; etc. could be better addressed or avoided, if quality statistical information was available. The monitoring and evaluation of international agendas also demand quality and comparable data across countries and over time in order to ensure all participating countries move together, and no one is left behind. This imposes the use of shared norms and standards in all areas of the common policies and international agreements.

The adoption of the revised SHaSA in 2018 came as a sigh of relief to address the poverty of quality data on the continent; as it provides a clear roadmap to provide comprehensive and comparable data across the continent. Over the past years, Africa has successfully generated harmonized data in several domains within the NSS and at regional, continental, and international levels. The SHaSA initiative is a unique initiative and has adopted best practices for statistical harmonization that are being emulated by other regions of the world that have embarked on regional integration. The triumph of the GPS-SHaSA initiative in providing the world with harmonized concepts and methodologies for GPS statistics has helped set the standards for the Global "Praia Group on governance statistics." SHaSA experiences and achievements can benefit academicians and researchers that are planning to conduct similar data harmonization.

Through the SHaSA initiative, comparable data are available in several areas, such as GPS, labor, migration, external trade, national accounts, HCPI, and ICP, even if some of them are not regularly produced and disseminated on time and do not cover all African countries. This shows that there is still a long way to go toward the production of a wide range of comparable and timely data on Africa. This situation will continue to affect the quality of policies and decisions on the continent and to encourage the use of international data sources to write Africa's narratives. Therefore, it is urgent to take necessary measures to address all impediments to statistical harmonization. These include:

(i) **Advocacy for the effective implementation of the Executive Council of the African Union Decision Dec. No. EX.CL/Dec.987(XXXII) to allocate 0.15% of the National Budgets to finance statistics.** An immediate action would be to create awareness and advocacy, and make sure that all African countries devote 0.15% of their National Budgets to finance statistical activities as per the decision of the Executive Council. Countries could also explore the possibility to allocate a percentage of the Official Development Assistance received from external development partners to statistical activities. These two ideas, if implemented, would provide African countries with some basic means to produce regular and harmonized data in key specific areas.

(ii) **Identification of strategic priority areas, including a key list of indicators by all data producers in collaboration with data users' communities, particularly decision-makers.** The SHaSA scope is too broad for African countries. With the limited resources, the continent does not have the capacity to produce harmonized data to cover all statistical dimensions. Therefore, focusing on a few domains and key indicators should be the starting point, like Eurostat did (Euro indicators).[14] Common norms, concepts, and tools for data collection and estimation and data validation processes can be developed and used in these areas. Expertise can be generated, peer learning, sharing experiences, and best practices and technical assistance can help African countries to produce comparable data in these areas. This will increase public trust and confidence in African data.

---

14 https://ec.europa.eu/eurostat/web/main/news/euro-indicators

(iii) **Development and exploitation of population and business registers, administrative records, and big data**. African countries have put too much emphasis on censuses and surveys, which are very costly, instead of harnessing administrative sources, including registration systems, the development of population and business registers, annual national enterprises balance sheets, and geographical maps. The exploitation of data in these records can improve data production in many areas and yield socioeconomic indicators at a lower cost, to respond to the demand for data. The integration of statistical and geospatial information and the explosion of "big data" generated by digital platforms and devices – which provide insights into human activities and environmental changes – offer enormous opportunities to generate and store data at a meager cost. NSOs should partner with private companies and data scientists to define scientific and professional principles, conceptual frameworks, and methodologies for their utilization.

(iv) **Promotion of statistical culture to raise more funding.** It is well-recognized that statistical operations demand substantial resources. The weak awareness of the importance of statistics in decision-making and evidence-based policy planning affects the demand for quality data significantly. Decision-makers and the general public are not sufficiently informed about the crucial role that statistics can play in achieving a better and more inclusive society. Hence, statisticians should market statistical products more effectively and make sure that all classes of the society understand and use statistics for their benefits by tailoring data in various formats in such a manner as to be understood by a variety of consumers. Data are a public good, and citizens have the right to get access to them. They must be open and available freely to everyone to use and republish as they wish, without restriction from copyright, patents, or other mechanisms of control. As the demand for comparable African data increases, the funding will increase for creating such data. Then statistics systems will have the means to produce quality statistics on Africa.

There are many research papers and strategic documents on statistical development that have led the foundation for the current statistical harmonization efforts in Africa. This chapter does not intend to cover all statistical development initiatives on the continent. It focuses only on the ongoing statistical harmonization activities of the producers of official data – Statistics Authorities and attempts to propose some solutions to statistical challenges on the continent. More research is needed in the area of statistical harmonization in Africa. A follow-up chapter will endeavor to assess the implementation of the proposed harmonization approaches and the recommendations of this chapter by African countries.

References

AFRICATRADE (2020). African Trade Statistics Yearbook, African Union Commission. https://au.int/sites/default/files/documents/39607-doc-af-trade_yearbook2020_v4_comp-compresse_1.pdf.

AU (2009). African Charter on Statistics, African Union Commission. Available at https://au.int/en/treaties/african-charter-statistics.

AU (2015). Africa's Agenda 2063: The Africa We Want. Available at https://au.int/Agenda2063/popular_version.

AUC (2018). An African Quality Assurance Framework for the African Statistics System, African Union Commission. https://au.int/sites/default/files/documents/32850-doc-quality_framework_for_the_ass_en_2016-03-17.pdf.

AUC, AfDB, UNECA, & ACBF (2018). *Strategy for the Harmonization of Statistics in Africa (SHaSA)*. African Union Commission https://au.int/en/ea/statistics/shasa.

Instituto Nacional de Estatistica (2015). The strategy for the harmonization of statistics in Africa, governance, peace and security statistics in Cabo Verde, AU AGENDA 2063 ASPIRATION 3: an Africa of good governance, democracy, respect for human rights. *Justice and the Rule of Law* (1).

UN (2015). United Nations Transforming Our world: 2030 Agenda for Sustainable Development. https://sdgs.un.org/2030agenda.

World Bank (2020). Purchasing Power Parities and the Size of World Economies Results from the 2017 International Comparison Program, World Bank Group. Purchasing-Power-Parities-for-Policy-Making-A-Visual-Guide-to-Using-Data-from-the-International-Comparison-Program.pdf.

Part II

Ex-post harmonization of national social surveys

9

Harmonization for Cross-National Secondary Analysis: Survey Data Recycling

Irina Tomescu-Dubrow[1,2,3], Kazimierz M. Slomczynski[1,3], Ilona Wysmulek[1,2], Przemek Powałko[1], Olga Li[2], Yamei Tu[4], Marcin Slarzynski[1], Marcin W. Zielinski[1], and Denys Lavryk[2]

[1] *Institute of Philosophy and Sociology, Polish Academy of Sciences, Warsaw, Poland*
[2] *Graduate School for Social Research, Polish Academy of Sciences, Warsaw, Poland*
[3] *Department of Sociology, The Ohio State University, Columbus, OH, USA*
[4] *Department of Computer Science, The Ohio State University, Columbus, OH, USA*

9.1 Introduction

Survey data recycling, SDR, is an analytic framework for ex-post harmonization of cross-national survey data. We started developing it in 2013, in response to (i) substantive research goals that called for a broader comparative data infrastructure than individual international survey projects (Tomescu-Dubrow and Slomczynski 2014) and (ii) the need for theory and methodology for the fast-growing field of ex-post harmonization (Dubrow and Tomescu-Dubrow 2016). In short, SDR (i) operationalizes aspects of survey quality and survey item properties so that methodological biases in the source data can be investigated systematically (Slomczynski et al. 2021) and (ii) provides a harmonization workflow (Wysmulek 2019) that maximizes retention of source information, ensures decision-making transparency, and enables flexible secondary use of the new, harmonized dataset.

This chapter discusses applications of the SDR approach for building a multicountry, multiyear database for comparative research on *political participation*, *social capital*, and *wellbeing* in the SDR project (wp.asc.ohio-state.edu/dataharmonization, NSF# 1738502). The resulting SDR database v.2.0 (hereafter, SDR2) has harmonized information from 23 international survey projects (cf. Table 9.1) for 4,402,489 respondents interviewed across 3329 national surveys that, taken together, span the period 1966–2017

Table 9.1 International survey projects selected for harmonization.

Acronyms	Project names
ABS	Asian Barometer Survey
AFB	Afrobarometer
AMB	AmericasBarometer
ARB	Arab Barometer
ASES	Asia Europe Survey
CB	Caucasus Barometer
CDCEE	Consolidation of Democracy in Central and Eastern Europe
CNEP	Comparative National Elections Project
EB	Eurobarometer
EQLS	European Quality of Life Survey
ESS	European Social Survey
EVS	European Values Study
ISJP	International Social Justice Project
ISSP	International Social Survey Programme
LB	Latinobarómetro
LITS	Life in Transition Survey
NBB	New Baltic Barometer
NEB	New Europe Barometer
PA1	Political Action (An Eight Nation Study)
PA2	Political Action II
PPE7N	Political Participation and Equality in Seven Nations
VPCPCE	Values and Political Change in Postcommunist Europe
WVS	World Values Survey

and 156 countries.[1] SDR2 provides both target variables and methodological indicators that store source survey and ex-post harmonization metadata (see Table 9.2).

Substantive target variables in SDR2 measure respondents' demographics (age, gender, education, civil status, income, household size, and place of residence), political behaviors (voting, engaging in demonstrations and boycotts, and signing petitions), political attitudes and opinions (interest in politics, trust in government, parliament, the legal system and political parties, and satisfaction with democracy), social capital (membership in political parties, in environmental and religious organizations and trade unions, interpersonal trust, and discussing politics), and wellbeing (happiness, life satisfaction, and health).

---

1 A first version of the SDR database (SDR1, described in Slomczynski et al., 2016) is deposited at Harvard Dataverse (Slomczynski et al., 2017). SDR1 spans the period 1966–2013 and 142 countries. It includes 1721 national surveys stemming from 22 survey projects with information from around two million people. SDR2 revises the approach to all SDR1 target variables and creates new ones using data from almost twice as many national surveys.

Table 9.2 SDR2 at a glance.

Time span	1966–2017
# countries	156
# respondents	4,402,489
# source survey projects	23
# source project-waves[a]	174
# source national surveys	3329
# source data files[a]	215
# source survey quality indicators	38
# technical target variables	9
# substantive target concepts	30
# harmonization controls[b]	74

[a] The difference between the number of source project-waves (174) and that of source data files (215) in SDR2 stems from the structure of the downloaded source data files. While generally a source data file includes all national surveys of a given project-wave, sometimes repositories provide separate files for each national survey of a project-wave, or one single data file for multiple project-waves.
[b] Capture properties of source items harmonized into a given target and certain harmonization decisions.

Methodological indicators are of two types. The first set captures aspects of source survey quality. The second set is harmonization controls, which capture properties of the source items harmonized into a given target variable and relevant SDR harmonization decisions. Scholars can use both types of indicators to partial out the effects of certain types of methodological bias, thus taking data recycling beyond repurposing old information for new research. These indicators can also serve as "filters" to redefine target variables and/or choose subsections of the harmonized data that more closely meet a researcher's needs (see Slomczynski et al. 2021, pp. 425–427).

The *scope* of harmonization in the SDR project, and thus, coverage in SDR2, are a function of theory-informed and pragmatic decisions regarding (i) the list of target concepts of interest, (ii) the selection of source survey projects, (iii) the operational definition of target concepts, and (iv) the operational definitions of methodological variability. The SDR Portal (http://cse-dnc215869s.cse.ohio-state.edu/), whose features we summarize in Section 9.4, enables users to check coverage in SDR2 and download the data in full or as a subset customized according to individual specifications. SDR2 and its documentation are also available from Harvard Dataverse (Slomczynski et al. 2023).

9.2 Harmonization Methods in the SDR Project

We constructed the SDR2 database following the harmonization workflow depicted in Figure 9.1. In developing this workflow, we broaden the definition of ex-post harmonization methodology to comprise both (i) the procedures of recoding and/or rescaling source variables from different national surveys into a common target, so that respondents' answers are comparable (Gunther 2003; Minkel 2004; Ehling and Rendtel 2006; Granda Wolf and Hadorn 2010, Fortier et al. 2011; 2017) and (ii) methods to account for

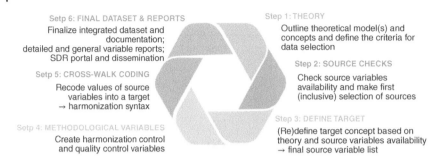

Setp 6: FINAL DATASET & REPORTS
Finalize integrated dataset and documentation; detailed and general variable reports; SDR portal and dissemination

Setp 5: CROSS-WALK CODING
Recode values of source variables into a target → harmonization syntax

Setp 4: METHODOLOGICAL VARIABLES
Create harmonization control and quality control variables

Step 1: THEORY
Outline theoretical model(s) and concepts and define the criteria for data selection

Step 2: SOURCE CHECKS
Check source variables availability and make first (inclusive) selection of sources

Step 3: DEFINE TARGET
(Re)define target concept based on theory and source variables availability → final source variable list

Figure 9.1 The SDR Harmonization Workflow. Source: WAPOR 2022 webinar "Survey Data Recycling and Data Harmonization" (https://wapor.org/resources/wapor-webinars/webinar-may-2022/.)

methodological variability due to unequal properties of the source surveys and to harmonization *per se* since such variability can weaken the validity and reliability of the target variables (Thiessen and Blasius 2012).

Before we discuss each step of the workflow, we note that, in practice, insights gathered later in the harmonization process can call for readjusting earlier decisions. Thus, the importance of thorough and on-the-fly documentation cannot be overstated (e.g. Fortier et al. 2023).

9.2.1 Building the Harmonized SDR2 Database

Step 1: *Theory* is key to the feasibility of ex-post harmonization projects (e.g. Granda and Blasczyk 2016), because it helps to set boundaries to the scope of data reprocessing, in terms of target concepts and their measurement(s), types of source data, and their time and geographic coverage. In the SDR project, macro–micro theories of individuals' political participation, social capital and wellbeing in comparative perspective, and the survey research literature, informed which source data we considered for harmonization.

We searched for survey projects that (i) maximize coverage of peoples' political behaviors and attitudes, social capital, health, wellbeing, and main correlates in different economic and political contexts; (ii) are cross-national, and, preferably, multi-wave studies, warranting the assumption of within-project comparability; (iii) intend to have national samples representative of the adult population; (iv) have English language documentation (study descriptions, codebooks, and master questionnaires); (v) are noncommercial; and (vi) are freely available for academic use.

The 23 international survey projects listed in Table 9.1 meet these criteria. Although some projects, such as WVS, ISSP, and ESS among others, are ongoing, we decided to harmonize data up to and including 2017. We chose this cut-off point for practical reasons, to facilitate the timely completion of the SDR project, and knowing that for 1966–2017, SDR2 will include, as theory calls for, country–year observations for different economic and political development levels.

Altogether, 174 waves from the 23 projects held harmonization potential for SDR2. Since these source data and their documentation were available through a few repositories including GESIS, ICPSR, and the UK Data Service, we automated downloads. The SDR team's data science specialist developed a script to parse the target web pages and retrieve the required files.

Concluding Step 1, from the 23 projects we downloaded for harmonization a total of 215 data files storing 174 project-waves, and their corresponding documentation. The difference between source data files and project-waves numbers stems from the structure of the source files. As a rule, we selected wave-level source data files, where one file contains all national surveys of a given project's wave. Sixty seven percent of source files in SDR2 are of this type. If wave-level files were not available at the time at which we retrieved the data (January–February 2018), we downloaded either the separate national survey files of a given project-wave (64 national survey files for six project-waves) or files containing the survey data from multiple waves of a given project (six data files for 23 waves). The downloaded source files correspond to a total of 3,329 national surveys.

Step 2: *Source checks* involve the systematic review of all available English-language documentation – source questionnaire, survey codebook, data dictionary, and data records in a given file – for the selected source data files, to find all source variables that match the theoretical target concept, and, in the process, identify possible survey quality problems.[2] Source checks are intrinsic to ex-post harmonization, because they reveal (i) the range of possible operationalizations of a target concept in the pulled data, and (ii) shortcomings in source survey quality, such as processing errors (Olkesiyenko et al. 2018), that could affect a target variable.

To ensure the accuracy of source checks, two or three coders in the SDR team independently identify all source questions with harmonization *potential* for a given target concept. They record key information about these measures (i.e. source variable names, labels, values, question wording, and the name of the dataset the items originate from) in standardized Excel format in ***Detailed Source Variables Reports (DVRs)***.[3]

Inter-coder differences are discussed in team meetings. Once resolved, DVRs become the basis for the next steps of the SDR harmonization workflow – (re)defining the target concept, constructing methodological indicators, and crosswalk coding. As such, DVRs are key also to harmonization process transparency (see Section 9.3).

During *Step 3: Define target*, we reassess the definition of target concepts and their possible operationalization given, on the one hand, theory and methodological considerations, and on the other hand, *available* source measures. At this stage, we also decide which inter-survey differences in properties (e.g. scale length) of the *selected* source measures to store in target-specific harmonization controls (see Step 4).

Team meetings that review the detailed information in DVRs are key to deciding which source variables to harmonize into the target measure(s) of a given target concept. Preparing handouts that organize the available source variables according to item characteristics, such as types of scales (e.g. frequency scales versus agree–disagree), or question wording, greatly aids team discussions.

We renounce items that have technical problems (e.g. empty values), serious quality problems (e.g. lack of clear value labels), or whose face validity we consider weak. For example, CDCEE 1-2 contains items on boycotts. A closer inspection of these items revealed that they capture *opinions* about boycott as a way to protest, while our target measures boycott *behavior*. Consequently, we excluded the CDCEE

---

2 In rare situations, we also reviewed the documentation in the national language of a given survey.
3 Generally, we prepare one DVR per target concept. However, for more complex concepts such as education, each target variable has a corresponding DVR. See also Section 9.4.

Table 9.3 Self-reported voting behavior in the last elections harmonized to T_VOTED target variable, SDR2.

Variable description	Variable name	Variable values
Respondent's declared behavior in the most recent (last) national elections	T_VOTED	1 = Voted 2 = Voted but casted a blank/invalid ballot 3 = Did not vote 4 = Did not vote: not eligible (too young, noncitizen status) 5 = Did not vote: not registered 6 = Did not vote due to external pressure 7 = Did not vote inferred from other answer categories (e.g. do not remember, NAP)

1-2 items from harmonization. For later checks of our decisions, we stored "rejected" source variables, with an explanation for the rejection.

Since we seek to facilitate flexible analyses of the harmonized data, we define target variables inclusively and retain important source information that scholars can use to adjust a target's operational definition to their research needs. The harmonized measure of reported voting behavior, T_VOTED, illustrates this approach (Table 9.3). Rather than two categories ("voted"/"else"), this variable has seven categories (for details, including missing values, see VOTED IN THE LAST ELECTIONS – General Target Variable Report, GVR, on Dataverse, https://doi.org/10.7910/DVN/YOCX0M).

Additionally, many target concepts in SDR2 are operationalized with two target variables. To illustrate, for each institutional trust measure (trust in parliament, political parties, legal system, and government), we construct (i) one target measure of intensity of trust, expressed by *numerical values on the harmonized 11-point scale* (for the harmonization of rating scales in SDR2, see Section 9.4), and (ii) one target that measures the relative position of a respondent in the *distribution* of trust in a given national survey. The scores of this distributional target variable *are percentiles* within the national sample, which indicate what share of respondents reports the same or lower trust than the individual. The two targets can be used also jointly: since the meaning of a respondent's answer on the harmonized trust scale depends on the proportion of respondents who, in the national survey the respondent belongs to, chose the same or a lower intensity answer, users can multiply the two target measures and account for this difference.[4]

The target concept "education" is another example. We measure respondents' education with the harmonized variables "completed years of education" and "education categories" (for a generalized framework for ex-post harmonization of education, see Schneider 2022). First, from a theoretical point, these indicators deal with different conceptualizations of education (investment, for years; certification, for categories). Second, since 107 source data files (50%) provide information for both years and education categories, methodologically it becomes possible to analyze how the measures relate to each other and to other variables. Third, we retain in SDR2 information from the source files that provide only years of education (71 of the 215 source data files, 33%), or only education categories (36 source data files, 17%).

---

4 For example, the value 7 on an 11-point scale of trust in parliament carries a different meaning if 85% of a national sample rated trust in parliament as 7 or below, than if 50% of a national sample did so. When including the interaction term in a regression analysis, we recommend including also the two "main" effects of the rating scale target and the distributional one.

Regarding Step 3, it is important to note that *technical* target variables (e.g. the geographic identification of survey samples or weighting factors) must be defined and also require harmonization. For example, we define two units for where the samples were drawn from – T_COUNTRY_L1U and T_COUNTRY_L2U.[5] T_COUNTRY_L1U identifies the smallest country-like unit provided in the source data. Usually, it corresponds to a country (e.g. Poland), but sometimes the unit is smaller – a subset of a country (e.g. East Germany) or an ethnic group within a country (e.g. Arabs in Israel) – while other times it is larger (e.g. Czechoslovakia in 1992). T_COUNTRY_L2U is harmonized to correspond to a *de jure* or *de facto* independent country: using the earlier examples, now the sample unit is Germany (after integrating east and west German samples), Israel (with both Jewish and Arab communities), and the Czech Republic and Slovakia (which in 1992 were already in the process of separation).

The harmonization of sample units at the country (L2) level has consequences for weighting, since the source data providers constructed weighting factors for design weights, post-stratification weights or their combination, at the L1 level. We construct the target variable T_WEIGHT_L2U with weighting factors at a country (L2) level. While values of T_WEIGHT_L2U are generally derived directly from the source weight values, exceptions exist. We discuss the harmonization rules in WEIGHTS – GVR on Dataverse. For a general discussion of weights in the context of harmonized data, see Joye, Sapin and Woly (date of publication of this SDH volume).

Step 4: Methodological variables in SDR2 allow researchers to (i) systematically investigate the extent to which certain methodological biases and errors in the source data impact the harmonized data, (ii) minimize information loss in the harmonized database, and (iii) enhance harmonization transparency. We construct two sets of methodological indicators: the first set – data quality controls – comprises measures of source survey quality. The second set – harmonization controls – stores information about survey item properties and ex-post harmonization decisions.

Source data quality controls capture variability in source survey quality. Drawing on the total survey error and total survey quality frameworks (e.g. Biemer 2016; Biemer and Lyberg 2003), the SDR approach identifies three relevant dimensions of quality, conceptualized as quality variables: (i) the quality of surveys as reflected in the source documentation (codebooks, questionnaires, technical reports, etc.); (ii) the degree of consistency between the official description of the data with the actual data records in the computer files; and (iii) the quality of the survey records in the source data files (for details, see Slomczynski et al. 2021; Slomczynski and Tomescu-Dubrow 2018). Appendix 9.A lists the data quality controls available in SDR2.

Harmonization controls corresponding to specific target variables capture differences between surveys, stemming either from source measures themselves (e.g. differences in the wording of questions that measure the same concept in different surveys, differences in scale properties) or from our harmonization decisions (e.g. deriving *age* from *birth year*, when source surveys provide only respondents' year of birth). To illustrate, we return to T_VOTED, the harmonized measure of self-reported voting behavior in the last elections. Its harmonization control, C_VOTED_TYPE (Table 9.4), helps users understand what the most recent elections pertained to.

———————————

5 L1U and L2U stand for level-1 and level-2 sample units. Both T_COUNTRY_L1U and T_COUNTRY_L2U are typically coded with regular ISO 3166-1 alpha-2 codes (https://en.wikipedia.org/wiki/ISO_3166-1). In the cases of sub-country division of a sample, we use an up-to-3-letter extension, e.g. DE-E for East Germany. In rare (and documented) situations, we further extend the ISO 3166-1 standard. For details, see the SDR2 MASTER file overview on Dataverse, https://doi.org/10.7910/DVN/YOCX0M.

Table 9.4 Harmonization control for T_VOTED, SDR2.

Variable description	Variable name	Variable values
Type of last election that respondents are asked about	C_VOTED_ TYPE	1 = Last parliamentary elections
		2 = Last presidential election
		3 = Last parliamentary and presidential elections took place on the same day
		4 = Not identified if parliamentary/presidential elections[a]

[a] Neither the source documentation nor external sources about the country's type of political system provide sufficient information to identify the election type.

In *Step 5: Crosswalk coding* we (i) map, for each target variable, the correspondence between values – missing values included – of the selected source items and target values, and (ii) specify the values of the harmonization control variables that accompany a given target variable. We do so using Excel files, which we call Crosswalk Tables, CWTs. CWTs are the files where actual recoding from source to target values takes place.

The CWT for a given target variable is based on that variable's Detailed Variable Report, DVR. While in DVRs each line corresponds to the *wording* of a source question, the CWT has separate lines for source *values,* which is necessary to prepare harmonization codes. In CWT, lines show source variable names, labels and response *options* from data dictionaries, rather than source *questions* from questionnaires and codebooks (which we keep in DVR). All CWT source information (including response options and their labels, realized values and distributions) was retrieved automatically from the source data files.

Step 5 plays an important role in additional source data availability checks. By looking at the distribution of each source variable selected for harmonization (which we provide on the level of a national survey), we reassess the consistency of source variables selection and check the recoding schemas from source to target values (i.e. the harmonization codes).

During *Step 6: Final datasets and reports* we create the SDR2 database and tie up loose ends of the harmonization workflow, such as editing the SDR documentation and finalizing the SDR Portal. A key milestone for this step is publicly sharing the harmonized data and their documentation.

For internal purposes, we use an open-source MariaDB engine to store the SDR2 data in a relational database comprising three datasets:

(1) The MASTER file stores harmonized information for a total of 4,402,489 respondents. It includes 50 target variables, 74 harmonization controls (whose values generally do not vary across respondents of a given national survey), and seven source data quality controls measured at the respondent level (e.g. missing case ID). The SQL syntax for constructing the MASTER file is automatically created from CWT mappings, with the help of Visual Basic for Application macros and additional scripting in the Linux shell environment.

(2) PLUG-SURVEY is an auxiliary dataset containing controls for source data quality measured, generally, at the national survey level, and a set of technical variables needed for merging this file with the MASTER file.

(3) PLUG-COUNTRY file is a dictionary of countries and territories used in the MASTER file. It contains basic geographical information: alpha/numeric standardized ISO country codes, country names, and codes and names for micro- and macro-regions of the world.

Publicly, researchers can access the three SDR2 datasets (in SPSS and STATA formats) and the full SDR2 documentation on Harvard Dataverse (Slomczynski et al. 2023). The SDR2 data are also available in CSV format (with corresponding syntax for reading the files into STATA/SPSS) from the SDR Portal (http://cse-dnc215869s.cse.ohio-state.edu/), whose extraction utility offers the possibility to download them in full or customize data selection according to specific variables, survey projects, countries and years.

In sum, the SDR harmonization workflow maps the steps that we took to construct the SDR2 database such as to retain source information, capture methodological variability within and between survey projects, and promote transparency at each stage of the harmonization process. While different ex-post harmonization projects will have specific goals and needs, the logic of the SDR approach can be useful as inspiration or guidelines.

9.3 Documentation and Quality Assurance

We strive to provide as detailed documentation as possible, to enable scholars to understand what we did, trace and replicate our decisions and adjust with maximum possible flexibility the SDR2 data to their own research specifics, if necessary. When harmonizing ex-post information from the 174 project-waves in the 215 source files that underlie SDR2, we are secondary users whose understanding of the data depends on the source survey documentation and project descriptions. Our approach to documenting harmonization in SDR is largely shaped by this user experience and awareness that documentation links to data quality (AAPOR/WAPOR 2021 Task Force Report on Quality in Comparative Surveys).

The SDR2 documentation provides overviews of the SDR2 database and its three datasets – MASTER file, PLUG-COUNTRY file and PLUG-SURVEY file – and descriptions of variables in these files. The documentation is available on Harvard Dataverse (Slomczynski et al. 2023).

The bulk of documentation pertains to the substantive target variables, for which we provide:

a) The GVR (PDF) with the operational definitions of the target variable, harmonization controls and source quality checks (if applicable), and the harmonization rules and procedures for constructing each type of measure. The GVR also highlights special cases that required additional harmonization decisions, together with the reasoning behind these decisions.

b) The DVR (Excel) with systematized information about all source variables used for harmonization into a given target variable of SDR2 (see also Section 9.2, Step 2). Specifically, DVR files include source variable names, labels, values, question wordings, and the name of the dataset they originate from. We extract this information from three main source elements: the core (master) questionnaire, the core (master) codebook, and the data dictionary of variable labels and values stored in the source data files. Having ready-available and systematized documentation of all source variables used to build a target variable allows to verify – and, if necessary, change – the decision to harmonize or not a given source variable. DVRs serve an important role for transparency beyond the SDR team. Interested scholars can use the DVRs to both check our decisions and assess how well the operational definition of a given target variable in SDR2 matches their research needs.

c) The CWT (Excel) and harmonization syntax. CWTs show in detail how source variable values are recoded into values of the target variable, and what values harmonization controls take (see Section 9.2, Step 5). CWT files are the basis for the MariaDB compatible SQL syntax used for creating the MASTER

Table 9.5 Effects of the indexes of survey quality deficiencies on mean value of trust in parliament, for 2045 national surveys in SDR2.

Survey quality index	OLS regression		
	B	SE	Beta
Documentation deficiencies	0.085	0.014	0.125
Processing errors	−0.356	0.079	−0.099
Computer file errors	0.117	0.047	0.060
Constant	4.080	0.067	—
Regression fit	$F = 18.64$, df = 3, 2041, $R^2 = 0.025$		

file on top of 215 tables corresponding to the 215 source data files. The syntax is provided in a single text file and includes creating necessary MASTER table columns, corresponding to target and control variables, but also some additional, internally used, columns (variables) which are needed for more complex procedures, e.g. implementation of some harmonization rules, described in GVRs.

Analyses on the first SDR data version show that the effects of survey-level summary indices of data quality controls on national survey estimates of protest behavior are significantly different from zero (Slomczynski et al. 2021). In this paper, the summary indices capture (i) quality of documentation, (ii) quality of source data processing, and (iii) quality of records in national surveys. The aggregate measures of protest behavior cover participation in demonstrations and signing petitions in the last year, and in the last 10 years, respectively. We find that the higher the quality of documentation and of data processing, the lower the percentage of persons who declare having attended a demonstration. However, the same controls have positive effects on signing petitions. The effect of the quality of records in source surveys is negative for mean participation in demonstration and for mean petition signing. The three quality indices explain between 5% and 10% of the variance in the means of the two types of protest participation.

Preliminary analyses on SDR2 examine the relations of data quality control indices and national estimates of trust in parliament for the 2045 surveys that carry this institutional trust variable (Table 9.5). The summary indices measure *deficiencies* in (i) documentation (mean = 0.315, sd = 0.284), processing errors (mean = 0.315, sd = 0.284), and computer file errors (mean = 0.268, sd = 0.524). The effects, in terms of regression coefficients, are small, but not trivial. Altogether, quality controls account for 2.5% of variance of the dependent variable.

9.4 Challenges to Harmonization

Challenges appear throughout the harmonization workflow (Section 9.2); some are project-specific, while others have a more general character. We focus on the latter. At *Step 1*, a point worth raising is that, even with well-specified selection criteria, source datasets that largely meet these criteria may not make the cut for harmonization. In SDR2, this happened with the Comparative Study of Electoral Systems,

CSES, which met most of our selection criteria. A closer look at CSES' sample design revealed that its surveys are often fielded as a part of panel studies. Since most datasets with harmonization potential for our project are not panel studies, we decided to leave CSES out.

A further issue linked to selection criteria is that we did not specify a minimum combination of distinct target concepts per source survey. In consequence, target variables are unequally available in the SDR2 database, which restricts the scope of certain comparative analyses. We encourage researchers to use the SDR Portal (Section 9.5) to check the span of possible analyses for given combinations of target variables.

The weak standardization of the content and format of documentation of cross-national survey projects (AAPOR/WAPOR 2021 Task Force Report), and its sheer volume, make source checks (*Step 2*) difficult (Tofangsazi and Lavryk 2018). Having different team members review the information independently, performing inter-coder reliability checks and resolving problems in team meetings are key to minimizing errors. However, for us, a separate issue remains. Because we use the English-language master questionnaires and codebooks to construct methodological indicators, part of the variation these measures convey could be noise.

It is at *Step 3* that "traditional" challenges of ex-post harmonization appear, such as losing important source information through harmonization, and finding substantial inter-survey variability in the properties of source items measuring the same concept. The solutions in the SDR approach are to (i) define the target variables inclusively (see Table 9.3 for the example on harmonized voting) and (ii) assume a *standard formulation of the source question* and *standard scale for answers* and use harmonization controls to measure deviations from these standards across surveys.[6]

In our experience, this approach serves well the harmonization of attitudinal and opinion source questions. First, we developed a scheme of comparing their wording according to possible differences regarding the **object** (e.g. parliament), **attribute** (e.g. trusting), **criterion** (e.g. intensity), **format** (e.g. whether the question is formulated in the projection mode), and **qualifiers** (e.g. restrictions on time and/or space). Valuation of the "object" is referenced, explicitly or implicitly, concerning time and space, providing an appropriate context. A detailed discussion of this scheme and how we use it to construct harmonization controls is provided in Slomczynski and Skora (2020).

Second, we assume as the standard that the rating scales of attitudes/opinions are of a specified **length** (fixed), **direction** (ascending) and **polarity** (unipolar), and account for deviations from the standard with specific harmonization controls (for an alternative approach to the harmonization of rating scales, see Singh and Quandt (Ch.17) in this volume).

We adjust the original length n of a source scale to a fixed target scale length r (11-point or 5-point) using the following linear transformation: source values k ranging from 1 to n are recoded to target values $\frac{r-1}{n*2}+\left(k-1\right)*\frac{r-1}{n}$.[7] This transformation is applied for $n \neq r$, treating r as a standard to which other scales are adjusted. The transformation assures that, for each original scale, the mean value is the same and the differences between original scale points and transformed scale points are minimized. The consequences of this transformation are that, for any n, including for $n > r$, values k could not reach the

---

6 Generally, we take the most common (frequent) source item property (wording/scale type) as the standard.
7 In the SDR project, the 11-point target scale ranges from 0-10, while the 5-point target scale ranges from 0 to 4.

Table 9.6 Within-project effects of length, direction, and polarity of the source scale on mean trust in parliament: NBB, EB, and CNEP.

Harmonization control variable	Variable values	Mean value (se)	F df	Eta^2
	New Baltic Barometer, NBB			
Scale length	4-point scale	4.689 (0.022)	$F = 1375.8$	Eta^2 = 0.091
	7-point scale	3.422 (0.027)	df = 1, 13815	
	Eurobarometer, EB			
Scale direction	Descending	4.412 (0.004)	$F = 4709.0$	Eta^2 = 0.011
	Ascending	3.632 (0.010)	df = 1, 4012955	
	Comparative National Elections Project, CNEP			
Scale polarity	Unipolar	4.634 (0.030)	$F = 347.9$	Eta^2 = 0.036
	Bipolar	5.649 (0.044)	df = 1, 9244	

extreme points, or *r*. However, this transformation, together with the additional information on the length of the scale, allows researchers to reshape it using different linear or nonlinear equations.

The primary attributes of rating scales may have a potential impact on respondents' answers. Usually, this impact is not present in the entire dataset, since various factors contribute to the randomness of the relationship between the length, direction, and polarity of the scale on the one hand, and substantive variables on the other. For instance, in SDR2 the correlation between the harmonization controls measuring length, direction, and polarity of the source scales of trust in parliament, and mean "trust in parliament" is close to 0.

However, if we focus on particular international survey projects, the effects of the three harmonization controls are statistically significant. Table 9.6 presents the results for different versions of the rating scale used in the same source projects. We observe statistically significant effects in the mean value of trust in parliament for length of the scale, scale direction, and scale polarity. On average, the differences in the mean value of trust for contrasting versions of the scale – a 4-point versus 7-point scale, ascending versus descending, and unipolar versus bipolar – are close to (or exceed) 1 point on the harmonized 11-point scale. These differences are statistically significant. In addition, scale length, scale direction, and scale polarity explain from 1.1% to 9.1% of the variance in trust in parliament. The effects of scale properties could be even stronger in analyses involving different projects.

Another harmonization challenge at Step 3 arises when a source survey asks several questions about the same theoretical concept. In such instances, we use more than one source variable to create a single target, and record this decision in a corresponding harmonization control.

The target "Participation in demonstrations" is a good example. Some surveys provide one measure of participation in *lawful* demonstrations, and a second measure of participation in *unlawful* demonstrations. We take both and flag this with the harmonization control C_DEMONST_SET. In harmonizing the information from the two measures, we apply the rule that the strongest answer wins, meaning that if a respondent answered "yes" to participation in lawful demonstrations, but "no" to participation in

Table 9.7 Rules of recoding a set of two questions about participating in demonstrations for the same survey/country/year.

Answer to the 1st source question	Answer to the 2nd source question	SDR (re)code in target
Yes	Yes	Yes
Yes	No	Yes
Yes	Missing	Yes
No	No	No
No	Missing	No
Missing (e.g. DK)	Same missing (e.g. DK)	Missing (e.g. DK)
Missing (e.g. DK)	Other missing (e.g. NA)	COMBI (SDR missing code[a])

[a] For details on SDR2 missing codes see Table 9.8.

unlawful ones, we assign them the target value "yes," participated in a demonstration. Table 9.7 illustrates the implementation of this rule for a set of two questions.

Harmonizing missing codes is another issue to contend with during Step 3. Target variables often have a complex missing values structure. They feature missing values "inherited" from the source items, as well as missing values stemming from ex-post harmonization. Table 9.8 shows how we code missing values in SDR2.

A notable challenge regarding methodological indicators (*Step 4*) is their different levels of measurement, which makes statistical analyses difficult (see Section 9.6.2). Quality controls for source documentation and those indicating if a national survey exhibits errors in source data records are measured at the national survey level. The quality of source data processing was assessed on the basis of the source data files in SDR2 and later standardized to the wave-level.[8] However, there are also some quality control variables that are measured at the respondent level, such as the SDR quality control variable flagging age, birth year or household size outliers.

Harmonization controls store information measured at the source data file level, which for 65% of the 215 files is the project-wave level (see section 9.2.1, Step 1). While it is expected that the values of harmonization controls vary among national surveys of different project waves, it happens that they also vary among national surveys of the same project-wave, ex-ante harmonization notwithstanding. In some instances harmonization controls also store information measured at the level of the respondent's answer option.

---

8 Standardizing the measurement of processing quality indicators to the wave-level entailed (i) aggregating the indicators stemming from source data files containing a single national survey and (ii) disaggregating the indicators stemming from source files containing the data of multiple project-waves (see Section 9.2, Step 1 for details about source data files in SDR2).

Table 9.8 Codes for missing values in SDR2.

SDR Tag[a]	SPSS (STATA) Codes	Label	Examples of source labels
		Harmonized source codes for missing values	
DK	-1 (.a)	Don't know	*both; can't choose; difficult to say; don't know; forget; hasn't heard enough; haven't thought much about it; it depends; no opinion*
NA	-2 (.b)	No answer	*no answer; no response; not stated*
REF	-3 (.c)	Refusal	*declined to answer; refused to answer; refusal*
DU	-4 (.d)	Don't understand the question	*do not recognize; do not understand the question; not clear*
DNR	-5 (.e)	Any combination of DK, NA, REF, DU	*don't know / no answer; no answer / refused*
INAP	-6 (.f)	Inapplicable	*not applicable; not available*
NEC	-7 (.g)	Not elsewhere classified	*interviewer error; missing data; missing; unknown; NA; not relevant; other*
		SDR-created codes for missing values	
UNFIT	-8 (.h)	Source value does not fit to target	*WVS/2 Czechoslovakia, 1990: source 5-point INTEREST IN POLITICS realizes only two valid response options*
ERR	-9 (.i)	Errors in source data and undocumented source values	*<null>; undocumented*
COMBI	-10 (.j)	Different missing types are identified on two or more source variables used for the target construct	
CINAP	-11 (.k)	For control variables only: Inapplicable	
INSUF	-12 (.l)	For survey: Insufficiently defined response categories	
QNA	-20 (.t)	For survey: Question not available	

[a] Abbreviations for the labels corresponding to the SDR2 codes for missing values. These tags are used in the Crosswalk Table (CWT) files that accompany documentation of SDR2 target variables, published on Dataverse.

Thus, to avoid errors when using the SDR2 data (see also Section 9.6.2), it is crucial to check the documentation, the GVRs, especially. A remarkable example is working with the target variable "participation in demonstrations."[9]

One of its harmonization controls, C_DEMONST_YEARS, stores source information on the time span during which a respondent engaged (or not) in demonstrations (e.g. in the last year, the last two years, or ever). This control is generally measured at the level of national surveys, since the time span usually appears in

---

9 T_DEMONST takes values from 1 to 5, where 1 indicates no participation, 2 respondents' unwillingness to participate, 3 respondents' willingness to participate, and 4 and 5 refer to participation in demonstrations in "yes/would/would not" (coded 4), or "yes/no" (coded 5) questions.

the wording of the source questions (e.g. *"During the last 12 months, have you taken part in a demonstration?"*).

Exceptionally, however, C_DEMONST_YEARS also stores individual-level information, because (i) in the 2004 and 2014 waves of ISSP, the time span appears as one of the item's response options (*Not done, never do; Not done, might do; In more distant past done; In the past year done*) and (ii) the 2006 and 2008 waves of AMB ask all respondents two questions, each with a different time span: having *ever* participated in a demonstration, and having participated *in the last year*, respectively (we take both items using the "strongest answer wins" rule exemplified in Table 9.7, and flag this decision in another harmonization control, C_DEMONST_SET).

Consequently, C_DEMONST_YEARS also contains respondent-level information, and cannot be used directly as a filtering variable for the "last year" or "ever" time spans. Filtering requires additional recodes, depending if a user would like to select only demonstrated "last year" or demonstrated "ever."

Concluding this section, we note that SDR2 has a complex multilevel structure (see also Section 9.6.2). Next to hierarchical levels (e.g. respondents nested in national surveys) the data also feature cross-classified levels, since 92 of the 156 countries (59%) at the L2 sample unit level (cf. Section 9.2, Step 3) are covered by more than one national survey in the same year. This situation allows for interesting analyses, but also requires attention when specifying regression models (see also Durand in this volume).

9.5 Software Tools of the SDR Project

9.5.1 The SDR Portal

The SDR Portal (http://cse-dnc215869s.cse.ohio-state.edu/) with the integrated *SDR*Querier system is built to customize users' experience with SDR2. *SDR*Querier is a new visual analytical system equipped with multilevel information queries through data visualization and user-friendly interactions (Tu et al. 2023).

First, to facilitate the *understanding* of SDR2, the SDR Portal contains a static documentation description and a dynamic interactive part. The purpose of the descriptive part is to familiarize scholars mainly with the data quality indicators (see Section 9.2, Step 4). The *Survey Data Quality* section displays values of survey quality indicators within project-waves and project countries. *Quality Association* calculates and plots quality measures association coefficients within or between international survey projects.

Importantly, *SDR*Querier is designed to present the structure of the SDR2 database. It also allows users to query variables of their interests or keywords. The system searches the information from the source survey metadata (e.g. codebooks, survey questionnaires, and data dictionaries) and the harmonized data, and suggests the most relevant substantive variables available in SDR2, accompanied by their corresponding sets of methodological variables. It also shows the list of the source questions used to construct the harmonized variables. Thus, it reveals not only the structure of the dataset but also the transparency of the harmonization process.

Second, the SDR Portal helps researchers assess data availability. Once a desirable set of variables is identified, *SDR*Querier visualizes the information on data availability by target variable and joint availability by source survey project. The data records are displayed by year with each point representing sample size by countries or respondents.

Once the data availability satisfies users' needs, researchers can proceed to the next step – exploratory data analysis. It is possible to retrieve univariate or bivariate descriptive statistics and distribution tables for selected variables for particular years, survey projects and countries.

Finally, after users are familiar with the data, they can choose to extract SDR2 in full, or only the fraction of the dataset they are interested in. Data Extraction on the SDR Portal offers a number of filters to select variables, source surveys and waves, world regions or countries, and years that are necessary for a particular study. Substantive variables are already grouped with their corresponding methodological control variables to be downloaded together. Users can choose to download the data in either SPSS or STATA format.

To sum up, the SDR Portal helps researchers understand the structure, availability, and capacity of SDR2, and navigates them through relationships between source data and harmonized data, as well as between target variables, harmonization controls, and source quality indicators before downloading the data in an intuitive, user-friendly, yet effective manner.

9.5.2 The SDR2 COTTON FILE

The SDR2 COTTON FILE, *Cumulative List of Variables in the Surveys of the SDR Database,* is a comprehensive data dictionary, in Microsoft Excel format (Powałko 2019). Its main purpose is to facilitate the overview of 88,118 variable names, values, and labels available in the original (source) data files that we retrieved automatically for harmonization purposes in the SDR2 project. The order of variables, the variable response options and the wording of labels strictly mirror the respective information in the source data files.

To browse the COTTON FILE, one may use (i) Excel's built-in filtering functionality or (ii) a custom search box written in Visual Basic for Applications. We created the second option specifically for the COTTON FILE to overcome the limitation of Excel's built-in filtering, which does not properly handle lists exceeding 10,000 unique values. The SDR2 INTRO to COTTON FILE (PDF) describes features and provides tips for using the file.

9.6 Recommendations

9.6.1 Recommendations for Researchers Interested in Harmonizing Survey Data Ex-Post

Our general recommendation for new ex-post harmonization projects is to start with solid work on theory and specific research aims of the new data. The strong temptation to add more source files and create more target variables in anticipation of meeting other users' needs could jeopardize the feasibility of the harmonization project.

Furthermore, we encourage researchers to maximize user flexibility of the harmonized dataset. In our experience, to do so it is important to (i) construct methodological indicators that document differences in source data quality, (ii) provide target variables that are amenable to different target concept definitions, and (iii) construct methodological indicators that store source item properties and harmonization decisions useful for target concept redefinitions, and/or accounting for methodology-induced variability in the target variable.

Accurate documentation is intrinsic to the survey data harmonization process (Kallas and Linardis 2010). A detailed record of decisions taken during each step and of decision changes ensures transparency and enables ongoing quality checks within the team. It facilitates analyses of the harmonized data and replication of the study. We hope that the documentation templates developed in SDR, and the logic of the SDR harmonization workflow in general, can be useful to anyone planning to harmonize survey data ex-post.

Finally, we encourage researchers to collaborate across social science disciplines and fields, and for larger-scale projects, with data and computer scientists. Such cooperation, including feedback from survey research methodologists and scholars versed in data reprocessing, can lead to valuable insights and a more streamlined harmonization process.

9.6.2 Recommendations for SDR2 Users

For users of the SDR2 data, our basic recommendation is to get familiar with the documentation and the database structure before analyzing the harmonized data. The SDR Portal provides an easy first step into these necessary explorations.

Next, we encourage scholars to use the source survey quality and harmonization controls to examine how national surveys and target variables meet expected properties, and, if necessary, to sub-select SDR2 data that are closer to their research needs. Since we construct substantive target variables with the intent to minimize source information loss, users should examine – via harmonization controls – measurement variation before including SDR target variables in statistical models.

Moreover, methodological indicators in SDR can be used in regression analyses to assess and isolate their direct effects on the dependent variable. By including methodological indicators known (or suspected) to be related to both the independent and dependent variables, researchers can statistically account for their effects and obtain a more accurate estimate of the relationship of interest. Other threats to internal validity, such as selection bias, measurement error, and omitted variable bias, should also be considered when interpreting the regression results.

As chapter 19 in this volume explains, not using weights in data analyses of complex harmonized surveys can seriously bias the results, leading to erroneous conclusions (see also Lavallée and Beaumont 2015). We recommend that SDR2 data users examine closely the harmonized weight variable (T_WEIGHT_L2), together with its accompanying controls (see Section 9.2, Step 3), including the one providing information on the level of distortion (deflation or inflation) of the sample size that occurs after applying source weights to the data. We encourage researchers to also check how weights are treated in the statistical software they use. While STATA automatically rescales the weights before applying them to analysis, SPSS does not. Incorporating survey weights in R is straightforward, thanks to the "survey" package.

Last but not least, the complex multilevel structure of SDR2 – and of other datasets harmonized ex-post – should be examined carefully (see also Durand (ch. 20) in this volume). In SDR2, respondents are nested in national surveys, national surveys are nested in project-waves, and project-wave are nested in international projects. At the same time, for substantive analyses, some country–year characteristics are theoretically important to include – for example, general domestic product, the index of economic inequality, or a measure of democracy, among others. The level of these variables does not always match the national survey level, since some countries in a given year are covered by more than one survey. Therefore, using harmonization and quality control variables on the level of surveys and project-waves together with substantive variables for country–years makes multilevel modelling extremely difficult. We suggest conducting the analyses in steps, eliminating first the effects of methodological variability and then proceeding with the substantive analysis. It is important to remember to account for the specific assumptions and considerations of multilevel models, such as independence assumptions, model convergence, and appropriate sample sizes at different levels.

Acknowledgments

We thank Weronika Boruc, Nika Palaguta and Zuzanna Skora for their contributions to the SDR database v.2.0. We thank Riu Qiu for his help with finalizing the SDR Portal. Last but not least, many thanks to Han Wei-Shen and Spyros Blanas for their key input in designing the SDR Portal and overseeing its implementation.

References

AAPOR and WAPOR (2021). *AAPOR/WAPOR Task Force Report on Quality in Comparative Surveys*. https://wapor.org/wp-content/uploads/AAPOR-WAPOR-Task-Force-Report-on-Quality-in-Comparative-Surveys_Full-Report.pdf.

Biemer, P.P. (2016). Total survey error paradigm: theory and practice. In: *The SAGE Handbook of Survey Methodology* (ed. C. Wolf, D. Joye, T.W. Smith, and Y. Fu), 122–141. SAGE.

Biemer, P.P. and Lyberg, L.E. (2003). *Introduction to Survey Quality*. New York: Wiley.

Dubrow, J.K. and Tomescu-Dubrow, I. (2016). The rise of cross-national survey data harmonization in the social sciences: emergence of an interdisciplinary methodological field. *Quality and Quantity* 50 (4): 1449–1467. https://doi.org/10.1007/s11135-015-0215-z.

Durand, C. *Forthcoming*. "On Using Harmonized Data in Statistical Analysis: Notes of Caution" (chapter 20 in this volume).

Ehling, M. and Rendtel, U. (2006). Synopsis. Research results of Chintex - Summary and conclusions.

Fortier, I., Doiron, D., Little, J. et al. (2011). Is rigorous retrospective harmonization possible? Application of the DataSHaPER approach across 53 large studies. *International Journal of Epidemiology 40* (5): 1314–1328.

Fortier, I., Raina, P., Van den Heuvel, E.R. et al. (2017). Maelstrom research guidelines for rigorous retrospective data harmonization. *International Journal of Epidemiology* 46 (1): 103–105.

Fortier, I., Wey, T.W., Bergeron, J. et al. (2023). Life course of retrospective harmonization initiatives: key elements to consider. *Journal of Developmental Origins of Health and Disease 14* (2): 190–198.

Granda, P. and Blasczyk, E. (2016). Data harmonization. In: *Guidelines for Best Practice in Cross-Cultural Surveys*. Survey Research Center, Institute for Social Research, University of Michigan. https://ccsg.isr.umich.edu/chapters/data-harmonization/.

Granda, P., Wolf, C., and Hadorn, R. (2010). Harmonizing survey data. In: *Survey Methods in Multinational, Multicultural and Multiregional Contexts* (ed. J.A. Harkness, M. Braun, B. Edwards, et al.), 315–332. Hoboken, NJ: John Wiley & Sons.

Günther, R. (2003). Report on compiled information of the change from input harmonization to ex-post harmonization in national samples of the European Community Household Panel—Implications on data quality (Working Paper #19). Statistisches Bundesamt.

Joye, D., Sapin, M., Wolf, C. *Forthcoming*. On the Creation, Documentation, and Sensible Use of Weights in the Context of Comparative Surveys. (chapter 19 in this volume).

Kallas, J. and Linardis, A. (2010). A documentation model for comparative research based on harmonization strategies. *IASSIST Quarterly 32* (1-4): 12–12.

Lavallée, P. and Beaumont, J.-F. (2015). Why We Should Put Some Weight on Weights. In: *Survey Insights: Methods from the Field, Weighting: Practical Issues and 'How to' Approach, Invited article*. Retrieved from https://surveyinsights.org/?p=6255.

Minkel, H. (2004). Report on data conversion methodology of the change from input harmonization to ex-post harmonization in national samples of the European Community Household Panel— Implications on data quality. CHINTEX Working Paper 20. Statistisches Bundesamt.

Oleksiyenko, O., Wysmulek, I., and Vangeli, A. (2018). Identification of processing errors in cross-national surveys. In: *Advances in Comparative Survey Methods: Multinational, Multiregional, and Multicultural Contexts (3MC)* (ed. T.P. Johnson, B.-E. Pennell, I.A.L. Stoop, and B. Dorer), 985–1010. Wiley https://doi.org/10.1002/9781118884997.ch45.

Powałko, P. (2019). SDR 2.0 cotton file: cumulative list of variables in the surveys of the SDR database. *Harvard Dataverse 2*: https://doi.org/10.7910/DVN/6QBGNF.

Schneider, S.L. (2022). The classification of education in surveys: a generalized framework for ex-post harmonization. *Quality and Quantity* 56: 1829–1866. https://doi.org/10.1007/s11135-021-01101-1.

Singh, R. and Quandt M. *Forthcoming* "Assessing and Improving the Comparability of Latent Construct Measurements in *Ex-Post* Harmonization" (chapter 17 in this volume)

Slomczynski, K.M., Tomescu-Dubrow, I., Jenkins, J.C. et al. (2016). *Democratic Values and Protest Behavior: Harmonization of Data from International Survey Projects*. Warsaw: IFiS Publishers.

Slomczynski, K.M., Jenkins, J.C., Tomescu-Dubrow, I. et al. (2017). SDR master box. *Harvard Dataverse 1*: https://doi.org/10.7910/DVN/VWGF5Q.

Slomczynski, K.M. and Tomescu-Dubrow, I. (2018). Basic principles of survey data recycling. In: *Advances in Comparative Survey Methods: Multinational, Multiregional, and Multicultural Contexts (3MC)* (ed. T.P. Johnson, B.-E. Pennell, I.A.L. Stoop, and B. Dorer), 937–962. Hoboken, NJ: Wiley.

Slomczynski, K.M. and Skora, Z. (2020). Rating scales in inter-survey harmonization: what should be controlled? and how? *Harmonization: Newsletter on Survey Data Harmonization in the Social Sciences* 6 (2): 16–28.

Slomczynski, K.M., Tomescu-Dubrow, I., Wysmulek, I., Powałko, P., Jenkins, J.C., Ślarzyński, M., Zieliński, M.W., Skora, Z., Li, O., and Lavryk, D., (2023). SDR2 Database https://doi.org/10.7910/DVN/YOCX0M, Harvard Dataverse.

Thiessen, V. and Blasius, J. (2012). *Assessing the Quality of Survey Data*. SAGE Publications.

Tofangsazi, B. and Lavryk, D. (2018). We coded the documentation of 1748 surveys across 10 International Survey Projects: this is what data users and providers should know. *Harmonization: Newsletter on Survey Data Harmonization in the Social Sciences 4* (2): 27–31.

Tomescu-Dubrow, I. and Slomczynski, K.M. (2014). Democratic values and protest behavior: data harmonization, measurement comparability, and multi-level modeling in cross-national perspective. *Ask: Research and Methods* 23 (1): 103–114.

Tu, Y., Li, O., Wang, J. et al. (2023). SDRQuerier: a visual querying framework for cross-national survey data recycling. *IEEE Transactions on Visualization and Computer Graphics* 29 (6): 2862–2874. https://doi.org/10.1109/TVCG.2023.3261944.

Wysmułek, I. (2019). From Source to Target: Harmonization Workflow, Procedures and Tools. Introduction and Part 1. Cotton File and Detailed Variable Report. Presentation delivered at the international event "Building Multi-source Databases" (December 16–20). Warsaw, Poland. https://wp.asc.ohio-state.edu/dataharmonization/about/events/building-multi-source-databases-december-2019/

9.A Data Quality Indicators in SDR2

Table 9.A.1 SDR2: Q-type measures for the quality of documentation.

F_SAMPLEUNIT_SVY	Flag: Type of source sample
QD_UNIVINFO_SVY	Quality control documentation: Information on universe provided
QD_SAMPLEINFO_SVY	Quality control documentation: Information on sample type
QD_SAMPLETYPE_SVY	Quality control documentation: Type of sampling schema used
QD_TRANSLINFO_SVY	Quality control documentation: Information on translation methods provided
QD_TRANSLVALUE_SVY	Quality control documentation: Professional translation methods employed
QD_PRETINFO_SVY	Quality control documentation: Information on pretesting provided
QD_PRETVALUE_SVY	Quality control documentation: Survey instrument was pretested
QD_MODEINFO_SVY	Quality control documentation: Information on interview mode provided
QD_RRINFO_SVY	Quality control documentation: Information on response rate provided
QD_RRVALUE_SVY	Quality control documentation: Response rate, numerical value
F_RRAPPROX_SVY	Flag QD_RRVALUE_SVY: Approximated value in documentation
F_RRCALC_SVY	Flag QD_RRVALUE_SVY: Calculated using documentation-based info
QD_FCTRLINFO_SVY	Quality control documentation: Information on fieldwork control provided
QD_FCTRLVALUE_SVY	Quality control documentation: Fieldwork control was carried out
QD_DOCU_INDEX	Quality control documentation: Index (accuracy)

Table 9.A.2 SDR2: Q-type measures for the quality of source data processing.

QP_GENDER	Quality control processing error types: Gender
QP_AGE	Quality control processing error types: Age
QP_BIRTHYR	Quality control processing error types: Birth year
QP_EDU	Quality control processing error types: Education level
QP_EDU_YEARS	Quality control processing error types: Education schooling years
QP_TRPARL	Quality control processing error types: Trust in parliament
QP_DEMONST	Quality control processing error types: Participation in demonstration
QP_PER_INDEX	Quality control processing error: Index (errors)
QD_DOCU_INDEX	Quality control documentation: Index (accuracy)
QR_RECORDS_INDEX	Quality control data records: Index (errors)

Table 9.A.3   SDR2: Q-type measures for the quality of data records the computer files.

QR_AGE_SUSPECT	Quality control data records: Suspect age value
QR_BIRTHYR_SUSPECT	Quality control data records: Suspect birth year value
QF_AGE_BIRTHYR_DIFF	Quality flag: Difference between source age and age derived from source birth year
QR_HH_SIZE_OUTLIER	Quality control data records: Household comprises more than 20 members
QR_DUPLICATE	Quality control data records: Nonunique record (NUR)
QR_ID_NONUNIQUE	Quality control data records: Nonunique source case ID
QR_ID_MISSING	Quality control data records: Missing source case ID
QR_DUPLICATE_SVY	Quality control data records: Survey contains nonunique records
QR_ID_ERROR_SVY	Quality control data records: Survey contains errors in case IDs
QR_HIMISS_AGE_GENDER_ SVY	Quality control data records: Survey has over 5% missing on source age or gender
QR_WGHTTYPE_SVY	Quality control data records: Type of source weight
QR_WGHTDESC_SVY	Quality control data records: Status of source weight variable
QR_WGHTDISTORT_SVY	Quality control data records: Sample size change when source weights applied

10

Harmonization of Panel Surveys: The Cross-National Equivalent File

Dean R. Lillard[1,2,3]

[1]*Department of Human Sciences, The Ohio State University, Columbus, OH, USA*
[2]*Ohio State University, Deutsches Institut für Wirtschaftsforschung, Berlin, Germany*
[3]*National Bureau of Economic Research, Cambridge, MA, USA*

10.1   Introduction

CNEF harmonizes data in order to empirically measure a well-defined theoretical concept common across time and space. At the most basic level, CNEF harmonizes these data because people everywhere want to compare themselves to others. In practically every culture of the world, one finds some variant of the aphorism "the grass is always greener on the other side of the fence." Of course, CNEF harmonizes data for other, more important reasons. CNEF harmonizes because cross-national studies with properly comparable data potentially yield knowledge about cross-national differences in well-being and evidence about policies and practices that can improve it. In a shrinking geopolitical world, individuals, researchers, policy makers, and governments increasingly want data they can use to compare, over time and across countries, the social and economic status of residents of those countries.

Hereafter, the verb "harmonize" refers to the process of making data from different surveys comparable (across countries and over time). For the purposes outlined above, one needs harmonized data to accurately compare the status of people across countries and over time (Wolf et al. 2016).

While interest is intense, internationally harmonized longitudinal data are sparse. A Google Scholar search of the phrase "cross-national" or "cross national" yields more than 2 300 000 results. Adding the phrase "longitudinal data" or "panel data" reduces the number of "hits" to 20 300. Note that a search yields 19 700 "hits" for articles that contain the cross-national phrases plus "cross-sectional" or "cross sectional." Adding the words "harmonized" or "harmonized" reduces hits to 6800 for longitudinal data and 4520 for cross-sectional data. These results measure published studies that include these words or phrases so they reflect both whether researchers have access to the data and their success in publishing it.

The Cross-National Equivalent File (CNEF) project harmonizes household-based panel data. Because CNEF harmonizes longitudinal data, it harmonizes across two dimensions – countries and time. Users of data from long-running panel studies can attest that a given country's survey instruments and questions change over time – in content, emphasis, and wording. To validly compare data for a given survey over time, researchers need the data to be harmonized. Harmonizing survey data collected in different countries is an equally, if not more challenging task described in more detail below.

Longitudinal data harmonized across survey years and across countries offer many statistical advantages over harmonized cross-sectional data. Researchers can use more sophisticated models to understand what determines the level of and changes in social and economic well-being of individuals, families, and households. With harmonized longitudinal data from different countries, researchers can compare and understand cross-national similarities and differences in the level and evolution of individual and population education, employment, income, and health and social well-being.

The CNEF project was the first project to harmonize data drawn from ongoing longitudinal panel studies. It was inspired by one of the earliest large-scale harmonization projects – the Luxembourg Income Study (LIS). Launched in 1983 with data from seven developed economies, LIS currently harmonizes data from nationally representative microlevel cross-sectional household surveys administered in over 50 high-and middle-income countries (see https://www.lisdatacenter.org/about-lis). For more context see Burkhauser et al. (2001), Burkhauser and Lillard (2005, 2007), Frick et al. (2007), and Lillard (2013).

The first version of CNEF was created in 1993. Economists Richard Burkhauser and Tim Smeeding at Syracuse University and Richard Hauser at Goethe-Universität Frankfurt am Main created CNEF to study how women's household income evolved after a divorce. Their study, funded by the US National Institutes on Aging (NIA), was one of the first to compare life-course patterns using longitudinal data from two countries. They drew data on US women from the Panel Study of Income Dynamics (PSID) and data on German women from the relatively new Socio-Economic Panel (SOEP) (Burkhauser et al. 1991).[1] Burkhauser and Hauser promised NIA that they would produce a cross-nationally harmonized panel data set that they would make freely available to researchers worldwide. Since then, researchers across a wide swath of the social sciences use CNEF to study a wide variety of questions related to social and economic well-being. Even a brief perusal of the CNEF bibliography (see https://www.cnefdata.org/documentation/bibliography) shows the wide range of domains and topics scholars study with CNEF data. The list includes studies of survey methods, data quality, employment, income, inequality in earnings, income, and health, intergenerational correlations in social and economic outcomes, and economics, physical, and mental well-being.

10.2 Applied Harmonization Methods

CNEF was established to overcome three major shortcomings of the harmonized data in the LIS. First, CNEF harmonizes longitudinal rather than cross-sectional data. Harmonized longitudinal data offer researchers the possibility of identifying how changes in policies and other factors causally affect individual behavior. Second, CNEF researchers can directly access the harmonized data. Direct access is important because researchers are more likely to delve into the data and test the robustness of the harmonization algorithms. Third, CNEF researchers can easily access the parent survey data. Researchers

---

1 Twenty-seven years later, Burkhauser and Hauser and colleagues used a much longer time series in the still vibrant CNEF to revisit the same question (Bayaz-Ozturk et al. 2018).

need such access to test the harmonization algorithms. Such access is not possible for much of the underlying LIS data because the original data are from restricted administrative sources. CNEF structures itself to make it easier to tap the collective experience, creativity, and insights of the worldwide research community who sometimes find errors in the harmonization coding, often suggest ways to improve it, and use original parent survey data to create new harmonized variables.

From its inception, CNEF has followed an overarching philosophical and practical approach to the harmonization endeavor. Scholars do not harmonize just to have harmonized variables but rather to develop data one needs to answer substantive research questions. This focus requires that scholars pay attention to particular aspects of the harmonization process, each of which contributes to a higher-quality result. Among other things, scholars must pay attention to how the effort gets framed, the transparency of the method, and how one evaluates the quality of the harmonized variable. With funding from the United States National Institute of Child Health and Human Development, CNEF is starting to build data transparency pages for each member country (see https://www.cnefdata.org/documentation/data-transparency). There, researchers can link to the original variables CNEF uses to construct their data. While this project is in process, we will eventually also provide users access to the codes, we use to construct all but the variables that require restricted data (e.g. tax burden estimates). CNEF is also in the process of reworking and publishing its comparability measures. These measures gauge the extent to which a harmonized variable matches the conceptually ideal measure. When complete, the data transparency pages will specify both the conceptual target for each variable and evaluate the extent to which each country's variables meet it.

One must consider who harmonizes the data. When scholars harmonize data to answer a substantive question, they will more likely be experts in at least one of the domains involved in the substantive research question. Their training, knowledge, and experience in specific domains make it likely that these scholars will have insights that allow them to develop more scientifically sound harmonization algorithms. They will also be able to pursue answers to issues or questions that arise when they lack specific knowledge because they can easily contact and consult with colleagues and other scholars working in the same area.

When one harmonizes data to produce inputs needed to answer substantive research questions, scholars clearly define what theoretical concept they want to measure. This conceptual frame helps the harmonization process in several ways. First, it guides scholars as they seek empirical counterparts. If a good and single empirical counterpart exists, the conceptual frame indicates any simple adjustments that the scholar must make (e.g. converting height in feet and inches to centimeters). Second, when a single empirical measure does not exist, scholars use the conceptual goal to determine if they can approximate the conceptual variable by combining two or more empirical measures. Third, when the study adopts a cross-national design, the theoretical concept must be independent of culture and country-specific factors. Finally, the conceptual variable provides a yardstick against which a scholar can evaluate how well or how poorly the harmonized variable measures the theoretical ideal. As importantly, when scholars have a conceptual variable, it is easier for them to document the quality of the harmonized variable (relative to the theoretical ideal).

The scientific process has the added benefit that external scholars evaluate the harmonized data quality. Before CNEF adds a harmonized variable, it must pass the review of journal referees and editors. While imperfect, peer review and peer evaluation constitute the main method by which CNEF evaluates its harmonized variables. In addition, peer use of the harmonized data subject those data and the underlying algorithms to continual and ongoing scrutiny. Discerning scholars verify the data and discover errors in the algorithm. In many cases, when other scholars discover errors or lapses in logic, they suggest solutions that CNEF evaluates. Even when researchers only call CNEF's attention to the errors, they help to improve the underlying algorithm. This process highlights that the most widely-used harmonized data likely measure their targets with greater validity over time.

When undertaking the specific process of harmonizing variables, a researcher should start by categorizing conceptual variables into one of two types. The first category includes data from the physical world that have clear objective measures. The second category includes more abstract concepts about which there is some debate. For the latter, to harmonize one must clearly define the concept and be able, at least in principle, to measure it in objective units.

It is straightforward to harmonize variables of the first type not only because every panel survey collects the data one needs but also because the harmonization process is trivial. When any kind of manipulation is required, it usually only involves conversion from one metric into another (e.g. converting feet and inches into meters). Examples of such concepts include age, biological sex, height, weight, time, pregnancy, geographic location, and distance.

Harmonization of the second type involves more difficult tasks. First, one needs to have a clearly defined theoretical concept. Second, one needs to identify data that correspond to the theoretical construct. For example, the concept of income should theoretically include all resources available to a person or household. Resources are anything with value and include both money and resources in-kind such as food or housing received. The availability or even comparability of all components is often less straightforward than it is for physical concepts such as age, height, or weight that are objectively measurable. Some surveys include an abundance of such data. In other surveys, such data are sparse or completely missing. Finally, one must consider whether and how the theoretical concept shifts over time.

Consider the specific research question: "Among people who work, do annual earnings of women relative to annual earnings of men differ across country x and country y?" The null hypothesis is that relative earnings of women do not differ in country x and country y.

To test this hypothesis in its simplest form, one needs four data. One must describe the concepts of sex, country, work, and earnings. To focus the example on the challenges of harmonizing a variable to measure earnings, we will cavalierly dispense with the first two concepts, even though there are fascinating issues that surround the concept of gender roles and the idea of what constitutes a country (e.g. was a resident of Dresden prior to reunification of East and West Germany a resident of Germany?).

The concepts of work and earnings have very clear theoretical meaning in economics. A person engages in paid work if he or she spends even one hour in exchange for resources (usually money). Earnings consist of all resources that accrue to a person in exchange for time spent working (Saumik 2020). The theoretical goal, therefore, is to empirically measure earnings of women and men while they were living in country x and country y during one calendar year.

The next harmonization task is to identify relevant survey data from each country and evaluate whether (and how) one can use it to measure earnings as theoretically defined, i.e. in a comparable way across the two countries. Here, we will mention data from two-panel surveys used in the CNEF – the US PSID and the German SOEP.

Table 10.1 lists variables in each survey that measure resources that survey respondents said they had received. These (and most) surveys measure money received. Most surveys do not collect all data that economists would theoretically consider to qualify as "earnings." For example, most surveys do not ask detailed questions about "in-kind" resources a person receives in exchange for their labor. Surveys usually omit such questions either because it is difficult to collect and process the many types of in-kind resources

Table 10.1 Earnings measures in PSID and SOEP.

PSID	SOEP
Wages and salary:	Wages and salary:
• Up to four jobs	• Primary and secondary job
• Self-employment	• Self-employment
• Professional practice/trades	
Other earnings:	Other earnings:
• Overtime pay	• Pay for 13th month
• Commissions	• Pay for 14th month
• Bonuses	• Christmas bonus
	• Holiday bonus
	• Miscellaneous bonus
	• Profit-sharing

PSID: Panel Study of Income Dynamics.
SOEP: Socio-Economic Panel.

or because people in a given country, time, or place rarely receive earnings in this way. Since a theoretically correct measure of earnings should include in-kind resources, a researcher should note their absence if he or she uses survey data from periods and/or countries where payment in-kind is more common. More generally, it is important to articulate why data are included or missing from the harmonized measure and how the inclusions or omissions meet or deviate from the theoretically indicated concept.

In Table 10.1, CNEF includes as part of earnings the pay German workers receive for a "13th" and "14th" month of work. The SOEP asks about these amounts to capture an anachronism of the German labor market. In Germany, employed people qualify for bonuses that have these labels. CNEF counts this money as earnings because a person only receives these bonuses if he or she works. Under the theoretical concept, they are earnings.

I should note that CNEF defines earnings as income (real resources) received in payment for labor performed over a year (12 months). Part of the harmonization process necessarily requires that one must also harmonize the period over which one measures variables such as earnings and income.

The next step in the harmonization process is to create the variable. This step simply totals the amounts reported for each category. The resulting variable measures money earnings from labor. It fails to include payments made in-kind and earnings received when a person worked more than four jobs in the United States and more than two jobs in Germany.

Note that CNEF leaves some decisions to researchers. Specifically, CNEF leaves all money variables denominated in national currencies and does not adjust for the purchasing power in each country. The decision to do so is partly because researchers still debate about how to put on equal footing the earnings, income, and wealth of people in different countries and in different years.

CNEF assigns, for every country, identical variable names, labels, and, in most cases, value formats. Variable names reflect the content – the first letter of the variable name represents the variable's category: demographic (D), employment (E), household composition (H), income (I), weighting (W), sample identifiers (X), location (L), medical or health (M), and psychological well-being (P). In the "long" data format, a year variable indicates the survey year from which the variable is drawn. In the "wide" format, the variable name includes the calendar survey year as a "suffix" that is appended to the basic variable name. CNEF's naming convention simplifies the coding for models run on many country data files. Researchers use one not multiple computer programs to analyze data in any or all panels.

Currently, CNEF harmonizes almost 100 variables across the above nine domains. The domains include variables that cover standard demographic characteristics (e.g. age, sex, marital status, education, and race/ethnicity); employment (e.g. work hours, employment status, occupation, and industry); inputs needed to construct equivalence scales (counts of household members of given ages); annual income (individual labor earnings, household income of different types, and estimated tax payments); place of residence; health (a wide range of medical conditions); and subjective well-being (SWB); sample weights and identifier variables. The documentation on the website lists price indexes for each country but these are not part of the data. It is important to note that not all variables are available in all countries. Sometimes a country simply does not collect the data needed to create a harmonized variable (e.g. some countries do not ask about race or ethnicity). In a few cases, CNEF separately lists some country-specific variables (e.g. taxes paid computed with a notable tax simulation program or a sample weight that is linked to survey-specific design).

CNEF is notable among all harmonization projects worldwide because it sets out early to create measures of the conceptual variable "disposable income." This variable theoretically measures the resources a household has after it has paid all taxes and after it has received public and private transfers of resources. To derive this variable, CNEF partners use tax simulation programs to estimate the federal and state taxes households pay each year and CNEF tallies as best as possible the public and private transfers households receive. CNEF adjusts gross household income (adding the net inflow if positive and subtracting if negative) to create a measure of household income that approximately measures income "post-government, post-transfers." This variable is one of the key variables that researchers and organizations recognize as a substantial contribution to the CNEF project.

For example, the Organization for Economic Cooperation and Development (OECD) uses this measure in its 2001 publication that examines the dynamic nature of poverty (OECD 2001). The authors' state:

> The CNEF data are extremely valuable for providing long panels that enable more comprehensive and detailed analysis of poverty dynamics, both for pre- and post-fiscal income. These data enable comparisons of the effects of national tax and transfer systems by providing the appropriate income variables defined identically.

Other examples of notable findings using CNEF data include a 2005 study (Burkhauser et al. 2005) that examined the share of household income replaced after the male partner of a married couple died. Using CNEF data, they compared the income replacement in Canada, Germany, Great Britain, and the United States – four countries with quite different social support programs. Despite differences, they found that if one counted income from both public and private sources, the share of income replaced was almost identical across the four countries. CNEF data on SWB are also being used. For example, Chesters et al. (2021) use HILDA and SHP (Switzerland) SWB CNEF data to explore heterogeneity in trends in SWB for people in different birth cohorts – all of whom experienced volatile economic conditions. Their results point to similarities and differences. As in most cross-national comparisons, overall one observes in both countries a similar U-shaped pattern in SWB over the life-course. They also find differences that invite further research. For example, they find for one particular five-year birth cohort, lower levels of SWB in Australia but not in Switzerland. They also find differences in the association between education and SWB in the two countries (controlling for other factors). These differences, or stylized facts, are opportunities for researchers to use data from other countries to explore and perhaps explain.

Researchers can find, on the CNEF website (www.cnefdata.org), more details about the variables and a partial and growing bibliography of manuscripts, published articles, books, and government reports that use CNEF data at https://www.cnefdata.org/documentation/bibliography. Table 10.2 lists the

Table 10.2 Current and planned CNEF members, by Parent Data Source.

	Current		
Country	**Survey name**	**Acronym**	**Years**
Australia	Household Income and Labor Dynamics in Australia	HILDA	2001–2018
Canada	Survey of Labor and Income Dynamics	SLID	1992–2009[a]
China	China Family Panel Study	CFPS	2010–2020[b],[c]
Germany	Socio-Economic Panel	SOEP	1984–2018
Great Britain/UK	British Household Panel Study	BHPS	1991–2008
	Understanding Society, the UK Household Longitudinal Study	UKHLS	2009–2018[d]
Italy	Italian Lives	ITA.LI	2019–2020[b]
Japan	Japan Household Panel Study	JHPS	2009–2016
Russia	Russia Longitudinal Monitoring Survey	RLMS-HSE	1995–2016[e]
South Korea	Korea Labor and Income Panel Study	KLIPS	1998–2018
Sweden	Administrative records	tbd	tbd[b]
Switzerland	Swiss Household Panel	SHP	1999–2018
USA	Panel Study of Income Dynamics	PSID	1970–1996 1997–2019[c]
Planned			
Austria	Austria Socio-Economic Panel	ASEP	tbd[b]
Canada	Longitudinal and International Study of Adults	LISA	2012–2020[c]
Israel	Israel Longitudinal Study	ILS	2012–2019
Mexico	Mexican Family Life Surveys	MxFLS	2002–2019[f]
South Africa	National Income Dynamics Study	NIDS	2008–2021[g]
Taiwan	Panel Study of Family Dynamics	PFSD	1999–2020[c]

[a] Statistics Canada ended SLID-CNEF file in 2010. If funded, LISA will replace it.
[b] Expected in 2022.
[c] Fielded biennially.
[d] Understanding Society incorporated the extant BHPS sample in 2010.
[e] The RLMS-HSE was not fielded in 1997 and 1999.
[f] MxFLS fielded in 2002, 2005–2006, 2009–2012, and 2018–2019.
[g] NIDS fielded in 2008, 2010–2011, 2012, 2014–2015, 2017, and 2020–2021 (subsample of the 2017 sample).

countries and surveys of CNEF members that contribute files or that are scheduled to contribute files in the near future. CNEF is a joint effort of researchers at the Ohio State University (where it resides) and the institutions that administer and manage each country's panel survey. CNEF researchers collectively devote substantial time, intellectual, and financial resources to support and develop CNEF. It is truly a collaborative effort of researchers worldwide who support, develop, and analyze harmonized panel data. We have applied for funding that, if awarded, will add CNEF files for surveys that are or will be administered by Statistics Austria, Statistics Canada, the Israel Central Bureau of Statistics, the Iberoamerican University and Center for Economic Research and Teaching, and the University of Cape Town, and Academia Sinica. We briefly describe the main features of the sample represented by the planned or ongoing surveys. For more details, visit the homepage of each survey. Below Table 10.2, readers will find a very brief description of the survey or data for each current and planned CNEF member.

10.2.1 CNEF Country Data Sources, Current and Planned

The institutions that administer the country surveys currently in CNEF include the Melbourne Institute of Applied Economic and Social Research at the University of Melbourne (HILDA); Statistics Canada (SLID); Peking University and Princeton University (CFPS); German Institute for Economic Research (DIW) in Berlin (SOEP); Institute for Social and Economic Research at the University of Essex (BHPS and UKHLS); University of Milan-Bicocca (ITA.LI); Keio University (JHPS); the Carolina Population Center at the University of North Carolina at Chapel Hill, Demoscope, and the Higher School of Economics – Moscow (RLMS-HSE); Korea Labor Institute (KLIPS); University of Duisberg-Essen (Swedish data yet to be named); Swiss Centre of Expertise in the Social Sciences at the University of Lausanne (SHP); and the Survey Research Center at the University of Michigan (PSID).

10.3 Current CNEF Partners

10.3.1 The HILDA Survey <https://melbourneinstitute.unimelb.edu.au/hilda>

The HILDA Survey is a household-based panel study first fielded in 2001. HILDA collects information about economic and SWB, labor market dynamics, and family dynamics of Australian households. Wave 1 includes 7682 households and 19 914 individuals. In 2011, HILDA refreshed the sample with 4009 individuals living in 2153 households. As of wave 17, HILDA retains 17 571 individuals from 9742 households. Each year HILDA interviews all adult members of each household (Wilkins et al. 2019). HILDA follows its panel members in the same way as in the other surveys.

10.3.2 The SLID <http://www.statcan.ca/start.html>

The SLID began in 1993 with a sample of about 15 000 households, containing approximately 30 000 adults. The SLID survey differs from the other surveys in that each of its panels lasts only six years. Statistics Canada chose to limit the length of the panels to keep the sample population representative of the national population because SLID also serves as Canada's labor force survey for the purpose of annual metrics. SLID added a second six-year panel in 1995. In 1998 SLID "retired" the first panel and launched the third panel that again overlapped with the second panel for three years. Thereafter, SLID retired old panels and introduced new ones using this procedure. The overlapping panels help to maintain continuity in the data. As in the other surveys, all current SLID families contain at least one member who was part of, or was born to one of, the initial household samples in each panel. Statistics Canada stopped contributing a SLID-CNEF file in 2010.

10.3.3   The CFPS <https://www.isss.pku.edu.cn/cfps/en>

With funding from the National Institutes of Aging, CNEF will create a file using data from the China Family Panel Studies (CFPS). CFPS is a household-based panel study first fielded in 2010 with a sample of almost 30 000 individuals aged 9 and older living in about 15 000 families. It is administered jointly by Princeton University and Peking University. CFPS fields its survey every other year.

10.3.4   The SOEP <https://www.diw.de/en/soep>

The SOEP is the English-language public-use version of the SOEP, a longitudinal dataset begun in 1984. The SOEP began with a sample of 6000 households living in the western states of the Federal Republic of Germany, including a disproportionate number of resident foreigners and non-German migrant workers. In June 1990, before Germany was officially reunited, the DIW fielded a survey of families in the eastern states and merged these data with the existing SOEP population to provide a representative sample of reunited Germany. DIW added new immigrant samples in 1994, 1995, 2013, and 2015 and two refugee samples in 2016. "Refreshment" samples were added in 1998, 2000, 2002, 2006, 2009, 2010, 2011, and 2012. The sample includes all household members, consisting of Germans living in the eastern and western German states, foreigners, and immigrants to Germany. In addition to gathering the core demographic and labor-related data, the DIW regularly adapts the survey to gather information on current social developments. The international version contains 95% of all cases surveyed (see 10.5684/soep. v34i) (see Wagner et al. 2007; Goebel et al. 2019).

10.3.4.1   The BHPS <https://www.iser.essex.ac.uk/bhps>
The BHPS began in 1991 with a sample of just over 5500 households containing approximately 10 000 individuals. Households were selected based on postal code of residence. The original BHPS sample represents the population of households with postal codes in England, Wales, and Scotland. The BHPS sample represents the UK population since the BHPS added 1500 new Scottish and Welsh households in 1999 and 2000 households in Northern Ireland in 2001.

   From 1994 to 2000 (waves 4–10), the BHPS separately interviewed youth aged 11–15. All current BHPS households contain at least one member who was either part of the original 5500 households, part of the first wave of booster sample households, or born to a member of one of these households. In total, BHPS fielded 18 waves of data. As part of wave 18, the BHPS participants were invited to join a new, much bigger survey called Understanding Society, the UK Household Longitudinal Study (UKHLS). Almost 6700 of just over 8000 BHPS respondents agreed to participate. BHPS sample members answered their first UKHLS surveys in 2010–2011 during wave 2 of Understanding Society. They are now regular members of the UKHLS panel.

10.3.4.2   Understanding Society, UKHLS <https://www.understandingsociety.ac.uk/>

The UKHLS is a panel study of individuals living in about 40 000 households that participated in the first survey wave in 2009. Each year, the UKHLS collects data on all household members. Individuals in households get interviewed annually in either face-to-face in-home interviews or via an online self-completion survey. UKHLS administers a separate questionnaire to youth aged 10–15. Individuals aged 16 and older complete a survey aimed at adults. The main survey sample consists of a large general population sample plus three targeted populations: an Ethnic Minority Boost Sample, the former British Household Panel Survey sample, and the Immigrant and Ethnic Minority Boost Sample. The sample includes households from all four countries of the UK.

10.3.5   The ITA.LI

The ITA.LI is a longitudinal quantitative and qualitative research project carried out by the Department of Sociology and Social Research of the University of Milan-Bicocca, within the scope of the Departments of Excellence project (Italian Law 232 of 11 December 2016). The project samples people age 16 and older in an initial sample of approximately 5000 families. The stratified sample includes more than 250 Italian municipalities. It uses a probabilistic sampling method in three stages developed in conjunction with the Italian National Institute of Statistics (ISTAT). The first survey went into the field in the fall of 2019 and was delayed for a few months because of COVID. Interviews began again in summer of 2020. ITA.LI expects an initial sample of 9000–9010000 individuals.

10.3.6   The JHPS <https://www.pdrc.keio.ac.jp/en/paneldata/datasets/jhpskhps>

The JHPS differs from the other household panels because the JHPS samples and follows 4000 men and women (not households). The initial sample was drawn using a two-stage stratified random sampling frame. In parallel with the 2004 Keio Household Panel Study, the JHPS asks sample members about their economic and employment status, education, and health/healthcare. When sample members are married, the JHPS asks their spouses the same survey questions.

10.3.7   The RLMS-HSE <https://www.cpc.unc.edu/projects/rlms-hse>

The RLMS is a series of nationally representative surveys designed to monitor the effects of Russian reforms on the health and economic welfare of households and individuals in the Russian Federation. The target sample size was set at 4000 households. A multistage probability sample of households was employed to get a nationally representative sample for the Russian Federation. Surveys were first conducted in 1992 but due to implementation problems and attrition, RLMS advises researchers to use data from the "Phase II"-period that begins in 1994. Although originally the sample was defined by geographic addresses, starting after round VII (1996), RLMS implemented the same following rules as the other household panel surveys (i.e. they follow individual household members when they move). RLMS attempted to recontact people who had moved out of the 1994–1996 sample dwelling units prior to the implementation of the new following rules. RLMS attempts to interview all household members aged 14 and older. Adults answer as proxies for children aged 13 and younger. Two institutions jointly run the project: the Carolina Population Center at the University of North Carolina at Chapel Hill, started by

Barry M. Popkin and now headed by Klara Peter, and the Demoscope team in Russia, headed by Polina Kozyreva and Mikhail Kosolapov (see Kozyreva et al. 2016).

10.3.8   The KLIPS <https://www.kli.re.kr/klips_eng/contents.do?key=251>

The KLIPS is a longitudinal survey of the labor market and income activities of households and individuals residing in urban areas. Launched in 1998, the KLIPS surveyed an initial sample of 13 321 individuals in 5000 households. KLIPS uses a two-stage stratified cluster sampling method. In selected households, KLIPS interviews individuals aged 15 and older. KLIPS attempts to interview household members who are not present for the in-person interview because they are temporarily absent (traveling abroad, hospitalized, studying overseas, etc.) using phone interviews. For adults, who are physically or mentally disabled, KLIPS conducts proxy interviews.

10.3.9   The Swedish Pseudo-Panel

With the same funding from the National Institutes of Aging that will create the CNEF file for China, CNEF is working with partners in Germany at the University of Duisburg-Essen and in Sweden to create pseudo-panel data with Swedish data from administrative registers. This innovative effort will break new ground by reconstructing the structure of the household-based sample of all the other countries using only administrative records. That effort is underway so the file has not yet been named. We expect researchers to get access to the file through Statistics Sweden in either 2022 or 2023.

10.3.10   The SHP <https://forscenter.ch/projects/swiss-household-panel/>

The SHP launched in 1999 with a sample of 12,931 individuals living in 5074 households. SHP added a refreshment sample of 6569 individuals living in 2538 households in 2004 and 9945 individuals in 4093 households in 2013. These three samples differ in the frame that generated them. The first sample (SHP_I) is a stratified random sample of private households whose members represent the noninstitutional resident population in Switzerland. In 1999, the methodology section of the Swiss Federal Statistical Office drew a simple random sample in each of the seven major statistical regions of Switzerland from the Swiss telephone directory. The 2004 refreshment sample was randomly drawn using the same method. SHP drew the 2013 refreshment sample from a register of residents of communes and cantons owned by the Swiss Federal Statistical Office.

   Like the other panel studies, SHP attempts to reinterview members of households that participated in an initial survey. In its early years, SHP followed individuals (as they formed new households) only if they were members of the original households. Other members of the household were interviewed as long as they co-reside with an original sample member. However, starting in 2007, SHP follows all household members as they form new households (without regard to their co-residence with a member of one of the original sample households). SHP interviews and releases new data annually (see Voorpostel et al. 2018).

10.3.11   The PSID <https://psidonline.isr.umich.edu/>

The PSID began in 1968 with a sample of 5000 families, with an oversample of lower-income individuals. The PSID follows all individuals from these original households and all of their offspring as they form new households. The PSID provides sample weights to represent the US population. Starting in 1997 the PSID began administering its survey every other year and no longer following every member or related member of families in the low-income oversample population. For a more complete discussion of the PSID see Hill (1992).

10.4 Planned CNEF Partners

CNEF is actively pursuing funding to add CNEF files for additional countries. We have secured funding for one country but do not yet know the final structure of the survey that will result. We have submitted grant applications to fund creation of CNEF files for five new countries and to continue the CNEF file for Canada with a different data set. Because we do not yet know what decisions will be made (and we do not want to raise hopes of researchers prematurely), we only provide brief details about the data from each country for which we hope to create CNEF files.

10.4.1 The ASEP

The Austrian Ministry of Science has allocated funds to create a household-based panel survey in Austria. The exact structure and content of the survey is yet to be determined. The proposal that the Ministry approved included a plan to create a CNEF file from the ASEP data. Stay tuned.

We have applied for funding to use data from each of the following surveys to create new CNEF files.

10.4.2 LISA <https://www.statcan.gc.ca/eng/survey/household/5144>

The Longitudinal and International Study of Adults (LISA), administered by Statistics Canada, began in 2012 and is administered biennially. LISA uses household interviews to collect information from approximately 34,000 Canadians aged 15 and older from more than 11,000 households. The LISA also contains information from several administrative data sources including the tax files, pension plan, and immigration landing records.

10.4.3 The ILS

The Israel Longitudinal Study (ILS) began in 2012 with a sample of almost 14 000 individuals aged 15 and older living in about 5000 families. ILS is administered by Israel's Central Bureau of Statistics. They field the ILS survey every year.

10.4.4 The MxFLS <http://www.ennvih-mxfls.org/english/index.html>

The Mexican Family Life Surveys (MxFLS) began in 2002 with a sample of approximately 35,600 individuals living in 8400 households in Mexico. Follow-up samples (2005–2006 and 2009–2012) include individuals who migrated within Mexico or emigrated to the United States of America and individuals in households that grew out from previous samples. The MxFLS-2 and MxFLS-3 relocated and reinterviewed almost 90% of the original sampled households. In 2018–2019, MxFLS fielded a fourth wave (MxFLS4) to a sample of 50%.

10.4.5   The NIDS <http://nids.uct.ac.za>

The National Income Dynamics Study (NIDS) began in 2008 with a sample of about 28,000 South African residents across the age distribution, living in approximately 7300 households. Survey participants and co-resident household members were reinterviewed approximately every two to three years in five waves of data collection (2008, 2010–2011, 2012, 2014–2015, and 2017). After a sample top-up in 2017, nearly 40,000 individuals participated. Between May 2020 and May 2021, a special telephonic follow-up panel survey was conducted on a subsample of 2017 NIDS household members, called the NIDS-Coronavirus Rapid Mobile Survey (NIDS-CRAM). Five waves of NIDS-CRAM were completed, with 7073 sample members aged 18 and older in the first wave and an additional 1084 added in the third wave.

10.4.6   The PSFD <https://psfd.sinica.edu.tw/V2/?page_id=966&lang=en>

The Panel Study of Family Dynamics (PSFD) began in 1999 with a nationally representative sample of approximately 1000 individuals born between 1953 and 1964 living in Taiwan. In subsequent surveys, the sample was expanded so that it now covers the population born from 1935 to 1991. The PSFD was administered annually from 1999 to 2012 and biennially since 2012. Since 2000, the PSFD also follows children from originally sampled families when the children reach age 16.

10.5   Documentation and Quality Assurance

To document its harmonization process, the CNEF project follows the principle of transparency. That principle and practices that flow from it, help researchers evaluate the quality of the harmonized data.

To be transparent, CNEF publishes the algorithm researchers use to harmonize each variable. Until recently, CNEF published the algorithm in a pdf document it posted on its website (https://cnef.ehe .osu.edu/data/codebooks). Those documents will soon be obsolete because CNEF has launched a new website (cnefdata.org) that will make it easier for users to navigate the harmonization algorithms, including an easy way for them to connect directly to the documentation for each original parent survey variable we use. At the time of this writing, we are still building the website, so the links are not yet active. When complete, researchers will be able to see CNEF's harmonization algorithm for each variable (in STATA code), the name of all original parent-survey variables the algorithm uses, and a "README" file that details any corrections/judgment calls researchers made. In the unusual cases that a particular survey changes variable names, we will reflect those changes. CNEF uses STATA to create its variables. We provide (or soon will provide) the option for researchers to get code to read the raw (comma-delimited ASCII) CNEF data into SAS, SPSS, and STATA. We hope to also provide R as an option but currently do not.

Above and beyond the clear documentation of harmonization algorithms, the principle of transparency yields multiple benefits. By being transparent, CNEF benefits from attentive researchers who find and help correct errors in the algorithm. Such errors are more readily apparent when researchers use a harmonized variable because those researchers tend to be subject matter experts and bring deep knowledge and experience. For the same reason, transparency also makes it easier for researchers to improve algorithms. Finally, transparency demands that researchers evaluate the quality of the harmonized data.

Since CNEF data follow individuals and households over time, CNEF evaluates data for internal and external validity. Evaluation of the data for internal consistency is part of the harmonization algorithm. When a new wave of data is added, CNEF researchers evaluate the time series distribution of each variable to check that it follows expected trends. This check is a first-order evaluation that assumes that the distribution of a particular variable changes because of a time-stable underlying process (e.g. the aging of the population). CNEF maintains this assumption if there are no obvious external factors at work (e.g. COVID mitigation policies). In addition to this check, CNEF relies on the panel nature of the data and the constant checking and rechecking of its algorithm as an assurance of the quality.

Beyond these checks, periodically, individual CNEF partners will evaluate the distribution of their harmonized variable against country-specific external sources. Such comparisons may validate the parent data or CNEF variables. For example, Kim and Stafford (2000) compare earnings in the PSID with earnings estimated using data from the March Supplement to the Current Population Survey (CPS). Though less frequent, this type of external validation helps to build confidence that the original data or the constructed variables capture the principal moments of the distribution of the intended variable – as it is reflected in independently drawn data from other external sources.

Finally, CNEF aims to evaluate how well the harmonization algorithm produces data that meet its target, that is, the theoretical concept that defines the desired variable. This evaluation, like the algorithms themselves, evolves and improves over time.

The effort sometimes fails. That is, it may be difficult to find ways to generate a variable in all countries that meets a theoretical construct. In that case, one must be ready to compromise. For example, one of my continuing challenges (and sources of intellectual dissatisfaction) is the harmonized variable CNEF labels as "Education with Respect to High School." When we joined the CNEF project, one of my first tasks was to harmonize data from the British Household Panel Survey to create CNEF variables. Anyone familiar with the British education system will know that their system does not follow the US system of a common curriculum through the end of secondary school. British secondary school students "complete" their education by taking a series of examinations in particular subjects. Students choose whether or not to take tests in one or more subjects. To complete high school in the United States, students must pass courses in a common set of subjects. We consulted with US university admissions officers to discover if they have a set of rules, some minimum set of tests passed in particular subjects, that they decide would qualify a British student as having education equivalent to a US high school graduate. After four months of trying, we gave up. A similar challenge exists when trying to define what counts as a "Year of Schooling."

For many years, we defended the absence of these two education CNEF variables for the BHPS by recounting the impossibility of finding a harmonization algorithm. My CNEF partners have been unsatisfied with my surrender. We recently have begun to rethink what concept those variables aim to measure.

While, we have not fully reconciled myself to a new concept. We are groping toward one, perhaps two. We are considering a distinction between a simple count of physical time spent in school or studies ("bodies in seats") and a much more ambitious variable that measures the theoretical concept economists developed that is labeled "human capital." In theory, human capital refers to any training, schooling, experience, or physical health that yields a positive return (in money or satisfaction). So CNEF measures of "education" and human capital are works in process. The continuing attempts to define and measure those variables across countries exemplify how the harmonization process evolves over time.

10.6 Challenges to Harmonization

CNEF, like every harmonization project, faces the usual challenges that arise when a given survey changes questions, categories, and content. CNEF has also faced challenges created when CNEF had to decide whether or not to add more countries. Some of these challenges have been resolved while others remain. For example, when deciding whether it made sense to include more countries in CNEF, the project had to evaluate the criterion that it uses to consider surveys "comparable." In CNEF's case, the household-based panel survey design forms the core criterion it uses to evaluate whether a new country's survey generates data that might be made comparable with data from existing CNEF data. To date, CNEF has not admitted new members whose parent surveys are not household-based surveys with the same (or closely similar) following rules. As noted above, CNEF does not rigidly adhere to our harmonization process. So it is possible that a new study could cause CNEF to revisit past decisions. So far, that has not happened. But it might.

Other factors come into play when deciding whether or not to admit new members. One of the key factors is the quality of the survey and data. We evaluate the sample design, following rules, and representativeness of the sample (relative to the national population). The candidate survey must have clear documentation and be committed to the scientific principle of data transparency. Briefly, the principle of data transparency requires that the survey provides clear documentation of its procedures and data construction so that researchers can understand what population the sample represents and how all derived variables are constructed. While not dispositive, CNEF gives preference to studies that are ongoing.

CNEF has also had to consider issues related not to adding another country but issues related to expanding the set of harmonized variables. Decisions about these issues involve questions such as what data might be harmonized and what overlap (in time or across countries) one needs to undertake valid comparisons. Decisions became less simple as CNEF grew. In keeping with its philosophy, CNEF allows and encourages researchers to conceive of and develop harmonized variables. As noted above, to admit a newly harmonized variable, CNEF not only evaluates the "success" of the proposed harmonization but also requires that a researcher publishes an article in a peer-reviewed journal that describes the harmonization algorithm and that earns the approval of the referees and editor.

CNEF's philosophy is that harmonization decisions should flow up from researchers so that a well-defined theoretical objective guides the harmonization process. To researchers contemplating whether or not to harmonize new variables for CNEF the CNEF team suggests a simple rule for researchers to consider. The rule is that researchers can harmonize any conceptual variable if the requisite data are available on at least two of CNEF's parent panel surveys. CNEF developed this rule after CNEF expanded to include data from four countries rather than the original two. Richard Burkhauser and Dean Lillard received US National Institutes of Aging funds to harmonize and add data measuring health using parent survey data from Germany, the United Kingdom, and the United States. They harmonized no additional health variables for Canada because both two health-related variables available in SLID (work limitations and self-reported health) were already included in CNEF.

CNEF faces ongoing challenges as it adds new members. In particular, CNEF has to be patient about the estimated taxes households pay because tax simulation programs are not available in all countries. Furthermore, the estimation of taxes requires special expertise and data. For most countries, researchers must measure not only levels of and changes in the tax code, income, employment, and earnings but also changes in marital status, births and deaths of household members, and residential locations. In most countries, CNEF uses estimated taxes from tax simulation programs written by experts in each country who are intimately familiar with their country's tax laws.[2] While it is feasible to simulate taxes paid in cross-sectional survey data, the longitudinal feature of the panel surveys in CNEF often means that the simulation programs yield more precise estimates because the panel surveys (usually) track a richer set of data on and changes in each individual's and each household's circumstances. To the extent that these demographic factors affect when and how much a country taxes individuals or households, the tax simulation programs can use that information to generate more precise estimates of taxes paid.

Because of the complexity of tax simulations, the simulated tax variables often are created some years after a new member joins CNEF. In some cases, e.g. Russia, we have not yet managed to create those variables. The task can be daunting because it requires attention to tax codes (which changed rapidly in some years) since the first year a given survey was fielded. This challenge and the challenge of incorporating changes in tax codes over time in countries where we have tax simulation programs remain as ongoing issues that need attention.

Although the inclusion of more panel surveys in CNEF raised the above challenges, it also led to positive engagement among CNEF members (pairwise and in different combinations) because the CNEF parent surveys began to adjust their content in order to more easily harmonize data. For example, during the process of harmonizing health variables, the PSID added more health questions, significantly expanding the set of harmonized health variables. Similar efforts added questions to collect data needed to harmonize smoking behavior and are being pursued to collect data to measure migration.

CNEF faced another type of challenge when researchers around the world clamored to include measures of life satisfaction that were available on most of the parent surveys. The dilemma these variables raise is that it is difficult to find a common scale on which to measure satisfaction that is independent of culture and country-specific factors. One can use these data for within-country comparisons but cross-country comparisons are more difficult. Despite these issues, CNEF added life satisfaction measures for every survey that collects those data but we reproduce the scales in toto. We do not attempt to harmonize them but rather make them available for researchers to study. As a side note, others have developed the method of "anchoring vignettes" to create a common point of reference that one could use to harmonize life-satisfaction measures (see Banks et al. 2004, and subsequent literature).

The expansion of CNEF also highlights challenges that arise specifically in longitudinal surveys. That challenge crops up when surveys change the design – dropping or adding questions or asking questions differently. These issues arise for researchers who harmonize both cross-sectional and longitudinal survey data. Such changes in longitudinal surveys involve specific advantages and disadvantages. The most

---

2 For the PSID we use the National Bureau of Economic Research tax simulation program written by Daniel Feenberg (see Feenberg and Coutts 1993). SLID appends administrative tax data for most of its sample. In all the other countries except Russia, researchers at each CNEF partner institution simulate taxes paid. For the HILDA see Goode and Watson (2006); BHPS and UKHLS see Bardasi et al. (1999); SOEP see Schwarze (1995). For the JHPS, KLIPS, and SHP consult the documentation or institution's scholars for details.

obvious disadvantage of dropping questions is that one stops being able to document the life-course evolution of a social or economic outcome for a given individual. Changing the wording of a particular question can be implemented on a longitudinal survey with less so-called "seam effects" if the survey asks the original question together with the newly worded question for two or more survey waves. Since longitudinal surveys need to recognize and collect data on new social issues, such changes are a reality that must be managed.

Other challenges arise when a particular study decides to field its survey less frequently (e.g. PSID), or stops its panel (SLID). These changes break the continuity that is one of the unique features of longitudinal data. Researchers should anticipate that they will have to face such events. In CNEF, harmonized data may not exist in some years for some countries and the comparability of variables may erode across surveys or for a given survey over time.

The other main challenge about harmonization per se and its documentation arises because of the hands-off and bottom-up approaches CNEF takes. As noted above, individual researchers collaborate to harmonize data. Such efforts may involve various configurations of researchers from two or more of the parent institutions, but the process is decentralized. In practice, this decentralized process has not resulted in many new harmonized variables in CNEF. CNEF must confront that challenge and find other ways to incentivize researchers to create new comparable variables.

10.7 Recommendations for Researchers Interested in Harmonizing Panel Survey Data

Researchers who want to harmonize panel survey data will benefit if they clearly articulate their goals. Most narrowly, researchers can consider their goal of generating data they want or need to answer research questions. To achieve that goal, researchers should harmonize the relevant panel data over time. Indeed, panel data offer analytical advantages over cross-sectional data. A primary advantage is that, with panel data, researchers can estimate models to identify the causal effects of changes in conditions and policies. Keeping that in mind, researchers should be careful to avoid a potential trap that arises when the project defends consistency at all costs. It can be that to defend the consistency of its data, a given survey becomes inflexible and fails to adjust its survey instrument to reflect changing conditions or to admit questions about newly emerging topics. Survey directors will always face this tension – between a desire to maintain temporal consistency in data and the desire to collect new data. The balancing act is an art form.

One of the challenges that every project faces, especially when harmonizing data from ongoing panel surveys, is securing funding. The funding challenge is the one that researchers should carefully consider. By its nature, it is difficult to plan for and control the flow of funding. However, researchers who want to harmonize panel data should, like the underlying panels, anticipate as much as possible how their project will evolve past the initial funding period. There is much to be gained by projecting various paths a project might follow and developing plans to rationalize further funding under each contingency.

CNEF confronted and endured through that period. In its early years, CNEF benefitted from an abundance of resources from the US National Institutes of Aging and from the German Institute for

Economic Research (Deutsches Institut für Wirtschaftsforschung or DIW). As those funding sources ended, CNEF faced the same crisis all projects face – how to finance its continuation. Richard Burkhauser and Dean Lillard managed to get some funding but with only sporadic success. They benefited from the largesse of the SOEP group – who donated significant resources. To try to be self-sustaining, CNEF does two things. First, it asks users to donate a modest amount when they get data. Second, tenured faculty running CNEF have always donated time to the project. CNEF ran on a shoe-string budget for many years.

Researchers interested in harmonizing survey data should anticipate and make contingency plans for periods when funding is either reduced or even completely unavailable. As importantly, one should have a vision to use in applications for funding from available sources.

In CNEF's case, that vision involved making it easier for researchers to find and use the data. With funding from the National Institute of Child Health and Human Development (NICHD), CNEF has developed a new web-based interface known as CNEF-Investigator. Researchers can use CNEF-Investigator to browse the data and view documentation and distributions of variables for all CNEF samples. To access Investigator, one requests a CNEF-Investigator account online at https://www.cnefdata.org/data/data-application. On CNEF-Investigator one can view distributions of variables not only for a single country in a single year but also for all countries side-by-side. For a subset of countries, approved researchers can select a basket of variables and download the data directly. CNEF is currently funded to upgrade and enhance the project's web-based platform with more metadata, links to documentation for each of the original surveys, and other features. For example, we have built on our web pages a feature that allows CNEF researchers to register working papers and published articles that use CNEF (https://cnef.ehe.osu.edu/bibliography/submissions). Researchers can also browse a bibliography that lists over 350 articles and books that use CNEF (see https://www.cnefdata.org/documentation/bibliography).

Researchers who want others to use their data should create a way to reach out to potential users. That outreach is easier with funding. For example, with funding, through a supplement grant to the PSID, we are promoting CNEF and the parent surveys of our CNEF partners. Until COVID intervened, the plan was to promote CNEF at three major social science conferences in 2020. But projects can also do outreach "on the cheap" by running information sessions free of charge whenever a researcher presents substantive findings at a university seminar.

In the United States context, the quest for funding is never-ending. On the one hand, the effort to secure funding never seems to end. Grant applications are more often rejected than accepted. That reality requires that researchers persist through difficult times. On the other hand, because one must explain to grant reviewers the scientific value of the project, the need to secure funding also helps to keep harmonization researchers sharp and thinking about what harmonized data offer social scientists and policy makers. On balance, the need to apply for and justify funding infuses vitality into the harmonization process. Rather than decry it, we recommend that researchers embrace and celebrate it.

10.8 Conclusion

Harmonization of panel data over time and across countries involves many challenges but also many rewards. The data are fundamentally important inputs to efforts to estimate causal effects of policies and changes in determinants of social and economic outcomes. The data are also fundamental to a deeper understanding of behavior that reflects a shared human condition and behavior that is shaped on the margins by culture and context.

Researchers who aim to develop such data face many challenges that are also opportunities. Funding will almost always be a challenge researchers will face. Despite those challenges, harmonized panel data from multiple countries offer exciting possibilities for social science and policy makers. The future looks bright.

References

Banks, J.W., Kapteyn, A., Smith, J.P., and van Soest, A. (2004). International Comparisons of Work Disability. *IZA Discussion Paper No. 1118*. https://ssrn.com/abstract=533807.

Bardasi, E., Jenkins, S.P., and Rigg, J.A. (1999). Documentation for derived current and annual net household income variables, BHPS waves 1–7. *Institute for Social and Economic Research Working Paper 99-25*.

Bayaz-Ozturk, G., Burkhauser, R.V., Couch, K.A., and Hauser, R. (2018). The effects of union dissolution on the economic resources of men and women: a comparative analysis of Germany and the United States, 1985–2013. *The Annals of the American Academy of Political and Social Science* 680 (1): 235–258.

Burkhauser, R.V. and Lillard, D.R. (2005). The contribution and potential of data harmonization for cross-national comparative research. *Journal of Comparative Policy Analysis* 7 (4): 313–330.

Burkhauser, R.V. and Lillard, D.R. (2007). The expanded Cross-National Equivalent File: HILDA joins its international peers. *The Australian Economic Review* 40 (2): 1–8.

Burkhauser, R.V., Duncan, G.J., Hauser, R., and Berntsen, R. (1991). Wife or frau, women do worse: a comparison of men and women in the United States and Germany after marital dissolution. *Demography* 28 (3): 353–360.

Burkhauser, R.V., Butrica, B.A., Daly, M.C., and Lillard, D.R. (2001). The Cross-National Equivalent File: a product of cross-national research. In: *Soziale Sicherung in einer dynamischen Gesellschaft, Festschrift für Richard Hauser zum 65. Geburtstag* (ed. I. Becker, N. Ott, and G. Rolf), 354–376. Frankfurt/Main: Campus Verlag.

Burkhauser, R.V., Giles, P., Lillard, D.R., and Schwarze, J. (2005). Until death do us part: an analysis of the economic well-being of widows in four countries. *Journal of Gerontology: Social Sciences* 60B (5): S238–S246.

Chesters, J., Simona, J., and Suter, C. (2021). Cross-national comparison of age and period effects on levels of subjective well-being in Australia and Switzerland during volatile economic times (2001–2016). *Social Indicators Research* 154 (1): 361–391.

Feenberg, D. and Coutts, E. (1993). An introduction to the TAXSIM model. *Journal of Policy Analysis and Management* 12: 189–194.

Frick, J.R., Jenkins, S.P., Lillard, D.R. et al. (2007). The Cross-National Equivalent File (CNEF) and its member country household panel studies. *Schmollers Jahrbuch (Journal of Applied Social Science Studies)* 127 (4): 627–654. (Also published in German as (2008) Die internationale Einbettung des Sozio-oekonomischen Panels (SOEP) im Rahmen des Cross-National Equivalent File (CNEF). *Vierteljahrshefte zur Wirtschaftsforschung, 77* (3), 110–129).

Goebel, J., Grabka, M.M., Liebig, S. et al. (2019). The German socio-economic panel study (SOEP). *Jahrbücher für Nationalökonomie und Statistik/Journal of Economics and Statistics* 239 (2): 345–360. https://doi.org/10.1515/jbnst-2018-0022.

Goode, A. and Watson, N. (eds.) (2006). HILDA User Manual – Release 4.0, Melbourne Institute of Applied Economic and Social Research, University of Melbourne, viewed January 2007, http://www.melbourneinstitute.com/hilda/doc/doc_hildamanual.htm.

Hill, M. (ed.) (1992). *The Panel Study of Income Dynamics: A User's Guide.* Beverly Hills, CA: Sage Publications.

Kim Y.S., and Stafford F.P. (2000). The Quality of PSID Income Data in the 1990's and Beyond. *PSID Technical Series Paper #00-03*, Survey Research Center, University of Michigan. https://psidonline.isr.umich.edu/Publications/Papers/tsp/2000-03_Quality_of_PSID_Income_Data_1990s_Beyond.pdf.

Kozyreva, P., Kosolapov, M., Popkin, B., and M. (2016). Data Resource Profile: The Russia Longitudinal Monitoring Survey-Higher School of Economics (RLMS-HSE) Phase II: Monitoring the economic and health situation in Russia, 1994–2013. *International Journal of Epidemiology* https://doi.org/10.1093/ije/dyv357.

Lillard, D.R. (2013). Cross-national harmonization of longitudinal data: the example of national household panels. In: *Understanding Research Infrastructures in the Social Sciences* (ed. B. Kleiner, I. Renschler, B. Wernli, et al.), 80–88. Zürich: Seismo.

OECD (2001). *Economic Outlook 2001.* Paris, France: Organization for Economic Cooperation and Development.

Saumik, P. (2020). Definition, measurement, and conceptual issues. In: *Hier fehlt der Herausgeber! Labor Income Share: Understanding the Drivers of the Global Decline*, 7–30. Singapore: Springer Singapore https://doi.org/10.1007/978-981-15-6860-2_2.

Schwarze, J. (1995). Simulating the federal income and social security tax payments of German households using survey data. *Cross-National Studies in Aging Program Project Paper no. 19*, Center for Policy Research, The Maxwell School, Syracuse University.

Voorpostel, M., Tillmann, R., Lebert, F. et al. (2018). *Swiss Household Panel User Guide (1999–2017)*, Wave 19 December 2018. Lausanne: FORS.

Wagner, G.G., Frick, J.R., and Schupp, J. (2007). The German socio-economic panel study (SOEP) – scope, evolution and enhancements. *Schmollers Jahrbuch (Journal of Applied Social Science Studies)* 127 (1): 139–169.

Wilkins, R., Laß, I., Butterworth, P., and Vera-Toscano, E. (2019). *The Household, Income and Labour Dynamics in Australia Survey: Selected Findings from Waves 1 to 17.* Melbourne: Melbourne Institute, Applied Economic & Social Research, The University of Melbourne.

Wolf, C., Schneider, S.L., Behr, D., and Joye, D. (2016). Harmonizing survey questions between cultures and over time. In: *Hier fehlen die Herausgeber! The SAGE Handbook of Survey Methodology*, 502–524. London: SAGE Publications Ltd. https://dx.doi.org/10.4135/9781473957893.

11

Harmonization of Survey Data from UK Longitudinal Studies: CLOSER

Dara O'Neill[1] and Rebecca Hardy[1,2]

[1]*Social Research Institute, University College London, London, UK*
[2]*School of Sport, Exercise and Health Sciences, Loughborough University, Loughborough, UK; Social Research Institute, University College London, London, UK*

11.1 Introduction

Longitudinal studies capture information on the same participants or households at different points of time. As a consequence, they are a key resource for furthering our understanding of how individuals and groups change over the life-course or across generations. The importance of this has been highlighted by the COVID-19 pandemic, with longitudinal studies uniquely equipped to help establish the antecedents and long-term impact of the disease and policy/healthcare response on different age groups and sectors of society (Settersten et al. 2020). There is an extensive history of longitudinal research in the United Kingdom (UK), with estimates suggesting that 1 in 30 UK residents were members of a study cohort in 2014 (Pell et al. 2014), a figure liable to have increased as new studies are commenced.

This growth in the number of UK-based longitudinal studies over the past 75 years and in the longevity of data collection among older cohorts raises both challenges and opportunities. The studies collectively provide us with the ability to expand our research focus across new populations and over longer periods of time, to explore new lines of cross-study and longitudinal comparison. These opportunities, however, concurrently require us to appreciate and address complex issues around measurement heterogeneity, changing survey practices, and the assessment of longitudinally dynamic characteristics.

Surmounting these challenges effectively and efficiently requires increased collaboration and coordination, between studies and more broadly with other stakeholders such as data curators and repositories. A recognition of this informed the foundation in 2012 of CLOSER (Cohort and Longitudinal Studies Enhancement Resources), a consortium of UK-based longitudinal studies and related organizations. CLOSER's broad remit has centered around maximizing the use, value, and impact of longitudinal studies and the data they collect, with a particular emphasis on the importance of data harmonization to

Survey Data Harmonization in the Social Sciences, First Edition.
Edited by Irina Tomescu-Dubrow, Christof Wolf, Kazimierz M. Slomczynski, and J. Craig Jenkins.
© 2024 John Wiley & Sons Inc. Published 2024 by John Wiley & Sons Inc.

achieving these aims (O'Neill et al. 2019). At its foundation, CLOSER brought together the UK Data Service, the British Library, and 8 longitudinal study partners:

- the Hertfordshire Cohort Study (HCS; Syddall et al. 2005);
- the Medical Research Council National Survey of Health and Development (NSHD; Kuh et al. 2011, 2016; Wadsworth et al. 2006);
- the 1958 National Child Development Study (NCDS; Power and Elliott 2006);
- the 1970 British Cohort Study (BCS70; Elliott and Shepherd 2006);
- the Avon Longitudinal Study of Parents and Children (ALSPAC; Boyd et al. 2013; Fraser et al. 2013);
- the Southampton Women's Survey (SWS; Inskip et al. 2006);
- the Millennium Cohort Study (MCS; Connelly and Platt 2014); and
- Understanding Society: the UK Household Longitudinal Study (UKHLS; Buck and McFall 2012).

The CLOSER consortium expanded in September 2020 to include 11 more longitudinal studies, but the harmonization work described in this chapter preceded this expansion and principally involved the original eight partner studies. These eight studies are varied in focus, design, and the period of time and generations covered, resulting in both opportunities and obstacles in making their data more validly comparable and integrable. As a consortium, CLOSER brings together researchers with diverse disciplinary knowledge, spanning both the social and biomedical sciences and with extensive longitudinal research experience. This expertise has helped guide CLOSER in identifying topic areas that could benefit from targeted harmonization efforts.

Accordingly, CLOSER has so far coordinated 16 harmonization-related work packages, each covering a specific research theme and led by researchers with established proficiency in the specific topic area (see Figure 11.1). The aims of these work packages have ranged in scope. Half have comprised efforts to

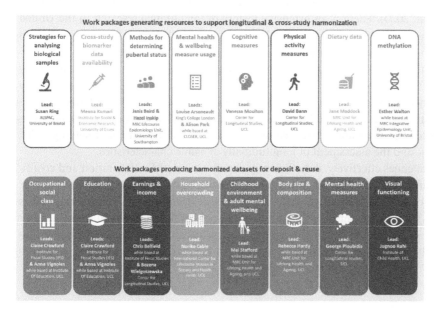

Figure 11.1 CLOSER's harmonization-related work packages.

document data availability and measurement comparability, culminating in the production of new research resources (such as online guides and interactive tools) that are intended to support cross-study comparisons and harmonization. A further eight projects have involved the actual retrospective/ex-post harmonization of study data, resulting in the creation of new harmonized datasets that have been or are in the process of being made available for wider reuse through deposit at the UK Data Service.

Section 11.2 will describe in more detail a selection of the work packages that have produced or are producing harmonized datasets to enable new cross-study research use. In so doing, we will endeavor to set out the diverse considerations and steps involved in the process of harmonizing longitudinal survey data.

11.2 Applied Harmonization Methods

CLOSER's different harmonization projects were intended from the outset to serve a shared purpose of developing knowledge on effective harmonization protocols and techniques, with earlier work packages helping establish and evaluate learning that in turn would inform later projects. As a result, these work packages were also linked to CLOSER's wider training and capacity-building efforts. The insights attained have in particular highlighted the challenges and importance of considering within-study differences in measurement as well as those between studies.

Given the broad range of topic areas addressed by the different CLOSER work packages, some adaptation and diversity in harmonization approach were required. The source data and their original measurement methods often differed notably across topics, ranging from study-specific questionnaire responses to more standardized clinical assessments. Some topics as a result posed greater challenges than others, with varying levels of data complexity and inherent subjectivity; yet as there were common steps across the work packages, this has enabled us to build a broad framework for the retrospective harmonization of longitudinal data.

We will begin this section with descriptions of exemplar work packages to illustrate techniques for harmonizing data from specific research areas, before identifying cross-theme commonalities and presenting a generalized protocol that summarizes CLOSER's retrospective harmonization process. The exemplar work packages selected have each focused on the original CLOSER partner studies in which a birth cohort approach was utilized, as these studies' similarity in design and research focus best facilitates joint analysis and data pooling (Davis-Kean et al. 2018).

11.2.1 Occupational Social Class

Concerns about social inequalities informed the decision to first establish the UK's longest running birth cohort, the NSHD (Wadsworth 2010), and this study, alongside the other subsequently established UK birth cohorts, has since become an important source of evidence on the determinants and consequences of such inequities (Kuh 2016; Pearson 2016). Occupational social class is a common indicator of socioeconomic position in UK longitudinal studies and is based on classification systems that estimate class using occupational details such as the type of job performed and the level of skill required for its undertaking (Elias et al. 1999). However, changing economic and employment circumstances over the seven decades currently covered by the UK birth cohort studies have resulted in modifications to official methods of occupational classification, hampering longitudinal comparisons. Additionally, the methods for capturing and encoding occupational detail during data collection (e.g. ascertaining the extent of supervisory responsibility in a role and the use of software to interpret free-text fields) have similarly varied over time (Dodgeon et al. 2019).

A CLOSER work package led by Claire Crawford (based at the Institute of Fiscal Studies) and Anna Vignoles (while based at the UCL Institute of Education), with additional support from Brian Dodgeon (Center for Longitudinal Studies at University College London (UCL)), sought to address these sources of heterogeneity and retrospectively harmonize occupational social class data collected by birth cohort studies in the CLOSER consortium (namely NSHD, NCDS, BCS70, ALSPAC, and MCS). Given the longitudinal nature of these studies, efforts were made to harmonize within studies to enable comparisons of childhood and adulthood measures, as well as across studies to enable cross-cohort research. Parental occupation data collected when the study members were aged 10–11 years were used as the childhood measure and the equivalent adulthood measures were then established using study members' own reported occupations in midlife (specifically age 41–43 years). Using existing lookup tables,[1] the occupational classifications were mapped where possible to the classification scheme introduced in 1990 (British Occupational Based Social Class, formerly known as Registrar General's Social Class; Galobardes et al. 2006), with this version adopted as the harmonization target due to its location at an approximate midpoint in the period spanned by the assessments from the different cohorts being harmonized. Challenges around the comparison of measures collected at different time points in a changing social context must necessarily inform the use and interpretation of such harmonized values, a point that will be addressed later in this chapter.

A number of additional practical challenges were encountered in the course of this work. No lookup table was identified that would enable mapping of the NSHD childhood values to the 1990 classification, so an earlier 1970 classification was used to derive a comparable but not fully equivalent version of social class for this variable. For NCDS and BCS70 childhood measures, some data substitutions were required for a subset of participants where earlier digitization of microfiche records had misread values and an alternative secondary source was available to now impute these values. Flag variables (i.e. supplementary categorical descriptor variables) were created to denote these in direct juxtaposition to the derived data and to inform subsequent usage and interpretation of these variables by ensuring research users understand the origins of the data and are equipped to exclude observations according to their preference or undertake sensitivity analyses where appropriate. As the MCS cohort had not yet reached adulthood, only childhood social class was harmonized for participants from this study. Additionally, this study captured occupational data on both parents in a manner that reported either or both parents' occupations, so the decision was made to harmonize whichever values were available and use the most advantaged value reported as the marker of childhood social class. Detail on father's employment among the ALSPAC cohort did not include sufficient information to derive social class (e.g. supervisory responsibility), so this study was omitted from the final set of harmonized data.

This final set of harmonized datasets (one per study), covering NSHD, NCDS, BCS70, and MCS, comprise either pseudonymized or original participant identifiers (depending on licensing restrictions of the

---

1 The lookup tables comprised matrices of cross-mapped classifications established by Paul Lambert and Ken Prandy (Cambridge Social Interaction and Stratification Scales) and David Rose and David Pevalin (Institute for Social and Economic Research).

original study data), childhood and adulthood measures, and flag variables. These datasets were deposited at the UK Data Service alongside user guide documentation that incorporates the original harmonization syntax. The impact of these datasets on longitudinal and cross-study research has already been demonstrated by work that has employed them in establishing how a more advantaged social class during childhood is associated in later life with better mental well-being (Wood et al. 2017) and lower body mass index (Bann et al. 2017). However, these other characteristics of study participants have often also necessitated harmonization to enable such comparative research, and therefore, have been the target of other CLOSER harmonization projects as discussed below.

11.2.2 Body Size/Anthropometric Data

Height and weight assessments have been a consistent feature of the UK birth cohort studies (Pell 2014) and as a result, these studies are important resources for furthering understanding of the rapid increase in obesity prevalence in recent decades (Power et al. 2013). However, in spite of the consistency of anthropometric data collection, divergences have arisen over time and between studies in their fieldwork and data handling methods. For example, changes have occurred between data collection sweeps/waves in the measurement scales utilized or in the reliance on self-reported versus clinical measurements. Differences have also arisen between studies in how some data cleaning steps have been implemented prior to the dissemination of data for research use. Consequently, the valid use of the data captured by these studies for longitudinal and cross-study evaluation first requires the retrospective harmonization of measurements collected using different study protocols (e.g. the use of different data collection modes) or recorded on different measurement scales.

A CLOSER work package led by Rebecca Hardy with support from Will Johnson (while both based at the MRC Unit for Lifelong Health and Aging at UCL) was undertaken to address this need. Following the identification of both clinically measured and self-reported variables from the target birth cohort studies (NSHD, NCDS, BCS70, ALSPAC, and MCS), a process of harmonizing the data definitions and validating the data values was begun (Hardy et al. 2019). Measurements on an imperial scale (e.g. stones and pounds) were transformed to a metric scale (e.g. kilograms). Missing clinical measurements were imputed with self-reported data from the same wave of data collection where such secondary sources were available. Similar imputation was performed for a small number of non-missing clinical measurements that had been previously winsorized during data cleaning (i.e. subject to cleaning to account for extreme outlier values), again where such alternative sources of measurement were available. For some data collection waves or sweeps during participants' midlife, height was only assessed for participants whose height had not been recorded at the prior sweep, and so the most recent available measurement was used instead to improve data completeness across all ages. Data were recoded as missing where they were deemed longitudinally unreliable, e.g. during sweeps where a female participant was pregnant, or where the values were biologically implausible cross-sectionally (e.g. height greater than 3 m) or relative to neighboring waves to support the longitudinal research usage of these data. The overarching aim across these steps was to ensure that the data across all waves and studies were handled in sufficiently equivalent ways where otherwise the data cleaning process would have varied. As with the occupational social class datasets, supplementary flag variables were appended to the body size datasets to denote pertinent characteristics of the source data and harmonization decisions (e.g. the precision of the original measurements and indicators of where data substitutions or transformation were performed) to inform subsequent analysis and interpretation of the harmonized data.

Like the social class data, the harmonized body size data have been deposited at the UK Data Service, alongside a user guide and the analytic scripts used to complete the harmonization steps outlined above. The utility of these data for both longitudinal and cross-study comparative research has been demonstrated by a number of publications to date (e.g. Bann et al. 2017; Bann et al. 2018; Johnson et al. 2020) including work in which the harmonized data were used to establish that younger generations in the UK are becoming overweight at younger ages (Johnson et al. 2015; see Figure 11.2). These findings have directly informed the UK government's childhood obesity strategy (Her Majesty's Government 2016), underlining both the policy and research impact of this harmonization work.

11.2.3 Mental Health

The UK cohorts are also a rich source of data on mental health over the life-course. However, variation in the assessment instruments used within and across these studies has informed calls for targeted harmonization efforts to fully realize their research potential (Department of Health 2017). A further CLOSER work package, led by George Ploubidis with support from Eoin McElroy (while both based at the Center for Longitudinal Studies at UCL), was undertaken to empirically evaluate the extent and consequence of measurement heterogeneity in this area. This project was conducted concurrent to another CLOSER work package, led by Louise Arsenault (King's College London) and Alison Park (then based at CLOSER), which completed initial scoping of mental health and well-being measurements across UK-based longitudinal studies.

Following the identification of relevant mental health measures in the CLOSER cohort studies, these measures were individually subject to psychometric validation including an examination of their underlying (latent) factor structures and measurement precision. Where the same assessment instrument had been implemented within or across the studies, these were then compared for measurement equivalence. This helped ascertain the extent to which these repeated applications of the same measure captured the same latent construct across different populations or time periods. Where different assessment instruments had been used to measure overtly similar aspects of mental health, supplementary work was undertaken to evaluate their comparability (i.e. to investigate the equivalence of data collected on

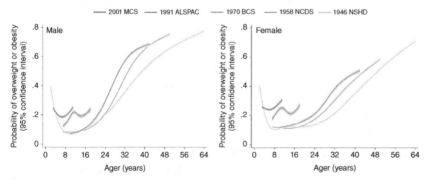

Figure 11.2 Illustration of cross-generational differences in the prevalence of overweight/obesity over the life-course, based on CLOSER harmonized data. Source: Reproduced from Johnson et al. (2015), https://doi.org/10.1371/journal.pmed.1001828.g003.

the same underlying symptoms but using different instruments). This involved a process of identifying potentially comparable measures by having two raters with relevant disciplinary expertise independently review and define the symptoms assessed by measurement items, with differences of opinion resolved by a third rater. For items deemed putatively comparable but where a different scale had been used, data recoding and transformations were then undertaken. Finally, the latent characteristics of any paired measures were examined to ascertain the measurement equivalence of these items (using multiple group confirmatory factor analysis) and to ensure there was a sufficient and quantified degree of similarity in the measures to support their combined usage in research. For childhood measures of mental health, this also took into consideration differences in respondents, which can pose challenges to the comparability of measurements (e.g. differences between parent and teacher reporting).

A detailed account of the above work has been published as an open-access CLOSER resource report (McElroy et al. 2020). At the time of writing, the harmonized data generated through this project are also in preparation for deposit at the UK Data Service for secondary research purposes.

11.2.4 Harmonization Methods: Divergence and Convergence

The harmonization approaches implemented by these exemplar work packages differed in some pertinent respects. This was to a large extent driven by differences in the source data they utilized, with these specific exemplars selected as they each targeted either questionnaire data, psychometric scales, or clinical assessments. The diverse modes of assessment used to originally collect the source data resulted in heterogenous sources of measurement imprecision and barriers to comparability that required consideration in how the harmonization was planned and how the resultant data outputs should be utilized. The work packages also differed in the number of assessment waves (study sweeps) included from each contributing study. The social class harmonization project, for example, targeted one period in both childhood and adulthood where the relevant measurements were captured at the same age across the studies being harmonized. In contrast, the body size harmonization work utilized all available data at every age regardless of whether there was cross-study alignment in ages. This variation was in part informed by the initial intended research use of the harmonized datasets, with the body size work seeking to establish datasets that could be used to model longitudinal trajectories of physical development. While each of the projects sought to harmonize across the CLOSER birth cohorts, variance in the availability of suitable data meant that some of the projects were able to include more studies than others in the end product datasets.

There were, however, also many similarities between the work packages, and indeed a common set of considerations and steps have underpinned these and the other harmonization projects led by CLOSER (as summarized in Figure 11.3). Each of CLOSER's harmonization projects commenced with work to explore study documentation and identify available data on the topics of interest, with searches typically focused on studies in the CLOSER consortium of similar design (e.g. birth cohorts). Following data access (which differed in complexity according to licensing restrictions), evaluations were completed of the data characteristics and content to establish the longitudinal and cross-study comparability of the study variables identified. All of the work packages utilized a form of data mapping or transformation to ensure the data were maximally standardized (either focused on observed or latent values) against the same target variable definition and measurement scale. These efforts were not always successful, and accordingly it was important to document where fully equivalent variables were not achievable, through

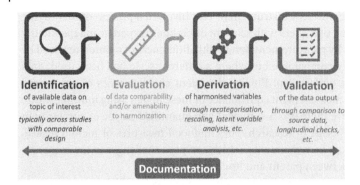

Figure 11.3 Generalized process of retrospective harmonization among the CLOSER work packages.

flag variables and dataset user guides. Output validation was performed to ensure that harmonized data were defined and created as intended and that they were of research utility, for example, through comparison of the derived measures with the source data, including through an examination of latent measurement properties, as well as verification of adherence to standardized metadata definitions across the harmonized outputs. All work packages had a longitudinal focus, which meant considering (i) changes in measurement practice and (ii) the meaningfulness of comparing assessments from different time points or contexts. These have implications for the valid research usage of the harmonized datasets, and so each of the work packages has identified and offered guidance on these considerations in the user guide documentation made available alongside the data. Indeed, a persistent thread throughout and across each harmonization project has been the importance of data documentation, in terms of both the source data and the harmonized outputs. This will be explored more in the Section 11.3.

The datasets are published separately for each work package by CLOSER at the UK Data Service. However, pseudonymized identifiers are included in all data files allowing cross-linkage of participant-level data between the different work package outputs, and facilitating linkage back to the source study data. While the harmonized data deposits are focused on specific topic areas, the inclusion of these identifiers equips researchers to jointly leverage the harmonized measures in conjunction with the wealth of additional data available from the CLOSER partner studies.

11.3 Documentation and Quality Assurance

Any effective retrospective harmonization endeavor is built on a full and valid understanding of the source data being utilized. Without knowledge of how the data were originally collected and processed, the risk of misinterpretation and mishandling during harmonization grows. Helpfully, the longitudinal studies in CLOSER have each developed effective strategies for capturing and sharing detail on their survey protocols and study data. Nonetheless, as some of the studies have been run over many decades, older documentation can be less accessible or offer differing levels of detail compared to that available for newer waves. Moreover, such differences in documentation are also commonly observed between studies.

The CLOSER harmonization work packages each involved a process of engagement with partner studies to draw on the expertise within the individual longitudinal studies included in the harmonization exercise. This helped address gaps in existing study documentation and provided vital insight to some of the idiosyncrasies of study coverage, assessment decisions, and data format/structures. Concurrently, CLOSER has been seeking to support and expedite such initial data scoping and evaluation through the creation of new resources that improve the detail and utility of documentation about the source study data, while also standardizing the content to make it more accessible and efficient to use. As mentioned above, a number of the CLOSER work packages are documenting data availability and comparability, and making this learning available through resource reports and online guides. These cover topics such as cognition, physical activity, and dietary measures, and are made freely available on the CLOSER website: www.closer.ac.uk/resources.

More broadly, CLOSER is also seeking to facilitate cross-study data discoverability and evaluation through the collation, standardization, and enrichment of metadata from across its partner studies. Metadata, often described as "data about data" (Gregory et al. 2010; Nadkarni 2011), documents the meaning of data (e.g. in the form of variable names, labels, and value categories/factor codes). The assessment instruments used to capture data from study participants are a further source of metadata (e.g. the specific survey questions asked), but this is typically stored separately to variable metadata, hindering their joint utility in clarifying the meaning of research data. CLOSER has established a process of automating the extraction and integration of these different sources of metadata from partner studies to document study data and its provenance according to an internationally agreed standard: the Data Documentation Initiative-Lifecycle (DDI-L; Vardigan et al. 2008). This information is made available as interactive documentation with advanced search functionality via an online platform, CLOSER Discovery: https://discovery.closer.ac.uk.

CLOSER Discovery and CLOSER's other resource guides serve the shared purpose of ensuring harmonization efforts are driven by an adequate and accurate understanding of the source data. The documentation of the harmonization process itself is of course equally vital in demonstrating the meaning and validity of the harmonized data outputs and ensuring the long-term utility of these data to new research needs. Through roundtable discussions and with expertise on existing dataset and harmonization documentation, CLOSER established a dataset user guide template to capture information consistently across the work packages and ensure that the decisions made and learning established is adequately recorded. This template comprises sections on the context of the harmonization work, an overview of the harmonization strategy, and detailed content on the source variables and specific derivation steps undertaken for each harmonized variable. To enhance that standardization further, and ensure the harmonized variables are documented in an equivalent way to the original study data, metadata regarding the harmonized variables are also being added to CLOSER Discovery. This enables the linkage of their metadata to that of the source variables from which these harmonized versions were derived, as well as to standardized documentation of the original assessment instruments. By capturing this provenance detail, CLOSER Discovery can support better understanding of the origins of harmonized data.

Quality assurance with the harmonized datasets was undertaken by the individual work package teams in conjunction with the CLOSER harmonization lead, as well as through collaboration with the original studies and external researchers. Code reviews (i.e. of the harmonization syntax commands) and data comparisons (between the source and derived datasets) were performed to ensure no unintended changes had arisen during the harmonization process. For some of the topics being harmonized, the longitudinal nature of the data offered a further basis for validation. For example, biologically implausible changes in body size across neighboring sweeps was used to screen for invalid measurements in the anthropometric harmonization work.

It was also important to consider changes in measurement practice across sweeps and wider contextual changes over time, as these can have a bearing on the valid comparability of data (e.g. changes in the colloquial usage of certain words could hinder the long-term comparability of some questionnaire items). Measurement equivalence testing, as used by the CLOSER mental health harmonization project, can help identify and screen for such effects. This approach was also used to pick up on the impact of cross-study differences in the recall period asked about by otherwise similar survey questions (e.g. "in the past week" versus "in the past month" . . .) or differences in the choice of respondents, further potential sources of measurement heterogeneity. During childhood, for example, different studies may rely more on parents/caregivers or teachers for assessing similar characteristics of study participants. By identifying the impact of such differences on data comparability, it was possible to ascertain the adequacy of the harmonization effort.

Given the datasets produced through CLOSER's harmonization work are being made available for secondary usage, it is important for new users to reflect on the validity of these data for their own research purposes. In this sense, validation is an on-going process and that is why detailed but accessible documentation is a key component of CLOSER's harmonization work.

11.4 Challenges to Harmonization

By its nature, retrospective harmonization is dependent on the availability of existing data that are amenable to comparison and integration with data from other sources. The process involved in successfully fulfilling the aims of retrospective harmonization, however, is both challenging and labor-intensive. Many of the potential hurdles associated with this process are heightened when dealing with longitudinal data, particularly as in the case of CLOSER's harmonization work when these data have been collected across broad time periods. Two of the cohort studies in CLOSER, NSHD, and MCS, for example, commenced 54 years apart. As previously identified in the applied methods' section above, contextual and societal changes can have sizable influence on the inherent comparability of data collected at different time points. An example of this is the widespread changes in employment types and conditions observed between the 1950s and 2000s in the CLOSER social class harmonization project, but also as seen in other CLOSER work packages on childhood circumstances and household overcrowding. Just because the categories used in variables can be recoded to be more overtly similar does not mean the data they represent can be interpreted in identical ways.

Survey assessment practices can similarly change over time, even within studies, as new technologies become available or scientific understanding undergoes development and revision. Efforts to minimize response burden and ensure continued engagement across study waves can result in longitudinal changes in survey protocols and content (Dillman 2009), e.g. in response to the inclusion of new assessment items of regional or national policy relevance. However, even where the same instruments and modes have been used across data collection sweeps, measurement equivalence cannot be assumed (Wang et al. 2017). This was a particularly relevant consideration in CLOSER's evaluation of data comparability in individual characteristics that underwent rapid development or change, for example, cognitive ability during childhood. The age sensitivity of measurements was also pertinent to cross-study harmonization, where the age at which participants were assessed differed between studies.

As a broad consortium, the studies within CLOSER have diverse designs and varied scientific aims, and these dissimilarities influenced their amenability to inclusion in CLOSER's targeted harmonization efforts. While many of the studies are based on a birth cohort design, prospectively assessing the same group of individuals across their life-course, other studies in CLOSER began with historical records (i.e. HCS) or utilize a household panel design (i.e. UKHLS). Furthermore, there has been some variation over the years in the disciplinary orientation of the CLOSER partner studies, with some supported by social science funding, and others biomedical. These disciplinary slants have shifted and intersected over time, but dissimilarities exist in the longer-term coverage and depth of detail on certain topics (e.g. biomarker and deoxyribonucleic acid (DNA) methylation data, two further topics addressed by CLOSER's harmonization-related work packages). Consequently, there were pertinent differences in design, sampling, and content that limited some studies' suitability to inclusion in the harmonization projects CLOSER has undertaken to date.

A related obstacle that can influence the harmonization process and its expediency is the fact that different studies' data may be made available under different access licenses. This can require adaptation in how harmonization steps are performed, e.g. the need to run analytic queries remotely for some studies and locally for others. Such licensing variations do also have implications for the ease with which harmonized data can subsequently be made available for secondary usage. In the context of the CLOSER harmonization work packages, these licensing considerations have meant that the data from each study are deposited at the UKDS separately to enable individual access requests for harmonized data according to the specific licenses of the different studies.

Retrospective harmonization typically involves a process of modifying data from different sources toward a point of convergence, a least common denominator. This transformation can involve a loss of information or precision. Establishing the extent and acceptability of that loss is crucial to the valid understanding and usage of harmonized data. This is why documentation of harmonization decisions in a way that facilitates both evaluation and amendment is so important to ensuring harmonized data are suited to addressing specific research needs. CLOSER has sought to maximize the accessibility and usability of that documentation, but ensuring sufficient recognition and support for the standardization of documentation remains a challenge given resource constraints and limited incentivization. However, CLOSER is collaborating with its study partners and the wider longitudinal research community to address these challenges.

11.5 Software Tools

Specialist tools were not typically used in the data processing stages of CLOSER's harmonization work, other than standard statistical software (such as Stata or Mplus). However, tools addressing data discovery and documentation were developed alongside or as part of these work packages. These have broad research utility in that they can assist with the usage of CLOSER's harmonized datasets as well as with planning new harmonization projects involving the CLOSER partner studies.

As mentioned, these tools include a cross-study repository of study metadata, CLOSER Discovery (https://discovery.closer.ac.uk). Metadata from CLOSER's harmonized variables are being incorporated into this resource to enhance their discoverability and provenance detail. As CLOSER Discovery adheres to the widely used DDI-L technical standard, its documentation of harmonized variables is being done in a manner that is machine-readable and interoperable with other cataloging platforms and software. This includes compatibility with other established and emerging standards, such as those for the translation and documentation of data processing code from different statistical software packages as developed by the C^2Metadata project (Song et al. 2019). CLOSER Discovery has additional new functionality for capturing and visualizing potentially comparable data collected longitudinally or across different studies that may not have yet been the subject of harmonization work. This new functionality could thus be utilized to explore and scope out new retrospective harmonization projects, as well as providing guidance to survey managers seeking to ensure new waves of data collection are harmonized prospectively with existing measures.

Another resource developed through CLOSER support is a searchable directory of mental health and well-being assessments used by UK longitudinal and repeated cross-sectional studies: the Catalog of Mental Health Measures www.cataloguementalhealth.ac.uk. This was part of a CLOSER work package (led by Louise Arsenault and Alison Park) that sought to improve the research use of the mental health data collected by UK studies. Detail on the harmonized mental health datasets being produced by CLOSER is now available on this site.

A further CLOSER work package (led by Esther Walton while based at the University of Bristol) developed a visualization tool for DNA methylation data collected by longitudinal studies, within and beyond the CLOSER consortium. This tool, available at www.ariesepigenomics.org.uk, seeks to characterize the human methylome across the life-course by bringing together data from multiple studies that collected data at different time points. The tool is built in R-Shiny and is linked to a related resource for researchers seeking to understand and undertake cross-study epigenetic research: Accessible Resources for Integrated Epigenomics Studies (ARIES; Relton et al. 2015).

11.6 Recommendations

CLOSER's harmonization work to date has been diverse in scope and methodology, yet these heterogeneous experiences also offer common points of learning that can inform recommendations for future longitudinal and cross-study harmonization efforts.

As discussed in this chapter, the effectiveness of retrospective harmonization is contingent on the availability of accurate and accessible documentation to assist data discovery and exploration. This can be a potentially challenging aspiration in the context of longitudinal studies that may have collected data across multiple decades, with inevitable changes in data management practices. Survey teams should be assisted in capturing and standardizing detail from older study sweeps and in adopting more effective documentation practices moving forward. These needs could be fulfilled through the wider provision of training on the efficient creation of standardized and interoperable data documentation in future survey sweeps, as well as through the development and promotion of relevant tools that support such efforts with existing data repositories and historical study documentation.

These documentation considerations are equally important during the harmonization work itself, even though this may occur later in the data lifecycle or for bespoke research purposes. By ensuring that harmonized datasets are well described in a manner that is compatible with the original study data, harmonization teams can improve the verifiability of their work and make it more easily compatible with other data from the source studies. Training and support for documentation best practice are therefore just as important for secondary data users planning ex-post data harmonization. Such harmonization teams should also be encouraged and assisted in publishing the harmonization syntax/code (or a variant thereof if disclosive material needs prior removal) to make sure all processing steps are captured in maximal detail. This openness helps toward ensuring the harmonization process is more fully reproducible, is amenable to evaluation and ultimately is adaptable to new research needs or new datasets. Admittedly, the utility of such code releases is currently dependent on the statistical software expertise and access of secondary users, and it is known that software usage can vary between and within disciplines (Pan et al. 2016), as well as cross-nationally (Dembe et al. 2011). Emerging technologies for the automated translation of data processing code (such as the DDI-L compatible tools from the aforementioned C^2Metadata project) can help further extend the utility of syntax sharing in ensuring data harmonization efforts better adhere to the concepts of Findability, Accessibility, Interoperability, and Reusability (FAIR; Wilkinson et al. 2016) that underpin effective science.

The documentation of retrospective harmonization work, particularly with reference to longitudinal studies, is also pertinent to planning prospectively (ex-ante) harmonized data collection. By establishing the points of commonality across earlier sweeps of the same study or across different studies, retrospective harmonization efforts can help those designing subsequent/future surveys access guidance on how to collect new measurements that are more directly compatible with existing study data. Prospective harmonization in itself is not without limitations. The inclusion of such broadly standardized instruments could reduce the use of more population-specific or novel measurement approaches, given the need to minimize assessment burden on participants. The move toward prospectively harmonized measures could similarly impinge upon the continued use of older measures, which has implications for backward compatibility of assessments within studies. An alternative strategy is to support the continued adoption and use of diverse assessment methods but ensure such approaches are amenable to calibration. CLOSER has led work in this area, by seeking to establish calibration metrics for multiple measures of blood pressure, lung function, and grip strength (Lessof et al. 2016). Additional support for and coordination of such calibration work going forward is required to identify and target areas potentially amenable to calibration. This is important to ensure individual studies are able to retain flexibility over their survey methodologies while also ensuring we better facilitate cross-study comparisons.

CLOSER has adopted a collaborative approach to its harmonization work, implementing a coordinated rather than solely centralized strategy in fulfilling its cross-topic harmonization aims. While this approach necessitates effort to foster and achieve cross-work package consistency where possible and as appropriate, it offers many benefits to such broad-based harmonization endeavors. It helps leverage relevant expertise and diversity of perspective (which is particularly pertinent given CLOSER's harmonization efforts span not only social science data but also biomedical), while also expanding available capacity to help expedite what is inherently a time-consuming process. A collaborative environment is also important to knowledge sharing and the collective creation and revision of best practice concepts.

Harmonization ultimately is about facilitating new avenues of comparative research and enabling the pooling of data to potentially estimate effects or associations with greater precision in order to enhance scientific insights and strengthen their impact. Within this simple summation, however, lies a profusion of assessment possibilities and methodologies. This chapter has provided an overview of the experiences and learning from CLOSER's efforts to retrospectively harmonize data from within and across different longitudinal studies. This chapter reflects CLOSER's wider aims to capture and share learning about our harmonization work, and through our diverse outputs provide guidance and support on similar endeavors. Harmonization is most accurately viewed as a dynamic process, with no single ideal strategy or solution. Fundamentally, effective practice in this area can be characterized by the open sharing of experiences, challenges encountered, and solutions identified, and a recognition that harmonized data should always be produced and utilized according to their appropriateness and suitability to addressing a specific research need.

For further information about CLOSER's harmonization work and additional guidance on accessing the resources discussed in this chapter, check out our website (www.closer.ac.uk), follow us on Twitter (@CLOSER_UK), or get in touch with us via email (closer@ucl.ac.uk).

Acknowledgments

The authors wish to acknowledge the key role played in this work by the harmonization work package leads and their research teams, as well as the important contributions of the wider study teams in the CLOSER consortium in sharing their valuable expertise and facilitating these harmonization efforts. Sincere thanks are also offered to the participants in those studies. CLOSER was funded by the Economic and Social Research Council (ESRC) and the Medical Research Council (MRC) between 2012 and 2017. Its initial five-year grant has since been extended to September 2024 by the ESRC (grant reference: ES/K000357/1). The ESRC took no role in the design, execution, analysis, or interpretation of the data or in the writing up of the findings described in this chapter.

References

Bann, D., Johnson, W., Li, L. et al. (2017). Socioeconomic inequalities in body mass index across adulthood: coordinated analyses of individual participant data from three British birth cohort studies initiated in 1946, 1958 and 1970. *PLOS Medicine* 14 (1): e1002214. https://doi.org/10.1371/journal.pmed.1002214.

Bann, D., Johnson, W., Li, L. et al. (2018). Socioeconomic inequalities in childhood and adolescent body-mass index, weight, and height from 1953 to 2015: an analysis of four longitudinal, observational, British birth cohort studies. *The Lancet: Public Health* 3 (4): e194–e203. https://doi.org/10.1016/S2468-2667(18)30045-8.

Boyd, A., Golding, J., Macleod, J. et al. (2013). Cohort profile: the 'children of the 90s' – the index offspring of the Avon longitudinal study of parents and children. *International Journal of Epidemiology* 42 (1): 111–127. https://doi.org/10.1093/ije/dys064.

Buck, N. and McFall, S. (2012). Understanding society: design overview. *Longitudinal and Life Course Studies* 3 (1): 5–17. http://www.llcsjournal.org/index.php/llcs/article/view/159/168.

Connelly, R. and Platt, L. (2014). Cohort profile: UK millennium cohort study (MCS). *International Journal of Epidemiology* 43 (6): 1719–1725. https://doi.org/10.1093/ije/dyu001.

Davis-Kean, P., Chambers, R.L., Davidson, L.L., et al. (2018). Longitudinal Studies Strategic Review: 2017 Report to the Economic and Social Research Council. London.

Dembe, A.E., Partridge, J.S., and Geist, L.C. (2011). Statistical software applications used in health services research: analysis of published studies in the U.S. *BMC Health Services Research* 11 (1): 252. https://doi.org/10.1186/1472-6963-11-252.

Department of Health (2017). *A framework for mental health research*. London, UK. https://www.gov.uk/government/publications/a-framework-for-mental-health-research.

Dillman, D.A. (2009). Some consequences of survey mode changes in longitudinal surveys. In: *Methodology of Longitudinal Surveys* (ed. P. Lynn), 127–140. Chichester, UK: Wiley https://doi.org/10.1002/9780470743874.ch8.

Dodgeon, B., Morris, T., Crawford, C. et al. (2019). *CLOSER Work Package 2: Harmonised Socio-Economic Measures User Guide (Revised)*. London, UK: CLOSER.

Elias, P., McKnight, A., and Kinshott, G. (1999). *SOC 2000: Redefining Skill – Revision of the Standard Occupational Classification (Skills Task Force Research Paper 19)*. London, UK: Department for Education and Employment http://dera.ioe.ac.uk/id/eprint/15135.

Elliott, J. and Shepherd, P. (2006). Cohort profile: 1970 British birth cohort (BCS70). *International Journal of Epidemiology* 35 (4): 836–843. https://doi.org/10.1093/ije/dyl174.

Fraser, A., Macdonald-Wallis, C., Tilling, K. et al. (2013). Cohort profile: the Avon longitudinal study of parents and children: ALSPAC mothers cohort. *International Journal of Epidemiology* 42 (1): 97–110. Retrieved from https://doi.org/10.1093/ije/dys066.

Galobardes, B., Shaw, M., Lawlor, D.A. et al. (2006). Indicators of socioeconomic position (part 2). *Journal of Epidemiology and Community Health* 60 (2): 95–101. https://doi.org/10.1136/jech.2004.028092.

Gregory, A., Heus, P., and Ryssevik, J. (2010). Metadata. In: *Building on Progress: Expanding the Research Infrastructure for the Social, Economic, and Behavioral Sciences*, vol. 1 (ed. German Data Forum (RatSWD)), 487–508. Opladen & Farmington Hills, MI: Budrich UniPress Ltd.

Hardy, R., Johnson, J., Park, A., and O'Neill, D. (2019). *CLOSER Work Package 1: Harmonised Height, Weight and BMI User Guide (Revised)*. London: CLOSER.

Her Majesty's Government. (2016). Childhood Obesity: A Plan for Action. London, UK. Retrieved from https://assets.publishing.service.gov.uk/government/uploads/system/uploads/attachment_data/file/546588/Childhood_obesity_2016__2__acc.pdf

Inskip, H.M., Godfrey, K.M., Robinson, S.M. et al. (2006). Cohort profile: the Southampton Women's survey. *International Journal of Epidemiology* 35 (1): 42–48. Retrieved from https://doi.org/10.1093/ije/dyi202.

Johnson, W., Li, L., Kuh, D., and Hardy, R. (2015). How has the age-related process of overweight or obesity development changed over time? Co-ordinated analyses of individual participant data from five United Kingdom birth cohorts. *PLOS Medicine* 12 (5): e1001828. https://doi.org/10.1371/journal.pmed.1001828.

Johnson, W., Norris, T., Bann, D. et al. (2020). Differences in the relationship of weight to height, and thus the meaning of BMI, according to age, sex, and birth year cohort. *Annals of Human Biology* 47 (2): 199–207. https://doi.org/10.1080/03014460.2020.1737731.

Kuh, D. (2016). From paediatrics to geriatrics: a life course perspective on the MRC National Survey of health and development. *European Journal of Epidemiology* 31 (11): 1069–1079. https://doi.org/10.1007/s10654-016-0214-y.

Kuh, D., Pierce, M., Adams, J. et al. (2011). Cohort profile: updating the cohort profile for the MRC National Survey of Health and Development: a new clinic-based data collection for ageing research. *International Journal of Epidemiology* 40 (1): e1–e9. https://doi.org/10.1093/ije/dyq231.

Kuh, D., Wong, A., Shah, I. et al. (2016). The MRC National Survey of Health and Development reaches age 70: maintaining participation at older ages in a birth cohort study. *European Journal of Epidemiology* 31 (11): 1135–1147. https://doi.org/10.1007/s10654-016-0217-8.

Lessof, C., Wong, A., Hardy, R., & CLOSER Equipment Comparison Team. (2016). Early findings from a randomised repeated-measurements cross-over trial to understand differences in measures of physiological function and physical performance. In *5th Panel Survey Methods Workshop*. Berlin. Germany: German Institute for Economic Research. Retrieved from https://www.diw.de/documents/dokumentenarchiv/17/diw_01.c.535244.de/psmw2016_lessof et al_2016_early findings.pdf

McElroy, E., Villadsen, A., Patalay, P. et al. (2020). *Harmonisation and Measurement Properties of Mental Health Measures in Six British Cohorts*. London, UK: CLOSER Retrieved from www.closer.ac.uk/wp-content/uploads/210715-Harmonisation-measurement-properties-mental-health-measures-british-cohorts.pdf.

Nadkarni, P.M. (2011). *Metadata-Driven Software Systems in Biomedicine: Designing Systems that Can Adapt to Changing Knowledge*. London: Springer London https://doi.org/10.1007/978-0-85729-510-1_1.

O'Neill, D., Benzeval, M., Boyd, A. et al. (2019). Data resource profile: cohort and longitudinal studies enhancement resources (CLOSER). *International Journal of Epidemiology* 48 (3): 675–676i. https://doi.org/10.1093/ije/dyz004.

Pan, X., Yan, E., and Hua, W. (2016). Disciplinary differences of software use and impact in scientific literature. *Scientometrics* 109 (3): 1593–1610. https://doi.org/10.1007/s11192-016-2138-4.

Pearson, H. (2016). *The Life Project*. London: Penguin.

Pell, J.P. (2014). *Maximising the Value of UK Population Cohorts: MRC Strategic Review of the Largest UK Population Cohort Studies*. Swindon, UK: Medical Research Council Retrieved from www.mrc.ac.uk/publications/browse/maximising-the-value-of-uk-population-cohorts.

Pell, J.P., Valentine, J., and Inskip, H. (2014). One in 30 people in the UK take part in cohort studies. *Lancet* 383 (9922): 1015–1016. https://doi.org/10.1016/S0140-6736(14)60412-8.

Power, C. and Elliott, J. (2006). Cohort profile: 1958 British birth cohort (National Child Development Study). *International Journal of Epidemiology* 35 (1): 34–41. https://doi.org/10.1093/ije/dyi183.

Power, C., Kuh, D., and Morton, S. (2013). From developmental origins of adult disease to life course research on adult disease and aging: insights from birth cohort studies. *Annual Review of Public Health* 34 (1): 7–28. https://doi.org/10.1146/annurev-publhealth-031912-114423.

Relton, C.L., Gaunt, T., McArdle, W. et al. (2015). Data resource profile: accessible resource for integrated epigenomic studies (ARIES). *International Journal of Epidemiology* 44 (4): 1181–1190. https://doi.org/10.1093/ije/dyv072.

Settersten, R.A. Jr., Bernardi, L., Härkönen, J. et al. (2020). Understanding the effects of Covid-19 through a life course lens. *Current Perspectives on Aging and the Life Cycle* 45: 100360. https://doi.org/10.1016/j.alcr.2020.100360.

Song, J., Alter, G., & Jagadish, H.V. (2019). C2Metadata: Automating the Capture of Data Transformations from Statistical Scripts in Data Documentation. In *Proceedings of the 2019 International Conference on Management of Data* (pp. 2005–2008). New York, NY, USA: Association for Computing Machinery. https://doi.org/10.1145/3299869.3320241

Syddall, H.E., Aihie Sayer, A., Dennison, E.M. et al. (2005). Cohort profile: the Hertfordshire cohort study. *Int J Epidemiol*. 34 (6): 1234–42. doi: 10.1093/ije/dyi127.

Vardigan, M., Heus, P., and Thomas, W. (2008). Data documentation initiative: toward a standard for the social sciences. *International Journal of Digital Curation* 3 (1): 107–113. https://doi.org/10.2218/ijdc.v3i1.45.

Wadsworth, M. (2010). The origins and innovatory nature of the 1946 British national birth cohort study. *Longitudinal and Life Course Studies* 1 (2): 121–136. https://doi.org/10.14301/llcs.v1i2.64.

Wadsworth, M., Kuh, D., Richards, M., and Hardy, R. (2006). Cohort profile: the 1946 National Birth Cohort (MRC National Survey of health and development). *International Journal of Epidemiology* 35 (1): 49–54. https://doi.org/10.1093/ije/dyi201.

Wang, M., Beal, D.J., Chan, D. et al. (2017). Longitudinal research: a panel discussion on conceptual issues, research design, and statistical techniques. *Work, Aging and Retirement* 3 (1): 1–24. https://doi.org/10.1093/workar/waw033.

Wilkinson, M.D., Dumontier, M., Aalbersberg, I.J.J. et al. (2016). The FAIR guiding principles for scientific data management and stewardship. *Scientific Data* 3: 160018. Retrieved from https://doi.org/10.1038/sdata.2016.18.

Wood, N., Bann, D., Hardy, R. et al. (2017). Childhood socioeconomic position and adult mental wellbeing: evidence from four British birth cohort studies. *PLOS ONE* 12 (10): e0185798. https://doi.org/10.1371/journal.pone.0185798.

12

Harmonization of Census Data: IPUMS – International

Steven Ruggles, Lara Cleveland, and Matthew Sobek

Institute for Social Research and Data Innovation, University of Minnesota, Minneapolis, USA

12.1 Introduction

IPUMS International is the world's largest collection of population microdata available for research. It is largely composed of census data in which variables are harmonized over time and across countries. The goal is to facilitate comparative and cross-temporal international research. In pursuing this goal, IPUMS does not reduce international differences to a set of least common denominators but aims to provide researchers access to the full detail of the original data (Minnesota Population Center 2020). All material is available at www.international.ipums.org.

Over 100 national statistical offices allow IPUMS to disseminate their census microdata for research and teaching (Meier et al. 2011; Ruggles et al. 2015). As of 2022, the database included roughly 400 censuses (Ruggles et al. 2011). IPUMS has recently begun adding household surveys to the database as well, to provide timely and more frequent data. Most countries contribute multiple censuses to the database, allowing the study of change over time. The median sample includes 10% of the national population and has 840,000 person records.[1] In total, the database includes individual-level information on over one billion persons (Cleveland et al. 2011; McCaa et al. 2011). The IPUMS samples are nationally representative and typically offer geographic detail to the second administrative level within countries, such as counties, districts, or municipalities.

Censuses are a national enterprise and are products of their times. Consequently, the source data IPUMS receives are inconsistent across censuses within countries and are thoroughly incompatible across countries. Over the decades, statistical offices have subscribed to varying degrees to international standards for census question wording, but the standards themselves have evolved (United Nations 2017). Documentation of the data is sometimes fragmentary – especially for older censuses – and is usually in the official national language. Many of the census microdata files were never edited for use outside the

---

1 Statistical offices only provide samples of their censuses in order to protect respondent confidentiality. See the sample design discussion in Section 12.2.

Survey Data Harmonization in the Social Sciences, First Edition.
Edited by Irina Tomescu-Dubrow, Christof Wolf, Kazimierz M. Slomczynski, and J. Craig Jenkins.

statistical office, and it is not unusual to encounter logical inconsistencies and other problems. In general, the older the data, the more common such issues are.

The size and complexity of the database have forced IPUMS to innovate. We must contend with both heterogeneity of the source material and the large scale of the data, which totals well over a terabyte. Conveying the resulting complexity to researchers without overwhelming them is a significant dissemination challenge. Efficiently filtering information and subsetting the database is essential to make it usable. For this reason, the data access system is an integral component of our approach to harmonization (Sobek et al. 2007; Sobek and Cleveland 2017).

Despite their unique strengths, census microdata have limitations. Censuses are taken only at intervals and tend to ask fewer questions than surveys, and these are largely demographic and socioeconomic. Some of the datasets are samples of individuals, not complete households, and therefore, lack critical family context for behaviors and outcomes. Confidentiality measures must be taken to make the microdata publicly available, and those come at a cost. Because the data are samples, small population subgroups can have too few cases for analysis. This limitation is amplified by the suppression of small variable categories that might identify individuals. The suppression of small area geography (places under 20,000 population) is the most burdensome measure for many researchers. The project's Frequently Asked Questions web page describes key features about the database, its strengths, and limitations (https://international.ipums.org/international-action/faq).

12.2 Project History

IPUMS International began in 2000 with a grant from the U.S. National Science Foundation. At the time the project was conceived, only a handful of census microdata samples existed outside the United States, and they typically required personalized arrangements to use. The published census results were incompatible and limited to basic tabulations. To make international comparisons required harmonization, and harmonization required microdata. Our prediction that reducing barriers to the microdata would unlock a flood of new work proved correct, as thousands of researchers ultimately flocked to IPUMS, and a series of global-level studies ensued on topics such as living arrangements of the elderly (Reher and Requena 2018), comparative internal migration (Bernard and Bell 2018), and educational homogamy (Esteve et al. 2012).

IPUMS aimed to democratize access to census microdata, granting free and equal opportunity to all researchers globally. Our novel vision required that the national statistical offices grant a third party – the University of Minnesota – the right to disseminate their countries' census data. Extensive negotiation, detailed memoranda of understanding, and modest license fees enable dissemination in perpetuity. Older census data, often stored on decaying media, injected urgency into the project. Archiving these files before they were permanently lost was a principal motivation.

Developing a consistent sampling approach was a key early challenge. IPUMS disseminates sample data, not full count census files. While some countries today create complex samples designed for public dissemination, most do not. Even fewer did in the early years of IPUMS, and almost none did so for their older censuses. We had to determine a method for drawing samples when we were entrusted with full count data or when we needed to instruct statistical offices drawing samples at our request. An individual researcher might prefer, and a statistician would typically recommend, that a sample design be optimized to suit the intended research. IPUMS samples, however, must serve thousands of disparate research projects. Statisticians with whom we consulted therefore recommended a more neutral, simple sampling strategy to ensure that data users could understand the design and estimate error terms accordingly. An additional consideration was that the sample be geographically representative of the level of geography provided in the data. We settled on a systematic sample of households (every nth household) after a random start, sampled across geographically sorted records. In essence, this amounts to fine geographic stratification. We sacrifice precision for any particular research question for the ability to efficiently deliver data suitable for the widest variety of applications. The simple design has the critical benefit of being easy to communicate and easy to implement by partner countries.

IPUMS International met its initial goals by 2004, harmonizing 29 censuses from 8 countries on 5 continents. By this point, we realized that our past methods, modeled largely on prior U.S. census harmonization, would not scale to meet our new commitments. Accordingly, in 2005 we halted data processing altogether, embarking on a year-long retooling from a craft process to an industrial one emphasizing division of labor and software development. We broke the harmonization process into discrete steps amenable to automation. Disaggregating the process offered the additional benefit of restricting the scope of decision-making to better take advantage of graduate student labor. Our new approach invested heavily in standardizing the material we received from partner countries up front, rather than leaping directly from the original material into harmonization. Under the new regime, we would develop comprehensive metadata describing each source data file and clean up data irregularities prior to the harmonization stage.

We developed a custom XML markup for the census questionnaires, enabling software to compile this text on demand any set of source variables. Thus, we could have this information at hand as we developed harmonized classifications. We also determined that the source variables were the place to document the universe of respondents (i.e. the population to which the underlying census question was addressed). To distinguish between missing and not-applicable cases, which are often conflated in the data, we developed a standardized procedure to empirically assess these two types of nonresponse. Where appropriate, we use programming to separate the missing and not-in-universe cases at the source variable stage. This universe-checking process is labor-intensive, requiring research staff to tabulate each source variable against one or more other variables. Though costly, it yields critical information for harmonization and serves as an important layer of data quality assurance.

The introduction of regularized and well-documented source variables made it much easier to create harmonized variables. Somewhat belatedly, we recognized that these new features we developed for internal processing could be offered to users through a modified web dissemination system. To this end, we added the function to view compiled cross-national questionnaire text for harmonized variables. This led in turn to changes in how we wrote variable comparability discussions. Instead of documenting all issues in detail, which was becoming increasingly impractical, we could note an issue in general terms and point the researcher to the questionnaire text to make their own determinations. By offering the unharmonized source variables to users through the dissemination system, we could also solve a usability conundrum: how not to lose any information from the original datasets without proliferating rare or unique variables. Instead of trying to harmonize every variable, we could focus on widely desired and difficult variables, and make others available in unharmonized form. We could also simplify harmonized variables in cases where incorporating all details from some samples would debilitate a classification that otherwise works well to capture key distinctions. With our new systems in place, we processed 35 censuses in 2006, after releasing a total of 29 censuses in the first 5 project years. Since then, we have averaged 20–25 samples per year, while regularly adding new capabilities to the web system. The complete list of samples and their attributes are described on the sample information page (https://international.ipums.org/international-action/sample_details).

12.2.1 Evolution of the Web Dissemination System

The key challenges for IPUMS web design are data discovery and the process of defining data extracts. Some users are inevitably confused by the concept of variables that span datasets, with documentation organized around variables. In our perennial attempts to improve the system, we carried out formal user testing, conducted surveys, observed users in workshops, and talked with researchers at international meetings.

The 2005 project redesign began to address the issue of information overload by filtering all elements of variable documentation based on a user's sample selections. This entailed reorganizing the comparability text into country-specific blocks denoted with XML tags. The compiled enumeration text for each harmonized variable could also be filtered. In addition, we made the source variables available for inclusion in data extracts. In 2008, we focused on improving data discovery: instituting dropdown variable menus to ease navigation, elevating the visibility of the source variables, and adding various viewing options to the variable browsing screen. In 2010, we simplified the extract process, merging variable browsing and extract selection while adopting a "data cart" metaphor. Instead of stepping users through multiple screens to define their extract, we converted those steps into an easily skipped set of options.

More recently, we reorganized the main data browsing screen, with the principal aim of encouraging users to select samples before exploring variables. We also added video tutorials in multiple languages to guide users through the process. Future usability enhancements include more ways to filter samples and interface changes to accommodate high-frequency survey data.

12.3 Applied Harmonization Methods

Data arrive from partner countries in many formats. They are converted into fixed-column ASCII files organized hierarchically, with each household record followed by a person record for each resident. A suite of utility programs deals with the most commonly encountered formatting issues, but preprocessing regularly requires sui generis solutions. A variety of checks are necessary to ensure the soundness of the household-person structure, which is critical for both technical reasons and substantive research purposes. Problems might include households without a householder, households with multiple householders, straggler person records disassociated from their household, or other issues. These irregularities must be rectified by logical inference, explicit identification of household fragments, or whole-household record donation.

After the data are consistently formatted, each file is subjected to a series of pre-harmonization steps to create fully specified input for the harmonization stage. This process centers on metadata development. A data dictionary is created for each dataset, recording its layout and the characteristics of its variables: column locations, labels, and codes. Non-English labels are translated, so all subsequent work can proceed in a common language. Variables are assigned unique names within the IPUMS system and subjected to a number of procedures to convert them into more standardized "source variables" that serve as input to the harmonization stage. These steps involve the addition of several additional fields in the data dictionary that further document and potentially recode the original variables. The goal is not to code the data into common classifications, but rather to rationalize and fully document each sample-specific variable. In developing the source variables, we consolidate stray values into designated missing value categories; label and numerically code blank values representing the not-applicable cases; and clarify the labels, which may be ambiguous or poorly translated. At the end of this stage of processing, every categorical value has a label and has been numerically coded. Meaningful categories in continuous variables, such as the not-applicable value, top-code, or missing values, are also labeled. A brief variable description is written to document each variable, in the process confirming that we understand its meaning.

Table 12.1 shows a small part of a data dictionary. The leftmost fields indicate the five variables being displayed (sex, relationship, marital status, etc.). Some rows contain variable-level information, while the intervening rows pertain to specific values. The "Original Variable" columns document the values in the data as provided to IPUMS. We translate the Spanish labels into English and collect frequencies using custom scripts, in the process identifying several undocumented values in RELATE and MARST. The "Standardized Variable" columns are used to convert the original variables into the source variables, cleaning up stray values and editing labels as necessary. For MARST, we use the "SCode" column to reassign the original undocumented values 7 through 9 to the value "9" with a label of "Unknown" in the source variable. The final column in the table documents the universe of people who should have a response for the variable (i.e. they were asked the census question from which the variable was derived). Other fields in the data dictionary (not pictured) describe additional features of each source variable and govern its display in the web dissemination system.

Data harmonization is the culmination of IPUMS processing, involving the development of variables that span countries and times. This requires determining which variables are conceptually the same across datasets. Those determinations cannot be made solely on the basis of variable names and labels but may require reference to codes, value labels, census question wording, and category frequencies. It can sometimes be a judgment call: weighing the value of user convenience against the possibility of misleading researchers by combining variables with differing shades of meaning or strikingly different population universes. If concepts or categories differ significantly, we create parallel harmonized variables to minimize the likelihood of user error.

The core activity of data harmonization is to equate variable codes and labels across samples so that each category means the same thing across all censuses (Esteve and Sobek 2003). The primary instrument for achieving this is a variable correspondence table like the one depicted in Table 12.2 for the variable "Class of worker." The columns on the left show the harmonized output values and their labels. Each column on the right represents an input dataset: in this case census samples from four countries spanning a 30-year period: Ecuador, Romania, Venezuela, and Tanzania. Note that the full correspondence table for this IPUMS variable includes over 300 samples. Each row in the correspondence table contains items that are conceptually the same and thus receive the same codes in the output data. In broad strokes, the process is as follows: a researcher identifies the source variables in the different samples, a program inserts the values and labels for those variables into the correspondence table from the appropriate data dictionaries, and a researcher then aligns the codes and assigns output codes and labels (the "harmonized codes" columns on the left). This sort of semantic integration is intellectual labor that

Table 12.1 Data dictionary example (selected fields).

| | | | | Original variable | | | Standardized variable | | |
Name	Column	Width	Variable label	Code	Original label	Translated label	Frequency	SCode	Output label	Universe
Sex	40	1	Sex							All persons
				1	Varon	Male	131.612	1	Male	
				2	Hembra	Female	140,478	2	Female	
Relate	41	1	Relation-ship							All persons
				1	Jefe	Head of household	48,508	1	Head of household	
				2	Conyuge (esposa)	Spouse	18,749	2	Spouse	
				3	Companera	Partner	15.989	3	Partner	
				4	Hijo	Child	138,276	4	Child	
				5	Cónyuge del hijo	Spouse of child	883	5	Spouse of child	
				6	Companera del hijo	Partner of child	946	6	Partner of child	
				7	Otro parentesco	Other relative	29,278	7	Other relative	
				8	Domestica	Servant	2,717	8	Servant	
				9	No pariente	Not related	16,733	9	Not related	
				0		[no label]	11	99	Unknown	
Marst	42	1	Marital status							All persons
				1	Soltero (nunca casado)	Single, never married	217,960	1	Single, never married	
				2	Casado	Married	46.122	2	Married	
				3	Viudo	Widowed	5019	3	Widowed	
				4	Divorciado	Divorced	1.665	4	Divorced	
				5	Separado legalmente	Legally separated	906	5	Legally separated	
				6	Anulado	Annulled	406	6	Annulled	
				7		[no label]	9	9	Unknown	
				8		[no label]	2	9	"	

Table 12.1 (Continued)

Name	Column	Width	Variable label	Code	Original label	Translated label	Frequency	SCode	Output label	Universe
				9		[no label]	1	9	"	
Cons	43	1	Con-sensual union							Persons age 15+
				1	Si	Yes	36,352	1	Yes	
				2	No	No	97.797	2	No	
				0	Menores de 15 anhos	Under age 15	137.941	9	NIU (not in universe)	
Lit	44	1	Literacy							Persons age 5+
				1	Si	Yes	141.514	1	Yes	
				2	No	No	79,384	2	No	
				3	No responde	Undeclared	6787	8	Unknown	
				0	No aplica	Not applicable	44,405	9	NIU (not in universe)	

is unlikely to be automatable in the foreseeable future. The work requires a holistic view of the universe of codes for each sample and consideration of the underlying questionnaire text (Sobek and Cleveland 2017).

Variable harmonization aims to retain all the detail in the original samples while providing a fully integrated database in which identical categories in different samples receive identical codes. We employ several strategies to achieve these competing goals. In cases where original variables are compatible and recoding is straightforward, we write documentation noting any subtle distinctions between samples. For some variables, it is impossible to construct a single uniform classification without losing information from samples with rich detail. In these cases, we construct composite coding schemes. The first one or two digits of the code provide information available across all samples. The next one or two digits provide additional information available in a broad subset of samples. Finally, trailing digits provide detail only rarely available.

The classification scheme for "class of worker" in Table 12.2 illustrates the composite coding approach. The first digit identifies four substantive categories consistently available in all samples: (i) self-employed, (ii) wage-salary worker, (iii) unpaid worker, and (iv) other. The Romania sample does not make the distinction between employers and own-account workers, so these categories are combined at the fully comparable first digit (codes 100–122). At the second digit, we distinguish employers and persons working on their own accounts, for samples that make that distinction. The third and final digit differentiates among types of own-account workers (agriculture and non-agriculture).

Table 12.2 Correspondence table: class of worker.

Harmonized codes		Input codes			
Code	Label	Ecuador 2001	Romania 2011	Venezuela 1981	Tanzania 2012
000	NIU (not in universe)	9 = Blank (N/A)	9 = Blank (N/A)	0 = Blank (N/A)	99 = Blank (N/A)
100	Self-employed		1 = Self-employed		
110	Employer	1 = Employer		7 = Owner with employees	1 = Employer
120	Working on own account	2 = Self-employed		8 = Own-account worker	
121	Own account, agriculture				4 = Own account, agriculture
122	Own account, non-agriculture				3 = Own account, non-agric
200	Wage/salary worker		2 = Wage or salary worker		2 = Employee
210	Wage worker, private employer	5 = Private sector employee			
211	Nonmanual worker, private			2 = Private sector professional	
212	Manual worker, private			4 = Private sector manual labor	
213	Domestic worker			6 = Domestic service	
220	Wage worker, government				
221	Federal, government employee	4 = State employee			
222	Local government employee	3 = Municipal employee			
223	Nonmanual worker, govt			1 = Public sector professional	
224	Manual worker, government			3 = Public sector manual labor	
300	Unpaid worker		3 = Unpaid worker		
310	Unpaid family worker	6 = Family worker		5 = Unpaid family worker	5 = Contributing family worker
320	Apprentice				6 = Apprentice
400	Other		4 – Other		7 = Other not specified

We use correspondence tables in a more limited fashion for continuous, as opposed to categorical, variables. Most continuous variables do have categorical information (e.g. not-applicable or missing values), and we recode and label those specific values in the correspondence tables. Sometimes we use the tables to recode binned values, such as income represented in hundreds of units, to the midpoints of the intervals they represent, producing pseudo-continuous data that can be harmonized into truly continuous variables. We note in the documentation when the data are transformed that way. In some cases, mathematical computations or complex programming logic are needed to harmonize continuous variables. In those instances, the correspondence tables serve only as metadata to identify the symbolic location of the constituent source variables.

The correspondence tables exemplify our metadata-centered approach (Sobek 2020). We do not write recode statements, except in exceptional circumstances. We write software to read metadata. Simply moving an item from one cell to another in the correspondence table accomplishes the recode. The benefits are significant: a researcher can readily interpret the coding decisions while seeing all the associated labels with their codes and frequencies. If a new code is needed to handle some variation introduced by a sample, the IPUMS researcher simply adds a row in the table and aligns the appropriate input codes to it. The correspondence tables also help with sustainability. Reorganizing the codes to accommodate a new sample is relatively easy compared to sifting through a mass of impenetrable logical assignment statements. IPUMS is a living project, and we can never know the full universe of labels and coding structures that will need to be incorporated into existing harmonized variables in the future. The correspondence tables provide a practical solution to this challenge.

The IPUMS Data Conversion Program (DCP), written in C++, reads the correspondence tables to produce the integrated output data. In some cases, however, conventional programming is required. The DCP supports modularized programming in which discrete variable-specific logic can be written by research staff without affecting the main structure of the application maintained by software developers.

12.4 Documentation and Quality Assurance

Harmonized data inevitably are more complex than the original. Composite coding and clear labeling can go only so far in conveying the compromises involved in combining items derived from unique questionnaires processed by dozens of organizations. IPUMS is a general-purpose research tool, and it is impossible to predict which differences in the underlying data might be critical for a specific researcher's analysis. Our solution is to write harmonized variable documentation that highlights comparative issues for the user. The documentation is intended as a component of an integrated data dissemination system and does not seek to exhaustively explain all potential issues. The aim instead is to write enough to alert the reader to the issue and point them to the metadata element where they can explore further for themselves.[2]

Variable documentation is initially drafted during data harmonization, noting any decisions or underlying differences that are not self-evident from the codes and labels. The comparability text notes

---

2 The best point of entry to explore IPUMS documentation is the variable browsing system accessed at https://international.ipums.org/international-action/variables/group.

differences over time within countries as well as cross-national issues. The text might note, for example, when additional detail is present in a source variable that could not be accommodated in the harmonized variable. Perhaps the reference period for the question differs between countries, or the wording of the question was unusual in one or more censuses, possibly meriting examination of the questionnaire text. One of the most persistent comparability issues that cannot be conveyed via codes and labels involves differences in the population universe for the question. For instance, if only persons aged 15 and above are asked the employment question in a census, that can distort comparisons to child labor rates in censuses that applied a lower minimum age. IPUMS staff systematically verify all variable universes. Significant universe differences always warrant mention in the variable description, but it is advisable for researchers to review that component of the metadata for all their key variables.

Another important step in the development of the data dictionaries involves connecting each source variable to the specific questionnaire text that pertains to it. To make this possible, the original census questionnaires and instructions are first converted into machine-actionable simple text files from their original formats (typically pdfs). XML tags are inserted to provide basic formatting for web display, and every distinct block of text is assigned an identification number. We enter in the data dictionary the text block number(s) associated with the census question(s) from which each source variable was derived. Using these tags, web software can compile questionnaire text on demand for users. Questionnaire wording is the most fundamental documentation for most variables and is invaluable during harmonization as well.

A variety of checks aim to ensure the quality of the output data. Certain types of errors are logged by the DCP as it runs: values not accounted for in correspondence tables, values created via programming that lack labels, or assigned values wider than the designated width of the output field. The program automatically tabulates frequencies of every variable in each sample, and these frequencies are used in the web interface as well as diagnostically. We collate the frequencies for each harmonized variable and compare distributions across countries and over time within countries. Most processes to resolve data issues are driven by metadata under the control of research staff, which minimizes time spent iterating with programmers.

The constructed variables created via programming require special attention. The most valuable of these are the family interrelationship variables identifying spouses and parents across the records within households (Sobek and Kennedy 2009). That programming is complex and often requires custom solutions because of idiosyncrasies in the source data. Not all samples contain the full set of relevant variables to link family members, and the categories identified within those variables can differ across censuses. For example, some samples combine children and children-in-law in the relationship-to-head variable, making links to parents ambiguous. Other censuses lack fertility information useful for distinguishing among potential mothers of a child. And many African samples include grandchildren in the "child" category of the relationship variable, requiring checks on proximity within the household and age differences to differentiate among potential parents. We construct a flag variable for the family interrelationships to indicate the specific programming rule under which each link was made.

If we discover that data for a variable are clearly erroneous, we suppress it from appearing on the web via a metadata switch. We can suppress entire variables or a single sample within a harmonized variable. We analyze universes and run logical checks to assess the integrity of household structures. We review frequency distributions for every variable and analyze the data when constructing more complex variables such as migration status, educational attainment (EDATTAIN), children ever born, and household interrelationship variables. In general, IPUMS does not edit the data, aside from consolidating obviously erroneous values – like impossible ages – into a single missing-value category. When a sample has a large proportion of missing values for a variable, we note that in the harmonized description. As the sole distributor of many of these files, we tend to be conservative about making changes to the data we receive from statistical offices.

12.5   Challenges to Harmonization

International census microdata present many harmonization challenges: language translation; undocumented data edits; global variation in census methodologies and coding schema; and country-specific geography, educational systems, and living conditions. IPUMS developed many of the techniques and approaches described in Sections 12.2 and 12.3 specifically to address harmonization challenges. We create *source variables* with associated documentation and quality checks to ensure that we fully understand the source material prior to harmonization. *Composite coding* allows us to create broad globally comparable categories while retaining details. We make *comprehensive documentation* and *enumeration materials* easily accessible via the web delivery system to aid data users in understanding the data. Where possible, we rely on *international standard classifications* to inform our harmonization approach. For particularly complex or place-specific variables, we create an additional set of *country-specific variables*, harmonized across census or survey rounds within a single country.

EDATTAIN presents a particularly challenging harmonization subject. Censuses typically include between one and five questions about the completed educational level of household members. Using the United Nations Educational, Scientific and Cultural Organization's (UNESCO 2012) International Standard Classification of Education (ISCED) as a reference, we focus on the highest level of education completed. Table 12.3 shows the composite coding scheme for EDATTAIN with example input from four census samples. The general ISCED classification system has six years of primary schooling, three years of lower secondary, and three years of higher secondary (6-3-3). As Table 12.3 illustrates, some countries report only the general level completed (Armenia 2011), some report number of years within levels (Bangladesh 2011), and still others report a combination of level, certification status, and years within level (Ecuador 2011 and Tanzania 2012). We rely on documentation from country educational organizations, UNESCO, and supplemental sources to understand a country's educational system.

Most countries follow the 6-3-3 standard. Difficulties arise in categorizing data from countries that do not. Table 12.3 shows how IPUMS assigns distinct detailed codes for countries that defy the standard. In this example, Bangladesh requires only five years of primary schooling, which IPUMS assigns a special code of 211 within category 2 – PRIMARY COMPLETED. In Armenia, four years of primary is considered complete, but it falls too short of a global primary completion standard. IPUMS, therefore, codes four years in Armenia to a special code within category 1 – LESS THAN PRIMARY. The detailed codes enable a researcher to reassign the category for their specific research agenda. Despite our composite coding approach, some country-specific details cannot be preserved in a globally harmonized coding scheme without the variable becoming unwieldy. Consequently, we create separate country-specific EDATTAIN variables harmonized across time that retain the details of each national education system. Researchers get the convenience of a roughly comparable international control variable as well as variables suited to detailed analyses within countries. As we create EDATTAIN and the country-specific

Table 12.3 Correspondence table: educational attainment.

Harmonized codes		Input codes			
Code	Label	Armenia 2011	Bangladesh 2011	Ecuador 2011	Tanzania 2012
000	NIU (not in universe)	99 = NIU frol (in universe)	99 = NIU (not in universe)	9999 = NIU (not in universe)	99 = NIU (not in universe)
100	Less than primary completed				
110	No schooling	1 = No elementary	0 = No class passed	100 = None 301 = Preschool 601 = Basic education, year 1	* = No schooling 0 = Nursery education
120	Some primary		1 = Class 1 2 = Class 2 3 = Class 3 4 = Class 4	201-203 = Literacy center, years 1-3 401-405 = Primary, years 1-5 602-606 = Basic, years 2-6	1 = Primary standard 1 2 = Primary standard 2 3 = Primary standard 3 4 = Primary standard 4 5 = Primary standard 5
130	Primary (4yr)	2 = Elementary			
	Primary completed, less than secondary				
	Primary completed				
211	Primary (5yr)		5 = Class 5 6 = Class 6 7 = Class 7		
212	Primary (6yr)			406 = Primary, year 6 501-502 = Secondary, years 1-2 607-608 = Basic, years 7-8	6 = Primary standard 6 7 = Primary standard 7 8 = Primary standard 8
220	Lower secondary completed	3 = Primary	8 = Class 8	503-505 = Secondary, years 3-5	9 = Secondary form 1

Code	Category			
	Secondary completed			
	General or unspecified track	9 = Class 9	609-610 = Basic, years 9-10 701-702 = Baccalaureate-mid. years 1-2	10 = Secondary form 2 11 = Secondary form 3 16 = Postprimary training 18 = Preform one education
311	General track completed	4 = Secondary 10 = Secondary School Certificate 12 = Higher Secondary Certificate	506 = Secondary, year 6	12 = Secondary form 4 13 = Secondary form 5 14 = Secondary form 6
312	Some college/ university		901-904 = Higher education, years 1-4	
	Technical track			
321	Secondary technical decree	5 = Preprofessional		17 = Postsecondary training
322	Postsecondary technical education	6 = Secondary professional	801-803 = Postbaccalaureate, years 1-C	
400	University completed	7 = Higher professional 8 = Postgraduate professions 9 = Candidate of science 10 = Doctor of Science 15 = Graduation (general) 16 = Graduation (honors) 18 = Masters and higher	905-908 = Higher education, years 5-8 1001-1006 = Postgraduate, years 1-6	15 = University and other related
999	Unknown/missing		9998 = Unknown	

versions of it, we document unique aspects of a country's educational system and the comparability issues they present.

Despite our best efforts, harmonization is not a perfect science. Conceptual and classification differences across countries and time periods sometimes vary so widely that they defy proper harmonization. When this happens, we can forego harmonization altogether, providing access only to source variables. Alternatively, we can assemble a variable with common coding across time and place, even though its categories are not fully amenable to harmonization. For example, many countries ask questions about materials used in the construction of walls, roofs, or floors of the dwelling. Censuses rarely include income or wealth information, so dwelling characteristics can be an important proxy for the household's standard of living. Common dwelling construction materials include brick, clay, wood, grasses, mud, tile, etc. However, the various terms for such materials and their unique combinations result in an unwieldy set of categories for a globe-spanning variable. In IPUMS, the wall materials variable (WALL) has 24 unique categories of brick, or brick combinations, and another 19 categories of adobe, clay, and mud combinations. Despite the proliferation of overlapping categories, a single variable consolidating this information provides users with the information needed to recode or reclassify to suit their analysis.

Geography poses a unique harmonization challenge. Census samples typically report first- and second-level subnational geography for place of residence specific to the data collection date. These subnational units – especially the more detailed second level – can merge, split, or change boundaries between censuses. The goal of harmonization is to construct units that share the same exact spatial footprint across census years (Kugler et al. 2017). The creation of spatially harmonized variables requires obtaining geographic information system (GIS) boundary files for the source data or digitizing old maps. The process involves overlaying the boundary files across censuses and combining units as needed until all boundary changes occur within the aggregated units and no changes cross their borders. The resulting geographies are stable over time, so researchers studying change can be assured that their analyses are not an artifact of differing spatial footprints. The original unaltered units are retained in separate census-specific geography variables, for researchers who require maximum detail at a specific point in time.

Long-lived projects sometimes require major revisions, as was the case about 10 years into IPUMS International, when we began spatially reconciling the geographic codes in the data with maps (Sarkar et al. 2016). Previously, we had harmonized geography purely on the basis of place name across time – treating geography like any other categorical variable. When a new unit split off from another, we used composite codes, so the constituent pieces nested together. In terms of spatial consistency, however, our approach was unsound. It did not guarantee that units with the same name occupied the same space across census years. Meanwhile, we realized that boundary changes were more common than we had assumed. We undertook a major overhaul, changing our approach to the one described above to create spatially consistent variables. Using GIS methods over a multiyear transition period, we completely replaced the geographic variables. Over the past two years, we have begun to extend this work on place of residence to migration and birthplace.

In retrospect, we lacked the geographic capacity and sufficient maps to undertake significant geographic processing in the early stages of the project. At the urging of our Advisory Board, we privileged data production over such costly improvements. This may have been the correct choice for the overall success of the project, but the transition to an entirely new set of spatially harmonized variables undoubtedly burdened early users of the database.

12.6   Software Tools

IPUMS has developed an extensive technical infrastructure dedicated to large-scale microdata harmonization.[3] The research staff perform data processing and manage the web dissemination system largely through manipulation of the metadata. We described above some of the key metadata components: data dictionaries describing the source datasets, and variable correspondence tables that recode the source data into harmonized classifications. Other metadata govern variable and sample display on the web, describe variable attributes for processing purposes, and drive specialized features of the dissemination system.

12.6.1   Metadata Tools

The IPUMS software development team has built a suite of programs supporting the development and maintenance of the metadata. Programs perform tabulations on the data, harvest frequencies and import them into the data dictionaries and correspondence tables, confirm the integrity of the different metadata elements, and generate reports. The metadata are stored as comma-separated-value format files, with XML used, where metadata structures are complex. The metadata are ingested into an SQL database. The DCP uses the database to recode the source files, and it generates some additional metadata as a byproduct, such as variable output frequencies. The web dissemination system utilizes the same metadata database, ensuring that the data and web interface remain in sync.

12.6.2   Data Reformatting

A set of utility programs are used in the early stages of data management, before the data are sufficiently systematized for the harmonization software to be applied. For example, utility programs are employed by research staff to convert rectangular datasets into hierarchical formats or replace leading blanks with zeros without corrupting string-character fields. The collection of utility programs has slowly grown over time and are mostly written in Python or R.

12.6.3   Data Harmonization

The IPUMS DCP is a C++ application that uses dictionaries, correspondence tables, and other metadata to produce a globally harmonized output file corresponding to each input dataset. Because every record is processed sequentially, there would be no advantage to using a database for this work. Output data can be produced as ASCII or parquet-format. Parquet is an open-source column-oriented format that organizes the data by variable rather than by row, yielding faster processing times for the extract system. Data production is a batch process distributed across a large cluster of processors. We normally produce a new iteration of the complete database once per year, adding new samples to the collection and modifying variables as necessary. Because variables are harmonized across samples, any change in the coding

---

3 A highly selective but growing collection of tools is available at github.com/ipums. Other IPUMS software is available upon request.

structure potentially affects all samples, necessitating reprocessing of the entire data collection. The complete harmonized database is archived annually and issued a distinct Digital Object Identifier (DOI).

12.6.4 Dissemination System

The IPUMS web system is a Ruby application running on a Rails framework. The system allows interactive browsing and specification of data extracts. User data requests are authenticated and subsequently invoke the extract program, written in Java and optimized for processing speed. Data requests are put in a queue and distributed across a number of dedicated processors. The system uses a data warehousing approach in which the extract program operates on the harmonized data files that were previously produced by the DCP, as opposed to harmonization occurring in real-time. The extract system works through each sample requested by the user, outputting only the selected variables while pooling the data into one file. The program calls on a separate routine to generate statistical package syntax corresponding to the variables in the extract. The syntax files provide the variable locations, variable labels, and value labels needed to read the extracted data into the user's preferred statistical package: SPSS, SAS, STATA, or R.

12.7 Team Organization and Project Management

The composition of the IPUMS International team fluctuates but has historically consisted of three to five PhDs in population-related fields, four to seven doctoral students, a varying number of undergraduate researchers, and the labor of three full-time-equivalent software developers spread across a large group of information technology professionals. An interdisciplinary strategic leadership group of six to eight faculty members meet regularly with team leaders to set policy, discuss allocation of resources, and offer guidance on specific challenges.

Since the second project phase, we settled into an annual processing cycle that culminates in a data release each summer. Adding new samples to the harmonized database requires reexamining our prior work, editing variable coding schemes and documentation, and updating static web pages. A batch process offers economies of scale. The staff works on all samples simultaneously at each stage of the process: data reformatting, metadata development, source variable creation, and harmonization – subdivided into roughly 80 discrete steps. By design, no one specializes in a particular sample or country. We want multiple people working on each dataset to increase the odds of uncovering irregularities and to ensure methods are applied consistently. To that end, we have an explicit second-person review at sensitive steps in the process. The work is organized in a check-out system so that team members can be self-sufficient in moving to available work. Thorough written instructions for each processing stage ensure that everyone uses the same general approach to formatting, documenting, and harmonizing the data.

Within our highly structured work process, graduate and undergraduate workers perform most of the labor-intensive tasks that cannot be automated. Senior staff check the work of the students and perform tasks requiring greater decision-making, such as composing comparability text, developing a new harmonized variable, or working with the software developers on a feature. All material is reviewed by senior staff before public release. Full-time staff meets weekly to confer about any issues or questions that arise, seeking first to apply established precedents from the project's long history. When consensus cannot be reached or policy considerations are involved, issues are raised with the strategic leadership group for resolution. The general expectation is that the staff bring a technocratic perspective to these meetings and the faculty represent research interests across social science disciplines.

Our emphasis on metadata development and full specification of source variables reduces the data harmonization stage to the final few months of the annual data cycle. It helps that many older harmonized variables already encompass most of the range of variation, we encounter in the new samples. Some variables, however, such as education and occupation, are exceptions that continue to take considerable specialized labor by a senior researcher every data release.

Variables are chosen for harmonization based largely on the frequency with which they are present in the source censuses. The IPUMS harmonized classifications are a logical arrangement of the universe of response categories available across all samples. The development of harmonized coding schemes is therefore empirically driven. It is only indirectly the product of international standards, to the extent that countries adhered to such guidelines. As the database grows, and a critical mass of censuses builds that include the variable concept, it becomes a candidate for harmonization. For example, we recently added a suite of variables reflecting the new international trend toward assessing functional disability. We do make exceptions for topics we deem of particular research interest, and in this area, the strategic leadership group's opinions are most determinative.

The issue of data confidentiality required high-level decision-making early in the project. Although the sample microdata always lacked names and other personal identifiers, further steps were necessary to adequately anonymize the data. The leadership group and research staff worked together to develop an approach that would satisfy the statistical offices and could be operationalized consistently. After much discussion, we arrived at minimum cell counts for categorical variables, principals for applying top and bottom codes to the long tails of continuous variables, small rates of swapping cases between geographic units, and a universal minimum population threshold for identifiable geographic units. These decisions then had to be embedded in the data processing workflow. In the process, we uncovered additional issues relating to migration and birthplace, which are person-level attributes that can include very small geographies. They do not pose the same confidentiality risk as place of residence but arguably warranted a different approach than other personal characteristics. This issue required another round of discussion between the research and leadership groups, weighing potential researcher confusion if we identified different units in household-level place-of-residence and person-level migration variables. Ultimately, we settled on consistency between these two types of spatial variables.

12.8 Lessons and Recommendations

We have learned many lessons since the inception of IPUMS International. Some of those lessons are implicit in the design changes to data processing and dissemination that we describe above. Some additional high-level observations warrant mention:

- Harmonized data pose a cognitive burden on users, and it is a challenge to make the data and the dissemination system intuitive. The issue extends beyond the sheer volume of material and comparability issues. It can be difficult for new users to comprehend an unorthodox system in which documentation and data access are oriented around variables and not datasets.

- In a project this complex, mistakes will be made, and it generates goodwill to acknowledge and correct them openly. Early in the project, we recognized the need to publicly record changes to the data for both transparency and reproducibility of results (https://international.ipums.org/international-action/revisions). We even engage in a kind of crowdsourcing, encouraging users to help us discover problems by mailing them an IPUMS mug if they point out a significant error.
- Spreadsheets and word processors – in our case Microsoft Excel and Word – are excellent tools for entering structured information and writing text, and they have many useful features built in. We have benefitted greatly by having the research staff use these common programs as the front end for our custom software. This approach forces the software team to devise robust validation routines, but we are most productive as a team when we minimize the technical burden on the research staff whose focus is on the meaning and soundness of the data.
- When other organizations have performed harmonization on behalf of IPUMS, it has generally not been fruitful. We attempted some distributed harmonization work early in the project and have shied away from it since. Harmonization is harder than it appears, and we have found that other groups are unlikely to make decisions consistent with our methods and philosophy.

The IPUMS organization has been harmonizing microdata for 30 years. Our processes evolved over that period to meet the challenges posed by numerous national and international census and survey data collections (Sobek et al. 2011; Ruggles 2014). The following recommendations derive from that long experience and our consistent aim to empower researchers with the necessary tools to capitalize on the power of harmonized data.

Data harmonization should be metadata-driven. A signature feature in the design of IPUMS has been the development of software-driven by metadata. Researchers manage the metadata, but the technical infrastructure built around the metadata components is maintained by a team of software developers. The metadata empowers the research staff to perform the functions that affect the subject matter of the population data they understand without becoming junior programmers. A metadata-driven system is also highly flexible, easing the addition of new samples to the database.

Data harmonization should retain as much detail as possible. Most data harmonization projects recode variables to the lowest common denominator available across datasets. This approach yields a substantial loss of detail that can sharply limit the usefulness of the harmonized database. The IPUMS composite coding approach offers a strategy for retaining detail available in a subset of datasets while also providing codes that are fully compatible. The creation of parallel variables offers an alternative way to retain detail when source variable coding schemes or conceptual differences are irreconcilable. Offering unharmonized source variables as well as the harmonized versions provide the ultimate backstop against information loss. One cannot predict all the uses to which researchers might wish to put the data, and it is important to provide maximum flexibility.

Highly heterogeneous input data should be standardized prior to harmonization. The challenge of harmonizing disparate source data can be mitigated by standardizing the variables prior to integrating them. In the case of IPUMS, with dozens of organizations contributing data of differing vintages and degrees of quality control, this extra processing step was essential. The aim of this intermediate stage is only to rationalize codes and develop sound metadata, not to harmonize codes across datasets. With source data issues resolved upfront, the harmonization stage can focus on substantive cross-national coding and conceptual challenges. The value of the standardization stage is one of the seminal lessons of the project.

Comparability issues should be fully documented. Harmonized databases should include full documentation of potential comparability issues. The creators of harmonized data should take advantage of all available information – such as questionnaire text, enumerator instructions, and comparative analysis of the results – to identify potential comparability issues. Perhaps the most critical issue is comparability in variable universes: the applicable population for the samples included in the variable. Universes resulting from complex skip patterns are often difficult for users to navigate, so clear documentation is essential. Providing ready access to the original questionnaire text is a necessary component of robust comparability documentation.

Documentation systems should be dynamic in large-scale harmonized databases. The IPUMS International documentation would total over 15,000 pages of text if it were published in conventional printed format. To make this material accessible and manageable, it is essential to limit the information displayed to only those elements relevant to a given research project, as defined by the user. This requires metadata encoded with tags for flexible display options, and documentation pages that are constructed dynamically. The capacity to filter information creates individually tailored documentation that highlights subtle problems of comparability without overwhelming users.

References

Bernard, A. and Bell, M. (2018). Educational selectivity of internal migrants: a global assessment. *Demographic Research* 39: 835–854.

Cleveland, L., Davern, M., and Ruggles, S. (2011). Drawing statistical inferences from international census data. Working Paper No. 2011-1. University of Minnesota, Minnesota Population Center. https://assets.ipums.org/_files/mpc/wp2011-01.pdf.

Esteve, A. and Sobek, M. (2003). Challenges and methods of international census harmonization. *Historical Methods* 36: 66–79.

Esteve, A., McCaa, R., and Lopez, L.A. (2012). The educational homogamy gap between married and cohabiting couples in Latin America. *Population Research and Policy Review* 32: 81–102.

Kugler, T.A., Manson, S.M., and Donato, J.R. (2017). Spatiotemporal aggregation for temporally extensive international microdata. *Computers, Environment and Urban Systems* 63: 26–37.

McCaa, R., Ruggles, S., and Sobek, M. (2011). IPUMS-International statistical disclosure controls. In: *Privacy in Statistical Databases 2010, LNCS 6344* (ed. J. Domingo-Ferrer and E. Magkos), 74–84. Springer-Verlag.

Meier, A., McCaa, R., and Lam, D. (2011). Creating statistically literate global citizens: the use of IPUMS-international integrated census microdata in teaching. *Statistical Journal of the International Association for Official Statistics* 27: 145–156.

Minnesota Population Center (2020). Integrated Public Use Microdata Series, International: Version 7.3. Data file and code book. https://doi.org/10.18128/D020.V7.3

Reher, D. and Requena, M. (2018). Living alone in later life: a global perspective. *Population and Development Review* 44: 427–454.

Ruggles, S. (2014). Big microdata for population research. *Demography* 51: 287–297.

Ruggles, S., Roberts, E., Sarkar, S., and Sobek, M. (2011). The North Atlantic population project: progress and prospects. *Historical Methods* 44: 1–6.

Ruggles, S., McCaa, R., Sobek, M., and Cleveland, L. (2015). The IPUMS collaboration: integrating and disseminating the world's population microdata. *Journal of Demographic Economics* 81: 203–216.

Sarkar, S., Cleveland, L., Silisyene, M., and Sobek, M. (2016). Harmonized census geography and spatio-temporal analysis: Gender equality and empowerment of women in Africa. Working Paper No. 2016-03. University of Minnesota, Minnesota Population Center. https://assets.ipums.org/_files/mpc/wp2016-03.pdf.

Sobek, M. (2020). The IPUMS approach to harmonizing the world's population census data. In: *Advanced Studies in Classification and Data Science. Studies in Classification, Data Analysis, and Knowledge Organization* (ed. T. Imaizumi, A. Okada, S. Miyamoto, et al.), 495–511. Springer.

Sobek, M. and Cleveland, L. (2017). IPUMS approach to harmonizing international census and survey data. Working Paper 31. United Nations Economic Commission for Europe, Conference of European Statisticians.

Sobek, M. and Kennedy, S. (2009). The development of family interrelationship variables for international census data. Working Paper No. 2009-2. University of Minnesota, Minnesota Population Center. https://assets.ipums.org/_files/mpc/wp2009-02.pdf.

Sobek, M., Hindman, M., and Ruggles, S. (2007). Using cyber-resources to build databases for social science research. Working Paper No. 2007-01. University of Minnesota, Minnesota Population Center. https://assets.ipums.org/_files/mpc/wp2007-01.pdf.

Sobek, M., Cleveland, L., Flood, S. et al. (2011). Big data: large-scale historical infrastructure from the Minnesota population center. *Historical Methods* 44: 61–68.

UNESCO Institute of Statistics (2012). *International Standard Classification of Education ISCED 2011*. UNESCO-UIS. ISBN: 978-92-9189-123-8.

United Nations (2017). *Principles and Recommendations for Population and Housing Censuses, Revision 3*. Department of Economic and Social Affairs, Statistics Division.

Part III

Domain-driven ex-post harmonization

13

Maelstrom Research Approaches to Retrospective Harmonization of Cohort Data for Epidemiological Research

Tina W. Wey[1] and Isabel Fortier[2]

[1]*The Maelstrom Research Platform, Research Institute of the Health Centre, McGill University, Montreal, Canada*
[2]*Department of Medicine, Division of Experimental Medicine, The Maelstrom Research Platform, Health Centre, McGill University, Montreal, Canada*

13.1   Introduction

Collaborative research initiatives implemented to understand the determinants of disease and inform public policies have become increasingly important in modern health sciences (Dalziel et al. 2012; Gallacher and Hofer 2011; Lesko et al. 2018; Thompson 2009). Long-term health outcomes are the result of many factors, involving the direct and interactive effects of individual and environmental risk factors. Understanding these outcomes requires high-quality epidemiological studies incorporating complex data, such as on genetics, lifestyle, social factors, and environmental exposures. Longitudinal observational cohort studies following individual participants over time can be powerful research tools toward this end (Burton et al. 2009; Manolio et al. 2006; Richmond et al. 2014). However, conducting such studies is complex, resource intensive, and time consuming, and single studies are limited in the size and diversity of the sampled population, types of data that can be collected, and longevity and generality of the research program implemented. Initiatives to synthesize and co-analyze data from existing cohort studies thus offer significant advantages (Gaziano 2010; Thompson 2009) for addressing research questions (e.g. increased statistical power, greater diversity of populations and exposures) and optimizing use of existing data resources and cost-efficiency of research programs (Burton et al. 2017; Roger et al. 2015).

Under such initiatives, data must be harmonized retrospectively (i.e. ex post) to ensure the comparability and inferential equivalence of measures from different sources. This presents significant scientific, methodological, and ethical challenges (as discussed in Chapter 1). Working with detailed individual participant data (IPD) from cohort studies, in particular, entails certain key challenges (Fortier et al. 2017; Lesko et al. 2018). Notably, organizational, ethical, and legal regulations on access to IPD, which may apply under specific conditions and often differ from one study to another, can strongly structure or limit options for processing source data and statistical analysis of harmonized data. Additionally, the complexity of cohort study data (typically integrating information about individuals and their environments from different data sources and at different time points) presents significant scientific and analytical challenges. Secure data infrastructure is required to manage, process, and analyze IPD both before and after harmonization, and depending on the access, management, and processing requirements of the project, the software and infrastructure needs can be very complex or specialized.

To address these challenges, the Maelstrom Research team brings together epidemiologists, analysts, statisticians, and computer scientists working to find and implement solutions to enhance the use of epidemiological study data. Our activities include developing tools and methods for data discovery, harmonization, documentation, coanalysis, and dissemination (Bergeron et al. 2018; Doiron et al. 2017; Fortier et al. 2017); performing data harmonization, documentation, and analysis (Bergeron et al. 2020; Doiron et al. 2013; Fortier et al. 2019; Wey et al. 2021); and providing expertise and support to diverse research entities. The core group is hosted at the Research Institute of the McGill University Health Centre (RIMUHC, Montreal, QC, Canada) and works with local and international partners, ranging from individual research groups with shorter-term targeted analyses to large consortia implementing broad research platforms for long-term use by many researchers. In all cases, we work closely with data users (investigators who will analyze harmonized data), data producers (participating studies who collected the source data), and a range of subject experts (e.g. biostatisticians, ethicists, and geneticists).

This chapter explores the application of the Maelstrom harmonization approach in two large initiatives that included data harmonization across multiple longitudinal population-based cohort studies for research on long-term health outcomes. CanPath (the Canadian Partnership for Tomorrow's Health, formerly the Canadian Partnership for Tomorrow Project) is an ongoing prospective multi-study Canadian cohort initiated in 2008 to support long-term broad research on the biological, behavioral, and environmental determinants of cancer and other chronic diseases (Borugian et al. 2010; Dummer et al. 2018). The MINDMAP project (2016–2020) is a collaboration among existing cohort studies across Europe, United States, and Canada to examine the causal relationships and interactions between individual determinants and the urban environment on mental well-being in older adults, within and across cities (Beenackers et al. 2018). Harmonization efforts in both projects have previously been described (Dummer et al. 2018; Fortier et al. 2019; Wey et al. 2021), and we use this chance to illustrate the applied Maelstrom harmonization process across projects and discuss the commonalities and differences in the approaches taken and challenges faced.

13.2 Applied Harmonization Methods

Maelstrom Research focuses on developing and applying a rigorous and transparent approach to retrospective harmonization, as detailed in published guidelines (Fortier et al. 2017). The generic approach includes a series of key, interrelated, and iterative steps (summarized in Table 13.1 and referred to throughout the chapter) to aid in anticipating and handling common challenges and decisions. However, the unique scientific, ethical, and organizational contexts of each project determine project-specific considerations and solutions. We thus present two illustrative harmonization projects that were comparable in scale of harmonization work (in complexity and number of datasets and variables to generate) that also presented key differences in project objectives and needs (Table 13.2).

Both the CanPath and MINDMAP projects harmonized data from cohort studies, meaning that the studies recruited and collected data on participants at an initial *baseline* data collection event (DCE),

Table 13.1 Summary of the Maelstrom Research guideline steps for rigorous retrospective harmonization.

Steps	Aims	Outputs
(0) Define research questions and objectives	Ensure project feasibility and reproducibility and to guide decision making	A clear and realistic protocol reflecting the potential and limitations of the project
(1) Assemble study information and select studies		
(1a) Document individual study designs, methods, and content	Ensure appropriate knowledge and understanding of each study	Repository of information and documents required to support study selection and data harmonization process
(1b) Select participating studies	Select studies based on formal criteria to ensure comparability and consistency	List of eligible participating studies
(2) Define variables and evaluate harmonization potential		
(2a) Select and define the core variables to be harmonized (DataSchema)	Define the set of variables needed to answer research questions and formal criteria for constructing each, to ensure content equivalence	List of DataSchema variables and relevant attributes and specifications for construction
(2b) Determine the potential to generate the DataSchema variables from study-specific data items (the source data collected by each study)	Determine the possibility for each participating study to construct each DataSchema variable based on collected data items and qualitatively assess the level of similarity	Repository of information on harmonization potential of each variable in each participating study
(3) Process data		
(3a) Ensure access to adequate study-specific data items and establish the overall data processing infrastructure	Ensure accessibility and quality (completeness, coherence, comprehensibility) of the data items from each study required to create the harmonized dataset	Adequate study-specific datasets (the ensemble of source data items collected by each study) to generate DataSchema variables. Suitable data infrastructure to hold and process study-specific and harmonized datasets. Information to assess the quality of study-specific data
(3b) Process study-specific data items under a common format to generate the harmonized dataset(s)	Create harmonized data and generate a single pooled dataset or multiple independent dataset(s) per study to be used for data analysis	One (or several) harmonized dataset(s) and associated documentation
(4) Estimate quality of the harmonized dataset(s) generated	Perform data quality checks, verify processing methods, and examine harmonized data content across studies to ensure adequate quality, understanding, and utility of the harmonized dataset(s)	Final validated harmonized dataset(s) ready for statistical analysis
(5) Disseminate and preserve final harmonization products	Implement a sustainable infrastructure to preserve and disseminate harmonized data, variable-specific metadata, and documentation of the harmonization approach	Accessible repository of harmonized data, metadata, and related documentation

Table 13.2 Overview of the CanPath and MINDMAP projects and an example of a harmonized dataset from each.

	CanPath	MINDMAP
Research network overview		
Research objectives		
Main outcomes of interest	Cancer and other chronic diseases	Mental well-being of older adults in urban environments
Individual covariates/risk factors	Sociodemographic, behavioral, physical, genetic, biomarker, environmental	Sociodemographic, behavioral, physical, biomarker, environmental
Environmental factors	Physical environment	Physical and social environment
End users (access to harmonized data for research)	Researchers approved by CanPath	MINDMAP researchers approved by MINDMAP committee
Study populations		
Geographic scope	National: 9 Canadian provinces	International: 11 countries across Europe, Canada, and United States
Target population	General Canadian population	Aging populations in cities
Timeframe	2008–ongoing	2016–2020
Participating studies	6	10
Example harmonization project		
Harmonized dataset	Baseline health and risk factor questionnaire	MINDMAP harmonized dataset – individual participant data[a]
Version	2.0 (October 2017)	2.0 (April 2020)
Studies	5	6
Subpopulations	12	10
Source study-specific datasets	12	30
Total participants	307 017	218 784
DataSchema (target) variables	694	993
Harmonization evaluations[b]	8328	4733

[a] Numbers for MINDMAP exclude variables derived from linked environmental data.
[b] Assessments of harmonization potential made for each DataSchema variable in each applicable source dataset.

before participants had developed outcomes of interest, and collected further data on the same participants at one or more *follow-up* DCEs. Each participating study could include multiple subpopulations (e.g. participants sampled from different locations or with different questionnaires). The research objectives required harmonizing diverse participant data from diverse subject domains (e.g. sociodemographic characteristics, lifestyle and behaviors, health history, risk factors, and biochemical measures) and linking these to environmental data. Additionally, both harmonization programs were undertaken as part of larger research consortia with broad research objectives, and the harmonized datasets generated were intended to be used flexibly by multiple research groups to address a range of questions.

Background and characteristics of each project influenced the harmonization process implemented. Collection of core data items was agreed upon prospectively by CanPath member cohorts, but some inevitable variation among studies and over time in participant sampling and data collection necessitated subsequent retrospective harmonization. Baseline data have been collected, and data collection is ongoing for follow-up events. In this chapter, we use examples from the harmonization of the baseline health and risk factor questionnaire (BL-HRFQ), which included self-reported participant information and represents a subset of the core information collected by CanPath (Fortier et al. 2019). The BL-HRFQ harmonization thus integrated data collected from multiple cohort-specific subpopulations of participants at recruitment (baseline), which had roughly similar start years among studies.

In the MINDMAP harmonization project, participant data from existing cohort studies were retrospectively harmonized and linked through geospatial information to area-level environmental exposures and social and urban policy indicators from publicly available resources (Beenackers et al. 2018; Wey et al. 2021). As most participating studies were completed at the time of harmonization, the MINDMAP initiative included all available baseline and follow-up DCEs from studies. Hence, source datasets had very heterogeneous temporal characteristics (start years, times between DCEs, and years covered) as well as differences in participant subsets and variables collected at each DCE even within studies. This heterogeneity was a key to the MINDMAP objective of examining individual and environmental effects across different urban populations but also created challenges for harmonization and analysis.

13.2.1 Implementing the Project

13.2.1.1 Initiating Activities and Organizing the Operational Framework

The initial stages of each project are critical for coordinated and efficient implementation of the subsequent harmonization work, starting with well-defined research questions, objectives, and protocols (guidelines Step 0). Teams must then develop a unified and detailed approach to address the research objectives, data harmonization framework, and data infrastructure needs. Both CanPath and MINDMAP implemented multicenter governance models, with centralized coordination of data harmonization, which allowed for standardization of work and pooling of expertise, while maintaining flexibility for individual cohorts.

CanPath includes several specialized task forces and working groups undertaking work related to ethical and legal issues to be addressed, as well as the selection of core data to be collected, the development of tools to support data collection, and others (Borugian et al. 2010). The Harmonization Standing Committee is one of these and includes cohort investigators, data curators, and representatives from the Maelstrom Research team. In CanPath, retrospective harmonization of all study-specific datasets was also conducted centrally by Maelstrom Research and supported by the Harmonization Standing Committee. Throughout the harmonization process, Maelstrom Research was in frequent communication with data managers from each study as needed and had monthly remote meetings with the

Harmonization Standing Committee to discuss scientific or methodological problems arising and undertake concerted decisions.

MINDMAP also had centralized work packages, including one for overall coordination and one for development of a conceptual model, methodological framework, and data harmonization platform. Maelstrom representatives were responsible for overseeing the harmonization process and customizing Maelstrom software, methods, and standard operating procedures to the specific project needs. The harmonization team included members from nine international groups, including Maelstrom Research and members of cohort studies with different domain expertise (Beenackers et al. 2018). Retrospective harmonization was centrally conducted by the harmonization team, but processing of specific variables was distributed based on domain expertise so that groups harmonized domain-specific subsets of variables across all studies. There were initial in-person meetings for agreement and training on harmonization methods, weekly remote meetings of the harmonization team, and communication as needed with study teams (data producers) and investigators (data users) as needed. Harmonized data were only accessible for use by approved MINDMAP investigators.

13.2.1.2 Assembling Study Information and Selecting Final Participating Studies (Guidelines Step 1)

In both projects, the harmonization teams assembled information about participating studies and data collected in a central data repository, working in close communication with study data managers to ensure high-quality source documentation. The prospective planning of CanPath and the involvement of Maelstrom Research during development and data collection meant that Maelstrom had close collaborations with data managers, access to detailed documentation, and ability to request clarifications about study-specific data from each participating cohort. After data collection in CanPath, BL-HRFQ questionnaires and data dictionaries were requested from cohort study teams and compiled and assessed by Maelstrom Research. While core data collection was coordinated, some variation across studies in applied methodology occurred, leading to some differences in sampling frames, data collection methods, and questionnaire wording, resulting in the presence of different subpopulations within studies. Thus, while all CanPath studies were included in harmonization in the BL-HRFQ v2.0 (October 2017), 12 cohort-specific subpopulations were treated as separate input and output datasets (Fortier et al. 2019).

In MINDMAP, gathering and assessing documentation from all studies was more challenging, requiring more time and effort to seek and understand information from study data managers. For example, while data dictionaries were provided by all studies, original questionnaires were not obtained for all. In addition, some data dictionaries were translated to English from other original languages to be understandable to the harmonization team. Another particular challenge in MINDMAP was to clearly document and understand complex longitudinal data structures (Wey et al. 2021). Study baseline years ranged from 1984 to 2012, and studies had different numbers of DCEs. Different sets of variables were collected, not just among studies, but also among DCEs within the same study. Of the 10 cohort studies comprising the MINDMAP network, a subset of six studies were included in v2.0 of the harmonized dataset (April 2020), based on availability of IPD for harmonization and analysis. Variations in recruitment or data collection methods within some studies resulted in a total of 10 subpopulations. Different DCEs within subpopulations were also treated as separate inputs, resulting in 30 datasets representing DCE/subpopulation combinations.

13.2.1.3 Defining Target Variables to be Harmonized (the DataSchema) and Evaluating Harmonization Potential across Studies (Guidelines Step 2)

In CanPath, the cohort PIs and harmonization working group held a series of workshops before data collection to discuss the scientific goals of the cohort and agree on core variables and standard questionnaires to collect data. After data collection, the harmonization team at Maelstrom Research reviewed and compared collected cohort data with the proposed variables to generate an updated DataSchema, which was reviewed by the Harmonization Standing Committee. The approach taken was to create as many harmonized variables as possible for broad usage. For example, variables targeting participant alcohol consumption included: ever drank alcohol, current frequency of alcohol consumption, current quantity of consumption on weekdays and weekends for several specific types of alcohol, frequency of binge drinking defined separately for men and women, and comparison of current alcohol consumption compared to heaviest intake. The harmonization team also assessed and documented the potential to generate each DataSchema variable, i.e. the harmonization potential, in each study-specific dataset. In general, for this step, a team member examines study data dictionaries and questionnaires in detail and deems harmonization potential as possible or impossible, based on availability and compatibility of variables in each dataset, and a second team member validates the initial assessment. Questions and differences are discussed and resolved within the harmonization team if possible, or flagged for follow-up with study data managers if needed, for example, in cases where the compatibility of study data collection methods was unclear from documentation. In CanPath, this step largely involved verification of variables collected against the prospectively planned variables and clarification with study data managers on differences, such as variable names or formats. Harmonization potential was generally high, but some heterogeneity among studies in specific variables collected created some limitations and special considerations. For example, different versions (long or short form) of the International Physical Activity Questionnaire (IPAQ) were used in different subpopulations, limiting coverage and compatibility of physical activity measures across subpopulations (Fortier et al. 2019).

In MINDMAP an initial target set of variables desired to meet the specific network research objectives was defined by investigators, and the final DataSchema was refined by the harmonization team based on detailed examination of documentation and evaluation of harmonization potential, with feedback from study teams. Team members with domain expertise were responsible for assessment and validation of the harmonization potential for relevant DataSchema variables, e.g. for cognitive function tests. Significant heterogeneity among studies and harmonizing data from baseline and follow-up events presented particular challenges for defining the MINDMAP DataSchema. To allow researchers to readily use different subsets of the harmonized data for their specific analyses, separate DataSchema variables were created for equivalent measures at each DCE (e.g. smoking status at baseline, smoking status at first follow-up, etc.), along with associated administrative variables about baseline year and data collection timeframes. Multiple versions of related variables were also created where possible, but options were generally more limited and variables less precise than in CanPath. For example, variables targeting alcohol consumption included: binary indicators of current alcohol consumption, ever drank alcohol, current frequency of binge drinking with the same definition for men and women, and current quantity of alcohol consumption per week for several types of alcohol. The lower precision of DataSchema variable definitions can also be seen in the variable "Highest level of education completed" (Table 13.3). In MINDMAP, this variable was defined in accordance with the International Standard Classification of

Table 13.3 Definition of DataSchema variable "highest level of education" in the two projects.

	CanPath BL-HRFQ		MINDMAP	
Variable name	SDC_EDU_LEVEL		sdc_highest_edu_0	
Label	Highest level of education completed		Highest level of education completed (ISCED 2011)	
Description	Highest level of education completed by the participant		Highest level of education completed by the participant in accordance with the International Standard Classification of Education (ISCED 2011) at baseline	
Categories	Name	Label	Name	Label
	0	None	1	Lower secondary or less (ISCED 0–2)
	1	Elementary school	2	Upper secondary (ISCED 3)
	2	High school	3	Postsecondary non-tertiary education or short-cycle tertiary education (ISCED 4, 5)
	3	Trade, technical or vocational school, apprenticeship training, or technical CEGEP	4	Bachelor, master, doctoral, or equivalent (ISCED 6, 7, 8)
	4	Diploma from a community college, preuniversity CEGEP, or nonuniversity certificate		
	5	University certificate below bachelor's level		
	6	Bachelor's degree		
	7	Graduate degree (MSc, MBA, MD, PhD, etc.)		

Education (ISCED 2011) (UNESCO Institute for Statistics 2012), and only four categories were created to accommodate as many studies as possible. In contrast, in CanPath, this variable was defined with eight categories to maximize precision of information.

13.2.2 Producing the Harmonized Datasets

13.2.2.1 Processing Data (Guidelines Step 3a)
In both projects, a central data server provides a secure repository to hold source datasets, harmonized datasets generated, and documentation and supporting files, which can be accessed remotely

by approved data harmonizers (see Software tools). Study data managers transferred study-specific IPD to the server, where data harmonizers could access the source data for assessment and processing. Users can also share files (datasets or documentation) directly with each other on the server.

The data harmonization teams first assessed source data received to verify that all necessary variables were accessible and comprehensible. This involved checking the list of variables received against the lists requested, verifying that adequate metadata were provided to understand the variables, verifying that the input data formats were compatible with data and analysis servers, and generating descriptive statistics to explore the content of study-specific variables. Questions about the data or requests for additional information were directed to study data managers as needed. In CanPath, the ongoing nature of the studies meant that dynamic and detailed feedback from all studies was possible, and in some cases, maturing versions of source datasets were sent and reassessed. In MINDMAP, feedback on source datasets was mostly restricted to clarifications about study-specific variables.

13.2.2.2 Processing Study-Specific Data to Generate Harmonized Datasets (Guidelines Step 3b)

Harmonization team members verified the previously documented harmonization potential of DataSchema variables (i.e. target variables) with the variables received in each study dataset, and, where harmonization was possible, wrote a processing script to generate each DataSchema variable from each source dataset. DataSchema variable definitions and specifications provided the standard reference for processing decisions, but more complex processing (either because of scientific content or algorithms applied) was generally discussed with the entire harmonization team and investigators. The final harmonization status for each DataSchema variable in each source dataset was documented as "complete" if source variables were the same or could be transformed to generate compatible DataSchema variables, or "impossible" if relevant source variables were not collected or were incompatible. This was an iterative process that could entail multiple rounds of discussion, questions to study data managers, updating harmonization status, or updating DataSchema variable definition and associated processing scripts for each study after all studies had been processed.

The majority of processing methods applied in both CanPath and MINDMAP were algorithmic or relatively simple transformations. In CanPath, almost half of the algorithms were direct mapping of source study variables to harmonized variables, reflecting the relatively standardized methodologies in data collection. In MINDMAP, the algorithms developed tended to be more heterogeneous among source datasets. For example, to generate the DataSchema variable indicating whether a participant had ever drunk alcohol at baseline, only direct mapping of study variables was needed in CanPath, whereas in MINDMAP, recoding of study variable categories and if-else conditional coding were needed in different datasets, and harmonization was not possible in some datasets (Table 13.4). Additionally, a prominent consideration in MINDMAP was that more complex algorithms were required for some variables to account for information from multiple DCEs (e.g. the variable "Ever drank alcohol" in follow-up events was created using if-else conditions accounting for prior information for a given participant, combined with recoding of study-specific variables as needed).

Table 13.4 Example of types of processing algorithms applied to generate the DataSchema variable "Ever drank alcohol" in CanPath and MINDMAP.

Variable description: Indicator of whether the participant has ever consumed alcohol Value type: Integer Categories: 0 (never consumed alcohol), 1 (ever consumed alcohol)					
Number Source Datasets	Study variable(s)	Study variable format(s)	Harmonization status	Status detail	Processing rule
CanPath					
12	Ever consumed alcohol	Categorical	Complete	Identical	Direct mapping
MINDMAP					
4	Ever consumed alcohol	Categorical	Complete	Compatible	Recode
2	(a) Do you drink alcohol? (b) Alcohol: drank before	Categorical Categorical	Complete	Compatible	If-else condition
4	[Unavailable]		Impossible	Unavailable	

13.3 Documentation and Quality Assurance

Documentation occurred throughout the harmonization process, with the philosophy that clear and detailed documentation is critical both internally for appropriate and consistent harmonization decisions to be made (guidelines Steps 1–3) and externally to preserve final harmonization products and help end users to properly understand data content (guidelines Steps 4 and 5). Internally, as Maelstrom works in a team and typically with remote collaborators, it is imperative that all harmonization team members and collaborators have easy access to necessary documentation. In both illustrative projects, information about participating studies and data collected was gathered and made available on the central data server. In the CanPath BL-HRFQ, harmonization documentation and processing scripts were held centrally and only updated by Maelstrom Research. In MINDMAP, because multiple groups simultaneously worked on generating harmonization processing scripts for different areas of information, working versions of the DataSchema and processing scripts were held on a GitHub, allowing for better version control and dynamic updating of scripts by multiple remote data harmonizers.

After generation of harmonized datasets, publicly available documentation was published for both harmonization projects on the Maelstrom Research catalog, including project descriptions, study descriptions, lists of DataSchema variables, and harmonization statuses. A web-based data portal was created for CanPath (https://portal.canpath.ca) for users to explore harmonized data content and simplify demands for access to data. In addition to the core information described above, the portal provides harmonization algorithms for each variable generated in the form of text markdown pseudo-script (Figure 13.1a) intended for easy understanding by users not involved with the harmonization, summary statistics, and procedures

Figure 13.1 Some of the information about harmonization processing documented for generating the DataSchema variable "highest level of education" in example datasets from (a) CanPath and (b) MINDMAP.

to apply for access to harmonized data. In the MINDMAP project, R markdown documents with applied harmonization scripts (Figure 13.1b) and any comments were preserved in the GitHub repository, which is also publicly accessible. Furthermore, summary statistics for MINDMAP variables are only available to approved data users through the secure server, as harmonized datasets are only intended for use within the MINDMAP network. Overviews of the harmonization process and outputs have also been published as open-access, peer-reviewed articles for both projects (Fortier et al. 2019; Wey et al. 2021).

Quality assurance of harmonization products is critical to ensure that they are adequate for statistical analyses (guidelines Step 4). Harmonized datasets produced were always validated within each study dataset by a different analyst from the initial harmonizer, with the intention of assessing both the technical process and the content of the harmonized variables. For every harmonized variable generated, the validator verifies the harmonization script, harmonization logic, and final harmonization status, and checks that the format of the variable produced matches DataSchema specifications, that the univariate distribution of the harmonized variable is logical, and that any conditional multivariate distributions are logical (e.g. skip patterns). A basic checklist for these validations was created for the CanPath BL-HRFQ harmonization and used to create a template that could be adapted for other projects. Comparisons of harmonized variable distributions across datasets were also performed to explore potential biases or flag issues that were less evident when checking within studies. For example, in CanPath, it became apparent that the average value for the variable "sitting height" differed systematically by a set amount in some subpopulations, revealing a difference in measurement methods that had not been documented and requiring adjustment to harmonization algorithms. In MINDMAP, one study had anomalous values for the variable "feels safe in the neighborhood" compared to other studies, revealing an error in the study-specific category labels originally sent. Each project might also have more specific criteria to be met. In CanPath, acceptable range limits for some variables were implemented, and values beyond these ranges were considered invalid (e.g. values above a possible maximum amount for alcohol consumed in a week). In addition to flagging potential issues, examining distributions across studies helped to understand the harmonized content, which includes basic variation in underlying populations. For example, the participant age and sex distribution across studies were more variable in MINDMAP than in CanPath, reflecting the more heterogeneous populations represented.

13.4 Challenges to Harmonization

Both projects shared common challenges in anticipating the needs of and implementing feasible work plans for large and complex harmonization projects. The research objectives required data items from diverse subject domains and data sources and a range of expertise. To address these needs, both initiatives established management structures built on close collaboration and extensive communication among working groups and task forces assigned to manage administrative, scientific, technical, and ethical aspects of the project, including a dedicated harmonization working group (Beenackers et al. 2018; Borugian et al. 2010). They also established a central harmonization team led by Maelstrom Research to achieve harmonization across all datasets, which required access to anonymized study-specific IPD for harmonization and working in collaboration with network researchers and cohort study teams. Both projects also delineated data sharing and publishing agreements prior to the start of harmonization.

Both projects also faced challenges in data transfer and assessment of study-specific datasets that resulted in delays. In CanPath, the ongoing nature of the studies, where harmonization work started while recruitment and data collection for some cohorts were still ongoing, presented both advantages and challenges. Centralized data use policies facilitated data access and transfer, and the harmonization team had extensive feedback from study teams and could receive updated versions of datasets. However, updating work for evolving versions of study-specific datasets and ensuring consistency among dataset versions created additional work for retrospective harmonization. Some of these issues were addressed for ongoing work by creating master participant ID reference lists, data preparation standards and checklists, and updated versions of the DataSchema. In MINDMAP, challenges arose in some studies from more restrictive data access regulations and delays in receiving data, affecting the overall project timeline. Significant time and effort were also spent by the harmonization team to understand study-specific documentation and make harmonization decisions.

As a prospective pan-Canadian cohort intended to create a research platform for wide usage over decades, CanPath faced challenges in planning data collection, coordinating work at national and study levels, anticipating future needs, and ensuring long-term sustainability (Borugian et al. 2010). On the other hand, although retrospective harmonization was necessary, the prospective coordination and unusually high ability for the harmonization team (Maelstrom Research) to work directly with study data managers on recently collected and documented study data resulted in high standardization and ability to resolve questions. In MINDMAP, there was much greater heterogeneity in study methodology, including sampling frames, data collection methods, and question formats. Including diverse populations was also of explicit interest to research objectives, and the composition of subpopulations in MINDMAP was greater (e.g. age ranges and sex ratios, as well as geographic areas represented, were more variable among populations). Different country and study regulations added challenges to understanding access levels and procedures for obtaining IPD. Notably, there was overall lower harmonization potential in MINDMAP than in CanPath, largely due to the prospective nature of CanPath. For example, 81.6% of the possible harmonized variables in the CanPath BL-HRFQ were complete, while 46.7% of the possible harmonized variables from IPD (i.e. not from environmental data) were complete in MINDMAP. In many cases, the precision of DataSchema variable definitions was also greater in CanPath due to the prospective harmonization and greater homogeneity of study populations (e.g. in the variable "Highest level of education completed," Table 13.3).

Ultimately, some limitations always remain in harmonized datasets. Notably, trade-offs between specificity of harmonized variables and number of studies across which the variables can be generated are often necessary. For example, in MINDMAP, generating harmonized variables on mental well-being and cognitive functioning was of focal importance. However, studies typically used different scales or tests. Limiting harmonization to only studies that used the same standardized test resulted in high precision but low harmonization potential (e.g. for a DataSchema variable of current cognition score measured by the Mini-Mental State Examination, only two out of 10 source datasets could be harmonized). On the other hand, creating a harmonized variable using rescaled values from different tests allowed for greater harmonization potential but loss of precision and potential biases (e.g. a DataSchema variable on immediate word recall rescaled to range from 0 to 10 allowed for harmonization of 7 out of 10 source datasets, but included study data collected with two different tests, one originally using a 15-point scale). In MINDMAP, these differences were clearly documented in comments about harmonization for each source dataset to allow researchers to decide how to use the harmonized variable. In-depth exploration of the best statistical methods to harmonize these types of measures to maintain integrity of content while minimizing loss of information and methodological bias are important and active areas of research (Griffith et al. 2016; Van den Heuvel and Griffith 2016). The approach taken in this case was to apply simpler harmonization methods (i.e. rescaling) and document the transformation, leaving investigators the flexibility to further explore and analyze the harmonized datasets as appropriate for their research questions.

Users of the harmonized data will also need to consider representativeness of the populations captured in the harmonized datasets and decide if and how to adjust for study differences if making inferences to a general population. Weighting variables were not addressed in harmonization for these projects, and users must refer to the original studies' data and documentation for this information. The papers describing the harmonization projects attempt to highlight these considerations, for example, providing some comparison of demographics of the harmonized populations against the general populations from which they were drawn (Dummer et al. 2018; Fortier et al. 2019; Wey et al. 2021), listing sources of study-specific heterogeneity in the harmonized datasets to consider (Fortier et al. 2019), and pointing users to individual study documentation where more information on weights to use for analysis should be considered (e.g. Wey et al. 2021).

13.5 Software Tools

To address outstanding needs of multi-study harmonization projects, Maelstrom Research has participated in developing the OBiBa suite of interoperable, open-source software (https://www.obiba.org;

Doiron et al. 2017). In particular, there was need for software geared toward data harmonization that would support both data interoperability across studies and dissemination of metadata to users. The software is designed and continues to be improved through close collaboration among software developers, epidemiologists, statisticians, and the scientific community to support all stages of harmonization projects in a secure, modular, customizable, and integrated way. All software is free to download and install for any users (https://www.obiba.org/pages/download) under a General Public License (GPL) v3, with free and paid support options (https://www.obiba.org/pages/support).

The core software applications of Maelstrom harmonization projects are Opal (v4.2.1 July 2021) and Mica (v4.5.8 July 2021) (Doiron et al. 2017) from the OBiBa suite, with secure integration with open-source R statistical software (R Core Team 2019). Both applications are written in Java, JavaScript, and PHP with web services and graphical user interfaces. Opal is the database application used to build centralized web-based data management systems. It can securely import and export a variety of data types for storage under a standardized format and supports data curation, quality control, summary, visualization, annotation, and query, as well as basic processing methods for harmonization. Mica is the application used to create websites and metadata portals with functionality to organize and publish information about studies and datasets. Both applications can establish secure connections with R, allowing additional advanced options for data manipulation (including harmonization processing), statistical analysis, and creation of dynamic reports. Additional functionalities can also be accessed through secure connections with other OBiBa applications, such as federated statistical analysis with DataShield (Gaye et al. 2014; Marcon et al. 2021), user account management with Agate, and data collection with Onyx. Used alone or together, the applications can accelerate the process of harmonization while reducing costs.

Maelstrom uses Opal and Mica internally across harmonization projects, but the configuration and specific functionalities can be tailored to each project. For each partnership, Maelstrom software experts can work with project investigators and programmers to establish and configure relevant servers and software settings. In both CanPath and MINDMAP, a central Opal server provided a data hub for gathering documentation and hosting source datasets provided by each study and harmonized datasets produced by the project, while a secure RStudio (RStudio Team 2016) connection allowed for performing more advanced data manipulation and statistical analysis with R software. Study-specific and harmonized datasets are hosted on the Opal server and accessible by approved users (e.g. data managers providing source study-specific data, analysts performing harmonization, and researchers accessing harmonized datasets), who can be given different levels of access and permissions. In the CanPath BL-HRFQ, harmonization processing was performed in Opal and RStudio, and documentation and versioning of the DataSchema and harmonization scripts were handled centrally in Opal. In MINDMAP, due to the need for multiple international groups to work simultaneously and flexibly on the harmonization processing and frequently evolving versions of study-specific harmonization scripts, scripts for harmonization processing were written and applied entirely through R markdown in an RStudio interface, and the DataSchema and R markdown versions were maintained and frequently updated in a GitHub repository. Information on the Opal server can be documented and disseminated with Mica. The Maelstrom and CanPath catalogs are both built with the Mica-Opal software toolkit (Bergeron et al. 2018) and illustrate a major advantage of the software integration, which allows metadata in the Opal server to be immediately available and readily updated in a connected Mica server. Cataloged study and variable metadata are free for all users to browse and search.

13.6   Recommendations

While every harmonization initiative is unique, certain strategies can promote rigor and efficiency across projects. The Maelstrom harmonization approach aims to provide generic but adaptable guidelines and open-source tools that address common needs of collaborative harmonization initiatives and promote FAIR data principles (Wilkinson et al. 2016). Essential components of the approach include implementing structured protocols and standardized tools, careful assessment of harmonization inputs and outputs, thoroughly and transparently documenting the harmonization process and products, establishing a collaborative environment with expert input, implementing an efficient and secured data infrastructure, and respecting ethical and legal considerations. In addition to CanPath and MINDMAP projects, the approach has formed the foundation for multiple other initiatives aimed at harmonizing epidemiological datasets for various research objectives (e.g. Doiron et al. 2013; Graham et al. 2020; Jaddoe et al. 2020; Pastorino et al. 2019; Sanchez-Niubo et al. 2019; Turiano et al. 2020).

Comprehensive planning and preparedness from the earliest project stages greatly facilitate all subsequent work, and structured protocols, data infrastructure design, and statistical analysis plans should be carefully delineated in advance. Where possible, prospective harmonization, such as in CanPath, can offer advantages for greater standardization of data collected from different studies, higher precision in definitions of harmonized variables, greater harmonization potential across studies, less time and resources spent in retrospective harmonization, and more power and flexibility for statistical analysis to address research questions. However, prospective harmonization is clearly not always possible or suitable for all research questions, and even with comprehensive planning, eventualities arise leading to unintentional heterogeneity in study methodology and requiring retrospective harmonization. Initial definition of the DataSchema, evaluation of harmonization potential, and requests for relevant study-specific data items are made based on documentation, which is challenging to comprehensively understand when harmonization is achieved by teams not familiar with the study-specific protocols and data content. Secure and effective data infrastructures must also be in place and ready to receive study-specific data and hold harmonized outputs before processing can take place. If specialized statistical methods are anticipated, any statistical expertise, software, or infrastructure needed must also be defined and implemented in advance.

Harmonization decisions, documentation, and assessments must be made with research objectives and project constraints in mind. For example, in CanPath, the approach was to define as many DataSchema variables as possible for broad future use, which was also possible due to prospective planning, whereas the strategy in MINDMAP was to generate variables targeted to the specific needs of network investigators. Different projects may also decide that different levels of heterogeneity are acceptable in constructing DataSchema variables, but the criteria should be clearly documented. For example, both CanPath and MINMDAP were interested in current smoking status of participants, but CanPath focused on cigarette smoking and only considered participants as past smokers if they had smoked 100 cigarettes or more, whereas MINDMAP included smoking of any tobacco products and did not specify other criteria for past smoking. These criteria were clearly noted in the DataSchema variable definitions and category labels, and any additional considerations for specific source datasets were flagged in harmonization comments. Harmonization is also an iterative process, and updates may need to be made to DataSchema definitions, processing decisions, and documentation as needed. For example, adjustments may need to be made to the final list of DataSchema variables and the evaluation of harmonization potential after all source datasets are received and assessed. Where appropriate (e.g. when maturing versions of harmonized datasets are generated), feedback from data users may be integrated into the harmonization process.

Ultimately, the utility of harmonized outputs is evaluated in the context of the intended research applications. It is critical to provide documentation that is as clear and comprehensive as possible about variable definitions, harmonization decisions, harmonization processing, and potential limitations to allow data users to make appropriate decisions in their analyses. Users of retrospectively harmonized data for secondary analysis need to carefully explore the data and documentation and ensure that they have adequate understanding of the content and limitations. Additionally, data users should carefully consider how study-specific heterogeneity in methodology might interact with the harmonization process to affect inferences drawn from analysis and test for potential systematic biases. For example, a researcher may examine if study differences in questionnaire formats are associated with heterogeneity in participant distributions of harmonized variable and, if so, if this can be adequately accounted for in pooled analysis and how it affects interpretation of any pooled estimates or subgroup analysis.

The Maelstrom guidelines were developed to promote a generic and rigorous approach for retrospective data harmonization and to encourage discussion and further advancement of standardized harmonization protocols. Applications to date have yielded valuable research resources and helped identify directions for continued improvement. Recent work at Maelstrom Research, for example, is exploring further use of R software functions to improve efficiency in the data processing step of harmonization without sacrificing rigor and quality in other steps. A general ongoing challenge is to make information on lessons learned from previous harmonization efforts more shareable and accessible for future efforts. This highlights an ongoing need for standardized guidelines for application and reporting of data harmonization as exists in other areas of health sciences (Little et al. 2009; Moher et al. 2009; Vandenbroucke et al. 2007). Publications and collaborations, such as the current volume, are significant moves toward broad sharing and dissemination of knowledge and ultimately developing more standardized harmonization approaches across the research community.

Acknowledgments

The data used in this research were made available by CanPath – Canadian Partnership for Tomorrow's Health (formerly Canadian Partnership for Tomorrow Project), and the MINDMAP Promoting mental well-being and healthy aging in cities project. The CanPath harmonization work was conducted by the Maelstrom Research harmonization team and coordinated by Anouar Nechba and Camille Craig (Maelstrom Research, RIMUHC) with the support of Tedd Konya (CanPath National Coordinating Centre, University of Toronto), and the regional cohort leaders: Dr. Philip Awadalla (Ontario Institute for Cancer Research), Dr. John McLaughlin (University of Toronto), Dr. Paula J. Robson (CancerControl Alberta, Alberta Health Services), Dr. Jennifer Vena (CancerControl Alberta, Alberta Health Services), Shandra Harman (Alberta's Tomorrow Project), Dr. Parveen Bhatti (BC Cancer), Dr. Philippe Broët (CHU Sainte-Justine Research Center), Dr. Trevor Dummer (University of British Columbia), Dr. Simon Gravel (McGill University), Jason Hicks (Dalhousie University), Dr. Guillaume Lettre (Montreal Heart Institute Research Centre), and Dr. Robin Urquhart (Dalhousie University). The MINDMAP harmonization work was conducted by the MINDMAP harmonization team and coordinated by Dr. Mariëlle

A. Beenackeers (Erasmus University Medical Centre, Rotterdam, Netherlands), Rita Wissa (Maelstrom Research, RIMUHC), and Dr. Dany Doiron (Maelstrom Research, RIMUHC).

References

Beenackers, M.A., Doiron, D., Fortier, I. et al. (2018). MINDMAP: establishing an integrated database infrastructure for research in ageing, mental well-being, and the urban environment. *BMC Public Health* 18 (1): 158. https://doi.org/10.1186/s12889-018-5031-7.

Bergeron, J., Doiron, D., Marcon, Y. et al. (2018). Fostering population-based cohort data discovery: the Maelstrom Research cataloguing toolkit. *PLoS One* 13 (7): e0200926. https://doi.org/10.1371/journal.pone.0200926.

Bergeron, J., Massicotte, R., Atkinson, S. et al., & on behalf of the ReACH member cohorts' principal investigators. (2020). Cohort profile: research advancement through cohort cataloguing and harmonization (ReACH). *International Journal of Epidemiology*, dyaa207. https://doi.org/10.1093/ije/dyaa207.

Borugian, M.J., Robson, P., Fortier, I. et al. (2010). The Canadian Partnership for Tomorrow Project: building a pan-Canadian research platform for disease prevention. *CMAJ* 182 (11): 1197–1201. https://doi.org/10.1503/cmaj.091540.

Burton, P.R., Hansell, A.L., Fortier, I. et al. (2009). Size matters: just how big is BIG?: Quantifying realistic sample size requirements for human genome epidemiology. *International Journal of Epidemiology* 38 (1): 263–273. https://doi.org/10.1093/ije/dyn147.

Burton, P.R., Banner, N., Elliot, M.J. et al. (2017). Policies and strategies to facilitate secondary use of research data in the health sciences. *International Journal of Epidemiology* https://doi.org/10.1093/ije/dyx195.

Dalziel, M., Roswell, J., Tahmina, T.N., and Xiao, Z. (2012). Impact of government investments in research & innovation: review of academic investigations. *Optimum Online: The Journal of Public Sector Management* 42 (2): http://www.optimumonline.ca/print.phtml?e=mesokurj&id=413.

Doiron, D., Burton, P., Marcon, Y. et al. (2013). Data harmonization and federated analysis of population-based studies: the BioSHaRE project. *Emerging Themes in Epidemiology* 10 (1): 12. https://doi.org/10.1186/1742-7622-10-12.

Doiron, D., Marcon, Y., Fortier, I. et al. (2017). Software application profile: opal and mica: open-source software solutions for epidemiological data management, harmonization and dissemination. *International Journal of Epidemiology* 46 (5): 1372–1378. https://doi.org/10.1093/ije/dyx180.

Dummer, T.J.B., Awadalla, P., Boileau, C. et al. with the CPTP Regional Cohort Consortium. (2018). The Canadian Partnership for Tomorrow Project: a pan-Canadian platform for research on chronic disease prevention. *CMAJ* 190 (23): E710–E717. https://doi.org/10.1503/cmaj.170292.

Fortier, I., Raina, P., Van den Heuvel, E.R. et al. (2017). Maelstrom research guidelines for rigorous retrospective data harmonization. *International Journal of Epidemiology* 46 (1): 103–105. https://doi.org/10.1093/ije/dyw075.

Fortier, I., Dragieva, N., Saliba, M. et al., & with the Canadian Partnership for Tomorrow Project's scientific directors and the Harmonization Standing Committee. (2019). Harmonization of the health and risk factor questionnaire data of the Canadian Partnership for Tomorrow Project: a descriptive analysis. *CMAJ Open* 7 (2): E272–E282. https://doi.org/10.9778/cmajo.20180062.

Gallacher, J. and Hofer, S.M. (2011). Generating large-scale longitudinal data resources for aging research. *The Journals of Gerontology: Series B* 66B (suppl_1): i172–i179. https://doi.org/10.1093/geronb/gbr047.

Gaye, A., Marcon, Y., Isaeva, J. et al. (2014). DataSHIELD: taking the analysis to the data, not the data to the analysis. *International Journal of Epidemiology* 43 (6): 1929–1944. https://doi.org/10.1093/ije/dyu188.

Gaziano, J.M. (2010). The evolution of population science: advent of the mega cohort. *JAMA* 304 (20): 2288–2289. https://doi.org/10.1001/jama.2010.1691.

Graham, E.K., Weston, S.J., Turiano, N.A. et al. (2020). Is healthy neuroticism associated with health behaviors? A coordinated integrative data analysis. *Collabra: Psychology* 6 (1): 32. https://doi.org/10.1525/collabra.266.

Griffith, L.E., van den Heuvel, E., Raina, P. et al. (2016). Comparison of standardization methods for the harmonization of phenotype data: an application to cognitive measures. *American Journal of Epidemiology* 184 (10): 770–778. https://doi.org/10.1093/aje/kww098.

Jaddoe, V.W.V., Felix, J.F., Andersen, A.-M.N. et al. (2020). The LifeCycle project-EU child cohort network: a federated analysis infrastructure and harmonized data of more than 250,000 children and parents. *European Journal of Epidemiology* 35 (7): 709–724. https://doi.org/10.1007/s10654-020-00662-z.

Lesko, C.R., Jacobson, L.P., Althoff, K.N. et al. (2018). Collaborative, pooled and harmonized study designs for epidemiologic research: challenges and opportunities. *International Journal of Epidemiology* 47 (2): 654–668. https://doi.org/10.1093/ije/dyx283.

Little, J., Higgins, J.P., Ioannidis, J.P. et al. Studies, STrengthening the REporting of Genetic Association (2009). STrengthening the REporting of Genetic Association studies (STREGA): an extension of the STROBE statement. *PLoS Medicine* 6 (2): e22. https://doi.org/10.1371/journal.pmed.1000022.

Manolio, T.A., Bailey-Wilson, J.E., and Collins, F.S. (2006). Genes, environment and the value of prospective cohort studies. *Nature Reviews Genetics* 7 (10): 812–820. https://doi.org/10.1038/nrg1919.

Marcon, Y., Bishop, T., Avraam, D. et al. (2021). Orchestrating privacy-protected big data analyses of data from different resources with R and DataSHIELD. *PLoS Computational Biology* 17 (3): e1008880. https://doi.org/10.1371/journal.pcbi.1008880.

Moher, D., Liberati, A., Tetzlaff, J. et al. (2009). Preferred reporting items for systematic reviews and meta-analyses: the PRISMA statement. *PLoS Medicine* 6 (7): e1000097. https://doi.org/10.1371/journal.pmed.1000097.

Pastorino, S., Bishop, T., Crozier, S. et al. (2019). Associations between maternal physical activity in early and late pregnancy and offspring birth size: remote federated individual level meta-analysis from eight cohort studies. *BJOG: An International Journal of Obstetrics & Gynaecology* 126 (4): 459–470. https://doi.org/10.1111/1471-0528.15476.

R Core Team (2019). *R: A Language and Environment for Statistical Computing*. R Foundation for Statistical Computing https://www.R-project.org.

Richmond, R.C., Al-Amin, A., Davey Smith, G., and Relton, C.L. (2014). Approaches for drawing causal inferences from epidemiological birth cohorts: a review. *Early Human Development* 90 (11): 769–780. https://doi.org/10.1016/j.earlhumdev.2014.08.023.

Roger, V.L., Boerwinkle, E., Crapo, J.D. et al. (2015). Strategic transformation of population studies: recommendations of the working group on epidemiology and population sciences from the National Heart, Lung, and Blood Advisory Council and Board of External Experts. *American Journal of Epidemiology* 363–368.

RStudio Team (2016). *RStudio: Integrated Development Environment for R*. RStudio, Inc. http://www.rstudio.com.

Sanchez-Niubo, A., Egea-Cortés, L., Olaya, B. et al. (2019). Cohort profile: the ageing trajectories of health – longitudinal opportunities and synergies (ATHLOS) project. *International Journal of Epidemiology* 48 (4): 1052–1053i. https://doi.org/10.1093/ije/dyz077.

Thompson, A. (2009). Thinking big: large-scale collaborative research in observational epidemiology. *European Journal of Epidemiology* 24 (12): 727. https://doi.org/10.1007/s10654-009-9412-1.

Turiano, N.A., Graham, E.K., Weston, S.J. et al. (2020). Is healthy neuroticism associated with longevity? A coordinated integrative data analysis. *Collabra: Psychology* 6 (1): 33. https://doi.org/10.1525/collabra.268.

UNESCO Institute for Statistics (2012). *International Standard Classification of Education: ISCED 2011*. Montreal, QC: UNESCO Institute for Statistics (UIS) http://dx.doi.org/10.15220/978-92-9189-123-8-en.

Van den Heuvel, E.R. and Griffith, L.E. (2016). Statistical harmonization methods in individual participants data meta-analysis are highly needed. *Biometrics & Biostatistics International Journal* 3 (3): 70–72. https://doi.org/10.15406/bbij.2016.03.00064.

Vandenbroucke, J.P., Elm, E. von, Altman, D.G. et al., for the STROBE Initiative. (2007). Strengthening the reporting of observational studies in epidemiology (STROBE): explanation and elaboration. *PLoS Medicine* 4 (10): e297. https://doi.org/10.1371/journal.pmed.0040297.

Wey, T.W., Doiron, D., Wissa, R. et al. (2021). Overview of retrospective data harmonisation in the MINDMAP project: process and results. *Journal of Epidemiology and Community Health* 75 (5): 433–441. https://doi.org/10.1136/jech-2020-214259.

Wilkinson, M.D., Dumontier, M., Aalbersberg, I.J.J. et al. (2016). The FAIR guiding principles for scientific data management and stewardship. *Scientific Data* 3: https://doi.org/10.1038/sdata.2016.18.

14

Harmonizing and Synthesizing Partnership Histories from Different German Survey Infrastructures

Bernd Weiß[1], Sonja Schulz[1], Lisa Schmid[1], Sebastian Sterl[1,2,3], and Anna-Carolina Haensch[1,4,5]

[1]GESIS – Leibniz Institute for the Social Sciences, Mannheim, Germany
[2]Freie Universität Berlin, Berlin, Germany
[3]Technische Universität Braunschweig, Institute of Psychology, Braunschweig, Germany
[4]University of Mannheim, Mannheim, Germany
[5]Ludwig – Maximilian University of Munich, Munich, Germany

14.1 Introduction

In the research project "Harmonizing and Synthesizing Partnership Histories from Different Research Data Infrastructures" (HaSpaD[1]), we harmonize and cumulate data of (longitudinal) partnership biographies from nine large German surveys via *ex post* (output) harmonization (Wolf et al. 2016). We use the term "partnership" as an umbrella term for any kind of close, intimate union. Our project focuses on the substantive area of family research. Nevertheless, the field of family research shares problems with other subfields of social sciences, which can be solved using harmonization efforts: even though family research in Germany can rely on large surveys, some family-related phenomena cannot be investigated because each survey by itself does not contain sufficient information, e.g. on special populations. The data harmonization and cumulation will, for instance, allow to analyze divorce risks across the full life-course or whether risk factors for separation have changed over time. In addition, most family researchers focus on a few surveys and exploit only one survey per research question. There are no or very few efforts to replicate results across surveys or investigate sources of heterogeneity between the surveys (for an exception, see Klein et al. 2013). Pooling all German research data creates a comprehensive data basis for research syntheses, which can provide better insights into the various explanatory factors for partnership events as compared to an analysis of a single survey.

---

1 Funded by the Deutsche Forschungsgemeinschaft (DFG, German Research Foundation) – Projekt number 316901171.

Survey Data Harmonization in the Social Sciences, First Edition.
Edited by Irina Tomescu-Dubrow, Christof Wolf, Kazimierz M. Slomczynski, and J. Craig Jenkins.
© 2024 John Wiley & Sons Inc. Published 2024 by John Wiley & Sons Inc.

Against this background, we initiated the HaSpaD project, which serves mainly two objectives: The *first objective* of the research project HaSpaD is to ease the use of harmonized partnership biography data in Germany. It does so by providing an open, accessible, and reusable infrastructure of customized Stata code to assist users in building their own harmonized and cumulated biography data set with additional user-defined harmonized variables. Biographical data on intimate partnerships encompass information such as data on the beginning and the end of a partnership, co-residing as a couple, and marriage. Additional user-defined variables in our harmonized data set are mostly sociodemographic variables of the respondents and their partners, such as age, sex, citizenship, and educational attainment. Within the project, we develop and provide users an online tool, the so-called *HaSpaD-Harmonization Wizard* (https://haspad.gesis.org/wizard), that generates Stata code used to create a harmonized data set, including customized variables and extensive documentation on the process of harmonization of data and variables. All in all, we were able to identify nine survey programs that were fielded between 1980 and 2020, which allow us to analyze partnership biographies that started between 1912 and 2020. The HaSpaD data contain information on more than 170,000 partnerships.

The *second objective* of the HaSpaD project is to investigate several methodological issues that arise when harmonizing and cumulating survey data. Specifically, we are among the first to address the issue of cumulating complex survey data with its accompanying survey weights for regression analysis (which is even an issue in single-survey research, see West et al. 2016, 2017). Furthermore, we investigate how to deal with systematically missing data in some of the data sets, e.g. with systematically missing partner data. Ideally, we mitigate some of these issues; at least, we highlight them and inform data users about possible consequences.

14.2 Applied Harmonization Methods

The entire research process within the HaSpaD project follows the logic of a so-called individual participant data (IPD) meta-analysis (Stewart and Tierney 2002), which acknowledges the fact that data harmonization might be part of a larger research endeavor that also includes a systematic and transparent data search. Accordingly, appropriate eligibility criteria were derived to guide the search for data sets suited for harmonization.

Harmonization took place in two ways: (i) harmonizing the data structure and (ii) harmonizing variables. First, we aim for data that meet the requirement of a longitudinal data structure with partnerships' biographical information (i.e. retrospective event history or prospective panel data). Second, we also performed harmonization procedures at the measurement level of variables, e.g. rescaling variables or creating new variables based on existing variables. These adjustments at the measurement level of variables were, for instance, applied to age, sex, citizenship, religious denomination, and educational attainment. If available, we conducted these procedures for both partners, so-called anchor respondents and their partners.

14.2.1 Data Search Strategy and Data Access

The HaSpaD project aims for complete coverage of survey data containing information on partnership biographies in Germany. However, in order to ensure data quality as well as feasibility within the project duration of three years, we introduced the following eligibility criteria for including surveys in our harmonization process:

1. The data should be publicly available for secondary data analysis to ensure the reusability of the data by the scientific community.
2. The survey should provide longitudinal data (i.e. event history or panel data) as we have a life-course perspective on family-related events.
3. The sampling design of the survey data is probability-based and targets the general German population.
4. The survey population is composed of respondents living in Germany (or West Germany and East Germany, respectively).

First, we screened all survey programs known to us (e.g. utilizing existing meta-analyses on divorce risks by Wagner and Weiß 2003, 2006) regarding the aforementioned eligibility criteria. Second, the HaSpaD team used the Data Catalog (DBK) and the Registration Agency for Social and Economic Data (da|ra), which are service infrastructures provided by GESIS – Leibniz Institute for the Social Sciences, to search for further data sets. Both databases offer the opportunity to search for metadata keywords. However, since the information on survey content is limited, data sets without a clear family scope that may contain information on partnership biographies are difficult to detect within these databases. Additionally, we searched the World Wide Web and known research data centers (for a list of research data centers, see https://www.konsortswd.de/en/datacentres/all-datacentres) to gather additional relevant data sources. Altogether, we identified 9 survey programs with 21 sub-studies that met the eligibility criteria for survey design and data accessibility. For an overview, see Table 14.1.

Table 14.1 Overview of the survey programs and sub-studies used in the HaSpaD project.

Short title	Survey title	Data citation
Pairfam	Panel analysis of intimate relationships and family dynamics (pairfam), Data Release 12.0	Brüderl et al. (2021)
ALLBUS-Cumulation	German General Social Survey (GGSS/ALLBUS) – Cumulation 1980–2016	GESIS-Leibniz-Institut Für Sozialwissenschaften (2018)
Family Survey I West I	Change and Development of Forms of Family Life in West Germany (Survey of Families)	Deutsches Jugendinstitut (DJI) (1992)
Family Survey I East I	Family and Partner Relations in Eastern Germany (Survey of Families)	Deutsches Jugendinstitut (DJI) (2018a)
Family Survey II	Change and Development of Ways of Family Life Second Wave (Survey of Families)	Deutsches Jugendinstitut (DJI) (2018b)
Family Survey III	Change and Development of Families' Way of Life third Wave (Family Survey)	Deutsches Jugendinstitut (DJI) (2018c)
Mannheim Divorce Study	Mannheim Divorce Study 1996	Esser et al. (2018)

(Continued)

Table 14.1 (Continued)

Short title	Survey title	Data citation
Fertility and Family Survey	German Fertility and Family Survey 1992	Bundesinstitut für Bevölkerungsforschung (2002)
GLHS West I	Courses of Life and Social Change: Courses of Life and Welfare Development (Life History Study LV-West I)	Mayer (2018c)
GLHS West II A –Personal Survey	Courses of Life and Social Change: The Between-the-War Cohort in Transition to Retirement (Life History Study LV-West II A – Personal Interview)	Mayer (2018b)
GLHS West II T – Telephone Survey	Courses of Life and Social Change: The Between-the-War Cohort in Transition to Retirement (Life History Study LV-West II T – Telephone Interview)	Mayer (1995a)
GLHS West III	Courses of Life and Social Change: Access to Occupation in Employment Crisis (Life History Study LV-West III)	Mayer (2018a)
GLHS GDR	Courses of Life and Historical Change in East Germany (Life History Study LV DDR)	Mayer (1995b)
GLHS East 71	East German Life Courses After Unification (Life History Study LV Ost 71)	Mayer (2004a)
GLHS East Panel	East German Life Courses After Unification (Life History Study LV-Ost Panel)	Mayer (2004b)
GLHS West 64/71	Education, Training, and Occupation: Life Courses of the 1964 and 1971 Birth Cohorts in West Germany (Life History Study LV-West 64/71)	Mayer and Kleinhenz (2004)
GLHS Panel 71	Early Careers and Starting a Family: Life Courses of the 1971 Birth Cohorts in East and West Germany (Life History Study LV-Panel 71)	Mayer (2014)
Generations and Gender Survey, wave 2005	Generations and Gender Survey, wave 2005 (Sub sample Germany)	Generations and Gender Programme (2019a)
Generations and Gender Survey, wave 2008	Generations and Gender Survey, wave 2008 (Subsample Germany)	Generations and Gender Programme (2019b)
SHARE	The Survey of Health, Ageing, and Retirement in Europe	The Survey of Health, Ageing and Retirement in Europe (SHARE) (2017)
SOEP	Socio-Economic Panel (SOEP), v35, data of the years 1984–2018	Liebig et al. (2019)

14.2.2 Processing and Harmonizing Data

The process of data preparation and harmonization was divided into two major tasks: harmonizing the biography data (e.g. start and end of a partnership) and harmonizing additional variables on respondent or couple characteristics. In addition to providing a harmonized data set at the level of (a) partnership(s) of a respondent, we also compiled a new metadata set describing the characteristics of the source data sets (e.g. survey area and method, research design, or sample type).

14.2.2.1 Harmonizing Partnership Biography Data

First, we created a data template on how to organize partnership biographies. We decided to follow the data structure of the survey program pairfam (Brüderl et al. 2017, p. 54ff), as it is familiar to many family researchers, well-documented, and other surveys can easily be adapted to it. We adopted the core structure of the pairfam data set in terms of the data units, the variable naming conventions, as well as parts of the missing and flag variable schemes. Thus, the HaSpaD data set template is organized as follows: the *data units* (i.e. rows in the data matrix) are partnerships of a respondent. Since a single respondent can have multiple partnerships, multiple rows in the data matrix may refer to one respondent (long format). If respondents did not report any partnerships, they are not part of the harmonized HaSpaD data set. In many source data sets, the data units are respondents, with information on partnerships stored in different variables (so-called wide format, e.g. ALLBUS, Fertility and Family Survey, Generations and Gender Survey, and SHARE). Wide-formatted data sets had to be reshaped to fit the target data template.

Pairfam's *missing scheme* (Brüderl et al. 2017, pp. 12, 54ff) was extended to include missingness due to survey design (i.e. values are systematically missing for an entire survey or specific subgroups of respondents). Moreover, some of the original missing codes in the source data were collapsed into one category (e.g. data inconsistencies or different sources of item nonresponse).

Furthermore, the HaSpaD data template contains seven *identifiers* for the source survey and sub-study, the individual, and the partner. It comprises eight variables for the *core biography*: beginning and end of a relationship (1 & 2 variable), of co-residing (3 & 4) and of being married (5 & 6), death of a partner (7), and date of interview (8). Time-related information was reorganized into century months. In some cases, we had to impute date information. For example, we imputed January for all "begin-variables" (start of a relationship, co-residing, and marriage) and December for all "end-variables" (separation, moving out, and divorce) for all surveys that only provide yearly information on events. For complete documentation of harmonization procedures, we introduced 16 flag variables that include information on whether imputation procedures were performed with respect to the time of an event, the original value of the corresponding date variable from the source data, data inconsistencies, and information from which level of institutionalization onward details on partnerships are available (e.g. relationship begin, cohabitation, or marriage date).

The heterogeneity of study designs and data structures posed challenges to data harmonization. The source studies SOEP and SHARE implemented a so-called multi-actor design, where both anchor respondents and their current partners could report their partnership biographies (producing duplicates for the current relationship of both partners). For these studies, we needed to develop a solution that ensured high comparability with other study designs and low data loss.

The Generations and Gender Survey (GGS) and certain sub-studies of the German Life History Studies (GLHS) followed a cross-sectional design (with a retrospective collection of biographies) with a partial follow-up wave. For the respondents taking part in the follow-up wave, the partnership biographies had to be harmonized across time. Due to the lack of a partnership identifier, we needed to formulate harmonization rules based on plausibility assumptions to identify and update identical partnerships reported in the follow-up.

Besides the structure and organization of information, the source surveys were diverse with respect to the surveyed information on partnerships and couples. The main differences in the assessment of partnership biographies between the source studies are as follows: (i) *Level of institutionalization*: While some studies collect information about living apart together (LAT) partnerships, unmarried co-residing unions, and marriages, other surveys only consider marriages. Additionally, the definition of partnership types can vary between surveys, as some surveys instruct respondents on which partnerships they should consider as relevant. (ii) *Definition of an event*: The surveys also vary in their definition of events that respondents should report. For example, in the case of a divorced couple, some surveys collect the date of the separation, the date co-residence ended, and the date of divorce; other surveys only cover one of these dates, mostly the date of divorce. (iii) *Number of partnerships respondents can report*: Many studies (such as ALLBUS or the Family Surveys I and II) restricted the maximum number of partnerships respondents can report and instructed the respondents on which partnerships are relevant for the survey program.

Our general guideline was to include all information on partnership biographies available in the source studies and to provide users of the HaSpaD data with the information they need about study-specific particularities. This enables users to decide on selecting biographies they would like to include in their analyses. In general, study-specific particularities are documented in the HaSpaD data manual. If information is systematically missing in a source study, the target variables are assigned the code "-10 Missing by study design." The flag variables provide additional information, such as from which level of institutionalization onward information on partnerships is available (variable "flag_beg") and whether the harmonization of partnership biographies across time leads to identification problems (variable "problem").

14.2.2.2 Harmonizing Additional Variables on Respondents' or Couples' Characteristics

The second task was to harmonize and cumulate additional variables such as respondents' and couples' characteristics that may predict partnership behavior and partnership outcomes. These variables are merged with the target data set after processing the biography data. Information on sociodemographic characteristics includes the date of birth, anchors' sex, regional context (East or West Germany), religious denomination, parental divorce, education level, and citizenship.

Since our source data sets vary in designs from (repeated) cross-sectional surveys to panel studies, we decided on the most integrative, most informative level for coding these variables. As most surveys are cross-sectional, we treat all predictor variables as time-invariant. We decided to use the last available and valid information in panel studies. That is, for religious denomination, we used the last available information on the respondent's denomination they stated. In general, we adopted established concepts, like the ISCED-97-scale, on educational level. Our choice of the ISCED-97-scale is because the data sets GGS and SHARE only provide this variable without detailed information on the highest school and vocational degree as source variables for ISCED-97.

As a brief example, we illustrate our approach to harmonizing respondents' religious denominations. We started by checking whether the information for the variable of interest is available in the data sources

Table 14.2 Target code and coding instructions for the harmonization process of the variable religious denomination.

Values	Labels	Coding instruction
[−10]	Missing by study design	For surveys that did not ask for the religious denomination of respondents
[0]	No Religion	Without religious denomination
[10]	Christian, not further specified	Indicating Christian denomination, but the exact religious group remains unknown
[11]	Christian, Roman Catholic	Roman Catholic
[12]	Christian, Protestant	Parishes belonging to the Evangelical Church in Germany (EKD), no congregational or new apostolic chapels
[13]	Other Christian Religion	The denomination is known, e.g. Russian Orthodox, new apostolic church, but groups not belonging to codes 11, 12
[20]	Non-Christian Religion, not further specified	Not further specified; exact non-Christian denomination unknown
[21]	Jewish	Jewish denomination
[22]	Muslim	Muslim denomination
[23]	Other Non-Christian	Other non-Christian denominations, groups not belonging to 21, 22
[30]	Other, not further specified	"Others" in general, not classifiable whether Christian or non-Christian

by examining questionnaires, codebooks, and the data files themselves. If available, we compared the coding (e.g. Roman Catholic, Protestant, Christian, or other) and the survey routines (surveyed once and repeatedly) as well as details of the variable coding schemes. Based on this information, we established a harmonizing approach to preserve as much information as possible. The lowest levels of our coding system (see Table 14.2) for religious denomination are subcategories that identify certain religious groups (e.g. 11: Roman Catholic and 12: Protestant) nested in more general categories of religious denominations (e.g. 10: Christian and not further specified). For surveys without detailed information on membership in certain religious groups, we rely on these parent categories; see the GGS data, for instance.

14.3 Documentation and Quality Assurance

14.3.1 Documentation

Documentation concerns the source data sets as well as the harmonized data set. Especially in a data harmonization project, proper documentation of the source data sets is of utmost importance and, unfortunately, sometimes neglected when publishing data. In the HaSpaD project, though, we only use data sets found in GESIS service infrastructures or in German Research Data Centers accredited by the German Data Forum (RatSWD), which ensures that the documentation adheres to a set of documentation standards. The documentation for the HaSpaD project has three core elements: (i) The data manual, which is available in German and English, provides an overview of the HaSpaD project (Schulz et al. 2021). For instance, it explains how to access the data, the data harmonization process, and the harmonized data itself. (ii) The actual harmonization process, especially with respect to variable harmonization, is documented in multiple HTML files that are dynamically created, adopting the idea of literate programming, i.e. mixing documentation and programming code (here: Stata code) (Knuth 1984, and for the technical implementation, see Section 14.5). These documentation files inform about the original source variables, the source data, the necessary steps to create the harmonized variables, and possible data idiosyncrasies. (iii) Finally, the HaSpaD-Harmonization Wizard is an online tool that allows users to select variables and surveys to include in the harmonization process (https://haspad.gesis.org/wizard). In addition, it also serves documentation purposes as it gives an overview of the availability, coding, and missing labels of additional target variables and provides general information on the surveys (such as number and type of partnerships, survey years, and links to the respective research data centers).

14.3.2 Quality Assurance

14.3.2.1 Process-Related Quality Assurance

Quality assurance measures have been implemented in the early stages of the project and in the process of data harmonization itself. For harmonization projects, it is mandatory to establish and maintain good working relationships with the data agencies, e.g. the research data centers. In the grant proposal stage, the principal investigators, Schulz and Weiß, informed their respective survey programs about the HaSpaD project and asked for letters of intent. This step was important since we had a multitude of specific questions about the respective source data sets, and fortunately, all research data centers fully supported us.

With respect to all Stata-related tasks, it should be noted that establishing the four-eye principle helps improve the Stata code quality. After one person has written a harmonization routine, a second person independently checks the validity of the code and the documentation. However, it also should be noted that this is a very labor-intensive approach that could not always be implemented.

14.3.2.2 Benchmarking the Harmonized HaSpaD Data Set with Official Statistics

To assess data quality, we compare administrative statistics on divorce risks with those estimated with the survey-based, harmonized, and cumulated HaSpaD data. Previous research has demonstrated the general feasibility of this approach (Boertien 2020; O'Connell 2007; Vergauwen et al. 2015). We decided to compare the duration-specific divorce rates for different marriage cohorts, i.e. the cumulative percentage of marriages ending in divorce by year of marriage and marriage duration in years. These can be calculated with both data sources. Our estimates include marriages of up to 25 years of marital duration.

Different mechanisms can lead to overestimating marriage stability in survey data: (i) divorcees are generally less likely to participate in surveys and provide a full account of their marriage histories (Bumpass and Raley 2006; Kalmijn 2021; Mitchell 2010). (ii) Interviewers and/or respondents might be motivated to underreport (past) marriages to shorten the interview (Kreyenfeld et al. 2013). (iii) Prospectively collected data on partnership biographies in panel studies are likely to *overestimate* union stability because it is more difficult to recruit individuals who are separated between panel waves for further interviews (disproportional panel attrition) (Boertien 2020; Müller and Castiglioni 2015).

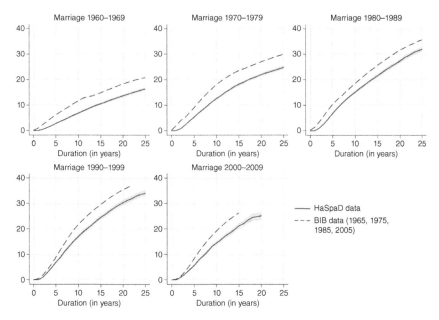

Figure 14.1 Cumulated marital dissolutions (in percent) by marital duration (in years) in the pooled HaSpaD data (note: without data from the Mannheim Divorce Study) and the German vital statistics BIB. Source: HaSpaD and BIB 2020, own calculations.

Figure 14.1 shows the cumulated percentage of marriages that have ended in divorce or separation by marriage duration and marriage cohort (marriage year for the administrative data). Estimates are from the harmonized and pooled survey data (solid lines) and official data from the BIB (2020) (dashed lines) in the respective panels of the plot. The results for the HaSpaD data are cumulated failure functions of the outcome "separation or divorce," with 95% confidence intervals (grayish areas around the colored lines) estimated with the life table method (Kaplan–Meier estimates). We excluded marriages formed before 1965 and after 2009 to compare mostly similar time frames. We also included the outcome "separation" to capture marriages that are about to divorce in the near future but after the interview. The exclusion of this outcome would lead to bias for the young marriage cohorts, for which calculations are mainly based on the panel studies pairfam and SOEP because, in these, separated marriages are more likely to drop out in future panel waves (and thus, before their legal divorce, see e.g. Müller and Castiglioni 2015). We excluded marriages from the Mannheim Divorce Study because its sampling procedure (approx. 50% of respondents married in their first marriage and 50% divorced or widowed after a first marriage) would lead to biased estimates of marriage stability compared to administrative statistics. The analysis is based on 78,736 marriages with full biographical information, of which 15,421 were dissolved at the time of the interview.

In Figure 14.1, we can observe an increasing marital instability across marriage cohorts in both the HaSpaD data and the vital statistics, which seem to level off in the two youngest marriage cohorts. We see that within all marriage cohorts, the dissolution risk is underestimated in the pooled survey data compared to the administrative data. There are few differences in this pattern across cohorts. In the two youngest marriage cohorts, the underestimation of dissolution risk is comparatively low within the first five years of marriage.

From our perspective, the survey data on Germany cumulated in the HaSpaD project produce quite similar trends in marital stability compared to the administrative statistics. The deviations that occur are mostly plausible given the different data generation mechanisms. Therefore, we are confident to draw meaningful conclusions concerning marital instability and its predictors over time from the HaSpaD data.

14.4   Challenges to Harmonization

The HaSpaD project aimed for complete coverage of survey data, containing information on German partnership biographies. Therefore, the first challenge was the scattered information on survey data across different research data infrastructures and the limited opportunities to search for information at the measurement or variable level. In the HaSpaD project, harmonization concerned the variable level and included harmonizing surveys with differing, often complex data structures. Another layer of complexity was introduced because surveys differed a lot with respect to the definition of partnerships, which the respondents should consider in their respective surveys. As these challenges were discussed in depth in Chapter 2, we chose to elaborate on methodological challenges of analyzing harmonized complex survey data in this chapter.

14.4.1   Analyzing Harmonized Complex Survey Data

Many of the surveys included in the HaSpaD data set offer survey weights to adjust for unequal sampling probabilities (design weights) or post-stratification weights to adjust for the difference between the (weighted) sample distribution and the population distribution for variables that are assumed to be related to key outcomes. The use of survey weights has often been described as "a mess" (Gelman 2007, p. 153) and does not become less complicated after combining several data sets. Zieliński et al. (2018) provide an overview of the use of sampling between 1966 and 2013, spanning 1035 national surveys from 22 international projects. Granda and Blasczyk (2016), Slomczynski and Tomescu-Dubrow (2018), as well as Tomescu-Dubrow and Slomczynski (2016), also give general overviews of the challenges of survey data harmonization.

An important first step in analyzing pooled (complex) survey data is to establish whether one is looking for design-based or model-based inference. A design-based perspective stresses the importance of randomization for inference to a well-defined, finite population, often through the help of survey weights. A model-based perspective focuses on the process that produced variation in the outcome. Potential HaSpaD data users are primarily interested in associations between variables, i.e. model-based inference, and typically fit general linear models or related models to the data. As already shown (Dumouchel and Duncan 1983; Kish and Frankel 1974; Solon et al. 2013), weighting survey data is not needed if the sampling is assumed to be non-informative, i.e. the sample distribution is the same as the population distribution. For instance, if the variables defining the sampling weights are included in the model and the model is correctly specified, design weighting is unnecessary, also in the case of pooled data (Joye and Wolf 2019).

However, the number of variables needed to account for different inclusion probabilities increases with the number of surveys included, even though a set of variables may be considered standard for German surveys (e.g. sex, geographical context such as East/West Germany). Since the set of covariates is limited and nonoptimal for estimating causal models (and some covariates are systematically missing for parts of the surveys), the assumption that the survey weights are unrelated to the dependent variable and conditional on the values of predictors becomes unlikely. The unweighted parameter estimators will then be biased with respect to the true model.

HaSpaD data users will probably focus on population-averaged estimates of the relationship between two variables or changes in these relationships over time, e.g. the estimation of the averaged influence of premarital cohabitation on the stability of one's marriage in Germany in 1980 compared to today. Such population-averaged (weighted) parameter estimates in regressions using complex survey data are usually derived from the pseudo maximum likelihood (Skinner et al. 1989) for variance estimates and the so-called sandwich estimator for survey-weighted variance estimates (Binder 1983).

In the case of survey-weighted regressions, an additional challenge arises if the pooled data is only a subsample of the samples of the source data sets, which is true for HaSpaD data. For correct standard errors (of survey-weighted estimates), it must be reflected that persons in partnerships are only a subsample of the surveys we included. This requires appending additional information about the non-included single respondents (i.e. number of observations or survey weights) to the HaSpaD data set. However, correct point estimates can be obtained without reflecting the subsampling in the weighting (West et al. 2008).

If applying the survey weights, the question is whether fitting separate regression models to every survey data set and then combining the estimates in a meta-analysis (a so-called two-stage meta-analysis, see, e.g. G.B. Stewart et al. 2012) or combining the pooled data and then fitting a multilevel model reflecting the heterogeneity between surveys (one-stage meta-analysis; see Asparouhov 2006; Carle 2009 for survey-weighted multilevel regressions in general) is more appropriate. As part of the project's methodological research, Haensch and Weiß (2020) showed that in two-stage analysis, bias could be introduced through survey weighting in the first stage, the single survey stage. They demonstrated that unless the coefficient of variance for the survey weights is small, the assumption of known within-study variances for two-stage meta-analysis is problematic and can result in biased point estimates (see also Burke et al. 2017; Stijnen et al. 2010). Fortunately, by avoiding the step of estimating point and variance estimates for the single surveys, weighted one-stage analysis remains approximately unbiased.

14.4.2 Sporadically and Systematically Missing Data

Furthermore, we faced the problem of sporadically (for some part of the observations in a survey) and systematically (for all observations in a survey) missing data. Not all the selected HaSpaD variables are provided in each of the harmonized surveys (see also Figure 14.3 for an illustration). While basic sociodemographic information like anchor's sex is available in all surveys, more specific information such as anchor's age at parental separation is not available in all surveys. Multiple imputation (MI) approaches for systematically missing data in covariates in generalized linear models have already been developed and tested (Audigier et al. 2018; Jolani 2018; Resche-Rigon et al. 2013).

HaSpaD data will mainly be used for discrete-time event history analysis. These analyses use data in a person-period format (or partnership-period format) (Singer and Willett 2003), a data format that leads to difficulties when using MI. With MI, it is crucial to include outcome information in the model for imputing partially observed covariates. This procedure is not straightforward in the case of discrete-time event analysis model, since we (i) usually have a partly observed (left- and/or right-censored) outcome, (ii) do not have a single outcome variable, but two: an event as well as a time-to-event indicator, and (iii) have to decide whether to impute data in the person format or in the person-period format (after transformation), especially if we look at time-invariant variables. This problem is highly relevant for HaSpaD data since most additional variables are constructed as time-invariant variables (see Section 14.2.2.2).

There is little guidance on how to incorporate the observed outcome information in the imputation model of missing time-invariant covariates for discrete-time event analysis. Therefore, Haensch et al. (2022) explored different methods using "fully conditional specification" (FCS) (Van Buuren et al. 2006) and a newer missing imputation approach, which is called "substantive model compatible fully conditional specification" (SMC-FCS) (Bartlett et al. 2015). These approaches vary in their complexity with which they incorporated the outcome into the imputation model, the respective FCS algorithm used, and the data format used during the imputation. Haensch et al. (2022) confirmed the results by White and Royston (2009) and Beesley et al. (2016) for continuous-time event analysis that imputing conditional on the (partly imputed) uncensored time-to-event yields high bias. A compatible imputation model for SMC-FCS MI (implemented as smcfcs.dtsam in the R package smcfcs, Bartlett and Keogh 2021) proves to be the key to imputations with good performance.

14.5 Software Tools

In the following, we focus on workflows and respective software tools we used that we deem interesting for the audience of this handbook. We will introduce our software stack for dynamically creating up-to-date documentation. The actual harmonizing work was done using Stata, i.e. executing mapping rules to restructure data sets to adhere to our HaSpaD data set template (see Section 14.2.2.1) and converting various variable scales from different survey programs into a unified, overall scale. This work results in a collection of Stata do-files for either harmonizing the data structure or harmonizing variables among survey programs.

As mentioned in Section 14.3.1, we adopted a literate programming approach to embed Markdown chunks in the Stata do-files. For this, we utilized "MarkDoc" (Haghish 2016), which is a general-purpose literate programming package for Stata. An example of a combination of markdown and Stata code can be found in Listing 14.1.

Listing 14.1 Example of a MwarkDoc Document

```
...
/***
# PROCEDURE
***/

/***
Target variable __denom__  is generated using the source variable
__rd01__  from the data set __ZA4586_v1 -0-0.dta__.
***/

use "'s102'/ZA4586_v1-0-0.dta", clear

// OFF
keep year respid rd01
numlabel rd01, add
//ON
...
```

After parsing this MarkDoc document in Stata, the documentation is extracted, resulting in an HTML file (see Figure 14.2).

With respect to disseminating the results of our harmonizing efforts and adhering to open science principles, it would be sufficient to put all Stata files and the respective documentation into a repository, such as GitHub, OSF (Open Science Framework), or GESIS' SowiDataNet|datorium, where users can download all materials. We provide our complete Stata code for all surveys and variables via the SowiDataNet|datorium for reuse (https://doi.org/10.7802/2317). However, since the code package includes the harmonization of all variables across all included data sets, users may prefer to select variables and data sets individually. Therefore, we decided to develop the so-called HaSpaD-Harmonization Wizard (https://haspad.gesis.org/wizard), an online tool that allows users to select survey programs and variables for harmonization (see Figure 14.3 for a screenshot).

After completing the selection process, the HaSpaD-Harmonization Wizard creates customized Stata code, which allows the user to merge the selected survey programs and variables from the source data

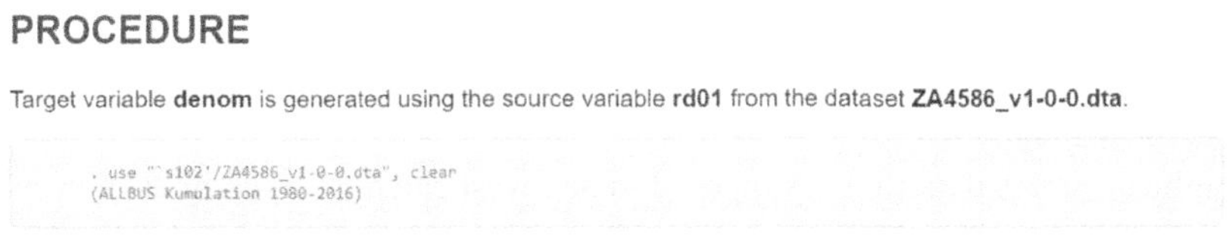

PROCEDURE

Target variable **denom** is generated using the source variable **rd01** from the dataset **ZA4586_v1-0-0.dta**.

```
. use "'s102'/ZA4586_v1-0-0.dta", clear
(ALLBUS Kumulation 1980-2016)
```

Figure 14.2 MarkDoc document after parsing in Stata.

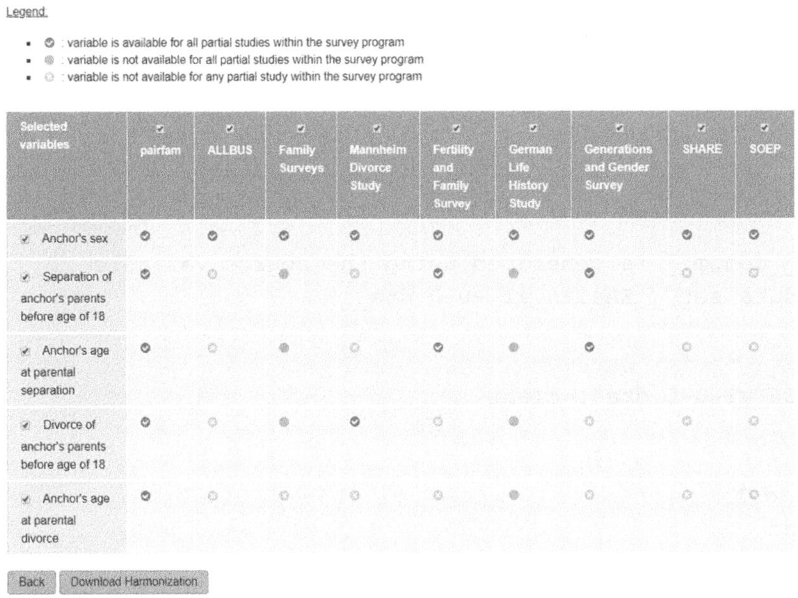

Figure 14.3 Screenshot from the HaSpaD-Harmonization Wizard illustrating the availability and systematic missingness of data for five selected variables over the nine survey programs.

sets into a so-called target data set containing the harmonized data. Altogether, we offer a ZIP file with all materials needed in a predefined, standardized folder structure that enables the user to start the data harmonization with the selected variables within the selected survey programs. The development of the HaSpaD-Harmonization Wizard included writing program code for the online user interface as well as extensive usability testing, which improved the usability of our tool considerably.

As mentioned above, a necessary precondition is that users must gain access to the respective data sets on their own. However, everything else, such as selecting surveys or variables, can be conveniently done via the HaSpaD-Harmonization Wizard, which provides customized Stata code. After executing these do-files in Stata, it will result in one harmonized and cumulative data file that can then be used for further research.

14.6 Recommendations

14.6.1 Harmonizing Biographical Data

14.6.1.1 Methodological Recommendations

When harmonizing biographical survey data that is supposed to be investigated with methods of event history data analysis, it is important to distinguish issues of the harmonization process related to the data structure and/or substantial variables. For harmonizing (manifest) substantial variables, we provide rules that map different source variables to a common target variable. Generally, our guiding principle was to transparently document all data processing and all decisions we made by clearly defined rules.

This includes carefully documenting data idiosyncrasies and inconsistencies in the data manual, as well as adding flag variables to the cumulative data set. In case we noticed inconsistencies in the source data, we decided against any attempts to "enhance" data quality by deleting those cases or by coding what seemed to be the "most plausible" answer (see Section 14.2.2.1). Thus, we recommend a critical use of the cumulated data sets and advise users to familiarize themselves with our data documentation before working with our HaSpaD data set.

14.6.1.2 Procedural Recommendations

Since we were dealing with large-scale and complex survey data sets, we aimed at developing good working relationships with the respective data providers. We recommend maintaining close communication with involved research agencies. It is important to reach an agreement to allow further users to obtain the exact same version of the source data used for *ex post* harmonization.

Furthermore, since programming – here in Stata – is a central part of the harmonization process, it is important to establish a basic quality assurance process, e.g. we established a four-eye principle to improve Stata code quality. We recommend keeping this rule, although it is very labor-intensive (see Section 14.3.2.1).

14.6.1.3 Technical Recommendations

Three principles are important when setting up the technical environment for data harmonization: automation, consistency, and flexibility. We recommend automating the harmonization process as much as possible, using existing or developing new programming tools. Our recommendation of consistency refers to variable naming schemes, file and folder names, increasing efficient file access, and programming to read and write data. It is also advisable to create a uniform documentation template of data processing, i.e. use automated HTML code documentation within code files to make documentation more efficient and time-saving if something must be changed. We recommend that the technical environment be as flexible as possible by avoiding fixed files and folder paths.

14.6.2 Getting Started with the Cumulative HaSpaD Data Set

The HaSpaD-Harmonization Wizard (see Section 14.5) provides customized Stata code to harmonize and cumulate the source data sets. As mentioned earlier, the respective source data sets must be obtained by the users themselves.

The "HaSpaD data manual" (Schulz et al. 2021) serves as a central entry point and helps to familiarize with the project. Overall, it contains information on the HaSpaD project and respective source studies. In addition, it describes the HaSpaD code and the underlying data structure. Chapter 1, "Quickstart," lists system requirements for users and provides brief step-by-step instructions on how to create the final HaSpaD-target data set. The following Chapters 2–6 introduce the HaSpaD project itself and provide more information on how to work with the HaSpaD-Harmonization Wizard, the Stata code, as well as the source data sets. From a conceptual and methodological standpoint, most important is Chapter 5 since it documents idiosyncrasies of the source data sets as well as the methodological challenges when working with the harmonized and cumulated HaSpaD data set. We discuss issues such as varying definitions of partnerships, differences in the maximum number of reported partnerships, or varying observation periods for partnerships. Furthermore, we illustrate the different reference populations, how to deal with survey weights, and the imputation of dates. Finally, we explain the process of working with panel surveys without partner-ID (the GLHS and the GGS).

To conclude, we recommend that users thoroughly read the HaSpaD data manual to familiarize themselves with the particularities of the included source data sets. Doing so helps to carefully evaluate the associated consequences with respect to their preparation of an analysis sample and make an informed decision concerning the respective model estimation, such as the inclusion of necessary controls.

Acknowledgments

The project "HaSpaD" was funded by the Deutsche Forschungsgemeinschaft (DFG, German Research Foundation) – Project number 316901171. Financial support is gratefully acknowledged. During the data preparation, all research data centers fully supported us by giving us important advice on preparing the source data. We are very grateful for this excellent support. Furthermore, harmonization efforts and data preparation profited very much from discussions with our scientific project partners Oliver Arránz Becker, Gerrit Bauer, Thomas Klein, Ingmar Rapp, and Michael Wagner. We would also like to thank the participants of the 18th Annual Meeting of the European Divorce Network in October 2020 for their fruitful discussion on the project procedures. Finally, we would like to thank the editorial team and anonymous reviewers who provided helpful comments on an earlier manuscript draft.

References

Asparouhov, T. (2006). General multi-level modeling with sampling weights. *Communications in Statistics – Theory and Methods* 35 (3): 439–460. https://doi.org/10.1080/03610920500476598.

Audigier, V., White, I.R., Jolani, S. et al. (2018). Multiple imputation for multilevel data with continuous and binary variables. *Statistical Science* 33 (2): https://doi.org/10.1214/18-STS646.

Bartlett, J.W., Seaman, S.R., White, I.R., and Carpenter, J.R. (2015). Multiple imputation of covariates by fully conditional specification: accommodating the substantive model. *Statistical Methods in Medical Research* 24 (4): 462–487. https://doi.org/10.1177/0962280214521348.

Bartlett, J. W., & Keogh, R., & Bonneville, R. & Ekstrøm, C. (2023). smcfcs: multiple imputation of covariates by substantive model compatible fully conditional specification. https://CRAN.R-project.org/package=smcfcs

Beesley, L.J., Bartlett, J.W., Wolf, G.T., and Taylor, J.M.G. (2016). Multiple imputation of missing covariates for the Cox proportional hazards cure model. *Statistics in Medicine* 35 (26): 4701–4717. https://doi.org/10.1002/sim.7048.

BIB (2020). *Anteil der geschiedenen Ehen der Eheschließungsjahrgänge 1965, 1975, 1985, 1995 und 2005 nach Ehedauer in Deutschland (Stand 2016)*. Bundesinstitut für Bevölkerungsforschung https://www.bib.bund.de/Permalink.html?id=10238144.

Binder, D.A. (1983). On the variances of asymptotically normal estimators from complex surveys. *International Statistical Review/Revue Internationale de Statistique* 51 (3): 279. https://doi.org/10.2307/1402588.

Boertien, D. (2020). The conceptual and empirical challenges of estimating trends in union stability: have unions become more stable in Britain? In: *Divorce in Europe*, vol. 21 (ed. D. Mortelmans), 17–36. Springer International Publishing https://doi.org/10.1007/978-3-030-25838-2_2.

Brüderl, J., Garrett, M., Hajek, K. et al. (2017). Pairfam data manual release 8.0. https://doi.org/10.4232/pairfam.5678.8.0.0.

Brüderl, J., Drobnič, S., Hank, K. et al. (2021). The German Family Panel (pairfam) (12.0.0) [Data set]. GESIS Data Archive. https://doi.org/10.4232/PAIRFAM.5678.12.0.0.

Bumpass, L. and Raley, K. (2006). Measuring divorce and separation: issues, and comparability of estimates across data sources. In: *Handbook of Measurement Issues in Family Research* (ed. S.L. Hofferth and L.M. Casper), 127–143. Lawrence Erlbaum Associates.

Bundesinstitut für Bevölkerungsforschung (2002). Deutscher Fertility and Family Survey 1992. https://doi.org/10.4232/1.3400.

Burke, D.L., Ensor, J., and Riley, R.D. (2017). Meta-analysis using individual participant data: one-stage and two-stage approaches, and why they may differ. *Statistics in Medicine* 36 (5): 855–875. https://doi.org/10.1002/sim.7141.

Carle, A.C. (2009). Fitting multilevel models in complex survey data with design weights: recommendations. *BMC Medical Research Methodology* 9 (1): 49. https://doi.org/10.1186/1471-2288-9-49.

Deutsches Jugendinstitut (DJI) (1992). Wandel und Entwicklung familialer Lebensformen in Westdeutschland (Familiensurvey). https://doi.org/10.4232/1.2245.

Deutsches Jugendinstitut (DJI) (2018a). Familie und Partnerbeziehungen in Ostdeutschland (Familiensurvey). https://doi.org/10.4232/1.13196.

Deutsches Jugendinstitut (DJI) (2018b). Wandel und Entwicklung familialer Lebensformen—2. Welle (Familiensurvey). https://doi.org/10.4232/1.13197

Deutsches Jugendinstitut (DJI). (2018c). Wandel und Entwicklung familialer Lebensformen—3. Welle (Familiensurvey). https://doi.org/10.4232/1.13198.

Dumouchel, W.H. and Duncan, G.J. (1983). Using sample survey weights in multiple regression analyses of stratified samples. *Journal of the American Statistical Association* 78 (383): 535–543. https://doi.org/10.1080/01621459.1983.10478006.

Esser, H., Gostomski, C.B., and von Hartmann, J. (2018). Mannheimer Scheidungsstudie 1996. https://doi.org/10.4232/1.13056.

Gelman, A. (2007). Struggles with survey weighting and regression modeling. *Statistical Science* 22 (2): https://doi.org/10.1214/088342306000000691.

Generations and Gender Programme (2019a). Generations and gender survey (GGS)—wave 1. DANS. https://doi.org/10.17026/dans-z5z-xn8g.

Generations and Gender Programme (2019b). Generations and gender survey (GGS)—wave 2. DANS. https://doi.org/10.17026/dans-xm6-a262.

GESIS-Leibniz-Institut Für Sozialwissenschaften (2018). ALLBUS/GGSS 1980–2016 (Kumulierte Allgemeine Bevölkerungsumfrage der Sozialwissenschaften/Cumulated German General Social Survey 1980–2016) (1.0.0) [Data set]. GESIS Data Archive. https://doi.org/10.4232/1.13029.

Granda, P. and Blasczyk, E. (2016). Data harmonization. https://ccsg.isr.umich.edu/chapters/data-harmonization.

Haensch, A.-C. and Weiß, B. (2020). Better together? Regression analysis of complex survey data after ex-post harmonization [preprint]. SocArXiv. https://doi.org/10.31235/osf.io/edm3v.

Haensch, A.-C., Bartlett, J., & Weiß, B. (2022). Multiple imputation of partially observed covariates in discrete-time survival analysis. Sociological Methods & Research, 1–39. https://doi.org/10.1177/00491241221140147.

Haghish, E.F. (2016). MarkDoc: literate programming in Stata. *Stata Journal* 16 (4): 964–988.

Jolani, S. (2018). Hierarchical imputation of systematically and sporadically missing data: an approximate Bayesian approach using chained equations. *Biometrical Journal* 60 (2): 333–351. https://doi.org/10.1002/bimj.201600220.

Joye, D. and Wolf, C. (2019). Weights in comparative surveys? A call for opening the black box. *Harmonization: Newsletter on Survey Data Harmonization in the Social Sciences* 5 (2): 2–16.

Kalmijn, M. (2021). Are national family surveys biased toward the happy family? A multiactor analysis of selective survey nonresponse. *Sociological Methods & Research* 004912412098620. https://doi.org/10.1177/0049124120986208.

Kish, L. and Frankel, M.R. (1974). Inference from complex samples. *Journal of the Royal Statistical Society: Series B (Methodological)* 36 (1): 1–22. https://doi.org/10.1111/j.2517-6161.1974.tb00981.x.

Klein, T., Kopp, J., and Rapp, I. (2013). Metaanalyse mit Originaldaten. Ein Vorschlag zur Forschungssynthese in der Soziologie. *Zeitschrift für Soziologie* 42 (3): 222–238.

Knuth, D.E. (1984). Literate programming. *The Computer Journal* 27 (2): 97–111. https://doi.org/10.1093/comjnl/27.2.97.

Kreyenfeld, M., Hornung, A., and Kubisch, K. (2013). The German generations and gender survey: some critical reflections on the validity of fertility histories. *Comparative Population Studies* 38 (1): https://doi.org/10.12765/CPoS-2013-02.

Liebig, S., Goebel, J., Schröder, C. et al. (2019). Socio-Economic Panel (SOEP), data from 1984-2018 (Version v35) [SAS, SPSS, Stata, SPSS, SAS]. SOEP Socio-Economic Panel Study. https://doi.org/10.5684/SOEP-CORE.V35.

Mayer, K.U. (1995a). Lebensverläufe und gesellschaftlicher Wandel: Die Zwischenkriegskohorte im Übergang zum Ruhestand (Lebensverlaufsstudie LV-West II T – Telefonische Befragung). https://doi.org/10.4232/1.2647.

Mayer, K.U. (1995b). Lebensverläufe und historischer Wandel in Ostdeutschland (Lebensverlaufsstudie LV-DDR). https://doi.org/10.4232/1.2644.

Mayer, K.U. (2004a). Ostdeutsche Lebensverläufe im Transformationsprozeß (Lebensverlaufsstudie LV-Ost 71). https://doi.org/10.4232/1.3926.

Mayer, K.U. (2004b). Ostdeutsche Lebensverläufe im Transformationsprozeß (Lebensverlaufsstudie LV-Ost Panel). https://doi.org/10.4232/1.3925.

Mayer, K.U. (2014). Frühe Karrieren und Familiengründung: Lebensverläufe der Geburtskohorte 1971 in Ost- und Westdeutschland (Lebensverlaufsstudie LV-Panel 71). https://doi.org/10.4232/1.5099.

Mayer, K.U. (2018a). Lebensverläufe und gesellschaftlicher Wandel: Berufszugang in der Beschäftigungskrise (Lebensverlaufsstudie LV-West III). https://doi.org/10.4232/1.13195.

Mayer, K.U. (2018b). Lebensverläufe und gesellschaftlicher Wandel: Die Zwischenkriegskohorte im Übergang zum Ruhestand (Lebensverlaufsstudie LV-West II A - Persönliche Befragung). https://doi.org/10.4232/1.13194.

Mayer, K.U. (2018c). Lebensverläufe und gesellschaftlicher Wandel: Lebensverläufe und Wohlfahrtsentwicklung (Lebensverlaufsstudie LV-West I). https://doi.org/10.4232/1.13193.

Mayer, K. U. and Kleinhenz, G. (2004). Ausbildungs- und Berufsverläufe der Geburtskohorten 1964 und 1971 in Westdeutschland (Lebensverlaufsstudie LV-West 64/71). https://doi.org/10.4232/1.3927.

Mitchell, C. (2010). Are divorce studies trustworthy? The effects of survey nonresponse and response errors. *Journal of Marriage and Family* 72 (4): 893–905. https://doi.org/10.1111/j.1741-3737.2010.00737.x.

Müller, B. and Castiglioni, L. (2015). Stable relationships, stable participation? The effects of partnership dissolution and changes in relationship stability on attrition in a relationship and family panel. *Survey Research Methods* 9: 205–219. https://doi.org/10.18148/SRM/2016.V10I1.6207.

O'Connell, M. (2007). The visible hand: editing marital-history data from Census Bureau surveys. In: *Handbook of Measurement Issues in Family Research* (ed. S.L. Hofferth and L.M. Caspar), 145–156. Lawrence Erlbaum Associates.

Resche-Rigon, M., White, I.R., Bartlett, J.W. et al. (2013). Multiple imputation for handling systematically missing confounders in meta-analysis of individual participant data. *Statistics in Medicine* 32 (28): 4890–4905. https://doi.org/10.1002/sim.5894.

Schulz, S., Weiß, B., Sterl, S. et al. (2021). *HaSpaD – Data Manual (September 2021) (GESIS Papers 021/12)*. GESIS – Leibniz Institute for the Social Sciences https://doi.org/10.21241/ssoar.75135.

Singer, J.D. and Willett, J.B. (2003). *Applied Longitudinal Data Analysis: Modeling Change and Event Occurrence*. Oxford University Press.

Skinner, C.J., Holt, D., and Smith, T.M.F. (ed.) (1989). *Analysis of Complex Surveys*. Wiley.

Slomczynski, K.M. and Tomescu-Dubrow, I. (2018). Basic principles of survey data recycling. In: *Advances in Comparative Survey Methods* (ed. T.P. Johnson, B.-E. Pennell, I.A.L. Stoop, and B. Dorer), 937–962. Wiley https://doi.org/10.1002/9781118884997.ch43.

Solon, G., Haider, S., and Wooldridge, J. (2013). What Are we Weighting for? (No. w18859; p. w18859). National Bureau of Economic Research. https://doi.org/10.3386/w18859.

Stewart, L.A. and Tierney, J.F. (2002). To IPD or not to IPD? Advantages and disadvantages of systematic reviews using individual patient data. *Evaluation and the Health Professions* 25 (1): 76–97.

Stewart, G.B., Altman, D.G., Askie, L.M. et al. (2012). Statistical analysis of individual participant data meta-analyses: a comparison of methods and recommendations for practice. *PLoS One* 7 (10): e46042. https://doi.org/10.1371/journal.pone.0046042.

Stijnen, T., Hamza, T.H., and Özdemir, P. (2010). Random effects meta-analysis of event outcome in the framework of the generalized linear mixed model with applications in sparse data. *Statistics in Medicine* 29 (29): 3046–3067. https://doi.org/10.1002/sim.4040.

The Survey of Health, Ageing and Retirement in Europe (SHARE) (2017). SHARE Waves 1, 2, 3, 4, 5 and 6. This paper uses data from SHARE Waves 1, 2, 3, 4, 5 and 6. https://doi.org/10.6103/SHARE.w1.600, 10.6103/SHARE.w2.600, 10.6103/SHARE.w3.600, 10.6103/SHARE.w4.600, 10.6103/SHARE.w5.600, 10.6103/SHARE.w6.600.

Tomescu-Dubrow, I. and Slomczynski, K.M. (2016). Harmonization of cross-National Survey Projects on political behavior: developing the analytic framework of survey data recycling. *International Journal of Sociology* 46 (1): 58–72. https://doi.org/10.1080/00207659.2016.1130424.

Van Buuren, S., Brand, J.P.L., Groothuis-Oudshoorn, C.G.M., and Rubin, D.B. (2006). Fully conditional specification in multivariate imputation. *Journal of Statistical Computation and Simulation* 76 (12): 1049–1064. https://doi.org/10.1080/10629360600810434.

Vergauwen, J., Wood, J., De Wachter, D., and Neels, K. (2015). Quality of demographic data in GGS wave 1. *Demographic Research* 32: 723–774. https://doi.org/10.4054/DemRes.2015.32.24.

Wagner, M. and Weiß, B. (2003). Bilanz der deutschen Scheidungsforschung. Versuch einer Meta-Analyse. *Zeitschrift Für Soziologie* 32 (1): 29–49. https://doi.org/10.1515/zfsoz-2003-0102.

Wagner, M. and Weiß, B. (2006). On the variation of divorce risks in Europe: findings from a meta-analysis of European longitudinal studies. *European Sociological Review* 22 (5): 483–500. https://doi.org/10.1093/esr/jcl014.

West, B.T., Berglund, P., and Heeringa, S.G. (2008). A closer examination of subpopulation analysis of complex-sample survey data. *The Stata Journal: Promoting Communications on Statistics and Stata* 8 (4): 520–531. https://doi.org/10.1177/1536867X0800800404.

West, B.T., Sakshaug, J.W., and Aurelien, G.A.S. (2016). How big of a problem is analytic error in secondary analyses of survey data? *PLOS One* 11 (6): e0158120. https://doi.org/10.1371/journal.pone.0158120.

West, B.T., Sakshaug, J.W., and Kim, Y. (2017). Analytic error as an important component of total survey error. In: *Total Survey Error in Practice* (ed. P.P. Biemer, E.D. de Leeuw, S. Eckman, et al.), 489–510. Wiley.

White, I.R. and Royston, P. (2009). Imputing missing covariate values for the Cox model. *Statistics in Medicine* 28 (15): 1982–1998. https://doi.org/10.1002/sim.3618.

Wolf, C., Schneider, S., Behr, D., and Joye, D. (2016). Harmonizing survey questions between cultures and over time. In: *The SAGE Handbook of Survey Methodology* (ed. C. Wolf, D. Joye, T. Smith, and Y. Fu), 502–524. Sage Publications https://doi.org/10.4135/9781473957893.

Zieliński, M.W., Powałko, P., and Kołczyńska, M. (2018). The past, present, and future of statistical weights in international survey projects: implications for survey data harmonization. In: *Advances in Comparative Survey Methods* (ed. T.P. Johnson, B.-E. Pennell, I.A.L. Stoop, and B. Dorer), 1035–1052. Wiley https://doi.org/10.1002/9781118884997.ch47.

15

Harmonization and Quality Assurance of Income and Wealth Data: The Case of LIS

Jörg Neugschwender[1], Teresa Munzi[1], and Piotr R. Paradowski[1,2]

[1]*LIS (Luxembourg Income Study) Cross-National Data Center in Luxembourg, Belval, Luxembourg*
[2]*Department of Statistics and Econometrics, Gdańsk University of Technology, Gdańsk, Poland*

15.1   Introduction

LIS – also known as the *Luxembourg Income Study* or *LIS Cross-National Data Center in Luxembourg* – is an independent, nonprofit cross-national microdata archive and research institute. One of the main reasons behind launching LIS is providing researchers with microdata on income (and later wealth) that facilitates comparative analyses across countries and time. Initiated at an international conference in Luxembourg in 1982, which saw the gathering of experienced researchers from various countries, the idea – rather innovative for the time – developed that "it would be possible to pool the knowledge and experience in these countries to create internally and externally consistent data sets for comparative studies, which are far superior to those currently in existence." (Smeeding et al. 1985). The main drivers for establishing the *Luxembourg Income Project* were relatively simple. Which welfare state policies work to protect risk groups, e.g. which public and private transfers provide income in old age? Even more importantly, what is needed to well compare income data across countries? Thus, LIS was founded based on the need to carry out conceptual harmonization that allowed the subsequent analyses to compare similar concepts, an objective that is still today a crucial element of LIS' day-to-day work.

Source data entering LIS differ substantially in terms of collection mode (surveys versus administrative data), type of information collected, level of detail, and structure of the data (Figure 15.1). Through the harmonization process, the *ex post* harmonized data are joined together in a common repository, where each set of data (or *dataset*, i.e. a group of individuals and households representing the total population in one year for one country) has the same structure and each variable has comparable contents. Researchers are able to access all datasets via a remote access system.

The harmonized LIS datasets are stored in two *databases:* the *Luxembourg Income Study (LIS) Database*, which includes income data, and the *Luxembourg Wealth Study* (LWS) *Database*, which focuses on wealth data. The LIS data target all individuals in a household unit (typically defined as a single person or a group of persons living in one dwelling and sharing a budget). Each dataset contains

Survey Data Harmonization in the Social Sciences, First Edition.
Edited by Irina Tomescu-Dubrow, Christof Wolf, Kazimierz M. Slomczynski and J. Craig Jenkins.
© 2024 John Wiley & Sons Inc. Published 2024 by John Wiley & Sons Inc.

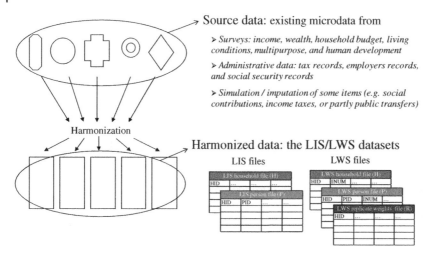

Source data: existing microdata from

➤ *Surveys: income, wealth, household budget, living conditions, multipurpose, and human development*

➤ *Administrative data: tax records, employers records, and social security records*

➤ *Simulation / imputation of some items (e.g. social contributions, income taxes, or partly public transfers)*

Harmonization

Harmonized data: the LIS/LWS datasets

LIS files   LWS files

Figure 15.1 *Ex post* harmonization at LIS.

household- and person-level data on labor income, capital income, pensions, public social security benefits (excluding pensions), private transfers, taxes and contributions and other nonconsumption expenditures, and consumption expenditures, as well as sociodemographic and labor market information. Besides the identical blocks of information from LIS datasets, the wealth datasets additionally provide information on assets and liabilities, contingent assets and liabilities, assets acquired in the past, and behavioral variables.

Concerning the frequency of microdata, LIS shortened over time the interval between two periods; data in the early years were provided every five years (until 2000), then every four years (from 2000 to 2004), and then every three years (from 2004 to 2016). Since 2019, various data are acquired annually, and country series are retrospectively extended following an annual frequency when feasible. As a result, LIS stores a quickly growing large pool of data, which at the time of writing consisted of around 850 datasets harmonized and documented for the scientific community, covering nearly 60 countries and, spanning over 50 years.

With this immense pool of harmonized data, nowadays researchers can easily analyze the data from a cross-national perspective, reflecting the initial idea since launching LIS. Scholars worldwide have used LIS datasets to compare economic and social policies and their effects on outcomes, including poverty, income inequality, employment status, wage patterns, gender inequality, family formation, child-wellbeing, health status, and immigration. In addition, the newer LWS datasets enable comparative research on wealth portfolios, assets, and liabilities, the relation between household income and wealth, and economic behavior.

The following sections shed particular light on the various harmonization efforts (Section 15.2) and quality assurance procedures (Section 15.3) at LIS, exemplified with practical examples. Section 15.4 focuses on core challenges of *ex post* harmonization. Section 15.5 describes quality assurance from a technical point of view. Section 15.6 concludes with key lessons learned from nearly 40 years of harmonization of microdata.

15.2 Applied Harmonization Methods

The *Luxembourg Income Study project* has been initiated by three main ideas: (i) take preexisting micro-data, (ii) establish a common conceptual framework for harmonized data, and (iii) provide comparable microdata files commonly accessible for researchers (Smeeding et al. 1985). These motives determine the scope of LIS' work. This section presents the main workflow of harmonization at LIS, which is defined by (i) acquisition of the data, (ii) the harmonization process, and (iii) data quality assurance. Section 15.2 focuses on the first two points; the latter is described in Section 15.3.

The central aspect of LIS' work revolves around the *ex post* feature of harmonization: LIS acquires preexisting survey and administrative data in their original shape. At the same time, LIS applies a strict policy to accept data for harmonization. There are various minimum entry conditions:

- The source data must contain income and wealth items that ensure a representative and comparable situation of well-being.
- The datasets need to contain the correct microlevel detail, i.e. information needs to be collected at the household and individual level.
- The unit of analysis should be the household; only in rare circumstances, LIS accepts alternative definitions, such as the tax unit, when there is no other dataset available in the country.
- Sampling has to guarantee coverage of the total population, and weighting[1] must ensure that the sample is representative of the total population and subgroups in the country, e.g. by age, sex, region, or total labor force.

In the first *opening* of the microdata, the LIS team assesses the quality of the source data. Particularly, the screening of the source data with respect to representativeness of income and wealth indicates the need for a conceptual framework that defines the ideal concepts underlying income and wealth items. This framework has been established gradually over many years of experience through involvement in international scientific projects and contribution to expert reports. More specifically, the conceptual framework of *disposable household income, assets, liabilities, and net worth* has been ensured by LIS' participation in the Canberra Group and other expert groups such as the OECD. The resulting publications (The Canberra Group 2001; UNECE 2011; OECD 2013a,b) have shaped into internationally acknowledged guidelines for producers and users of household income distribution statistics.

A second element of the *opening* is to clarify whether the income and wealth items are collected with a low degree of missing values. LIS has established as a rule of thumb that the percentage of missing values for *disposable household income* should not exceed 20%. As this situation is a result of accumulating person and item unit-non response patterns from the individual income sources, it means that the main individual income source is expected to have only about 5–10% of missing values. A higher percentage of missing values gets particularly problematic for representativeness of large households, where missing values are more likely to occur. On the other hand, when the source data were already imputed by data providers (and this is typically the case with wealth data, which suffers from high numbers of missing values), the entry criterion for inclusion into the databases focused on the quality of the imputation of missing data. Imputation has to be carried out utilizing well-documented statistical techniques, such as multiple imputation, in compliance with latest scientific standards.

Once a dataset passes the entry criteria, the actual harmonization work starts. Before presenting concrete examples of day-to-day tasks of harmonization work at LIS, it is crucial to clarify general guiding principles that have influenced the development of the conceptual framework, the workflow at LIS, and the technical infrastructure for the LIS and LWS Databases. Due to cross-national differences in the source data, LIS established *operational comparability* as a guiding principle for *ex post* harmonization. *Operational comparability* entails finding the right compromise between, on one side, creating variables following concepts that are purely comparable from the theoretical point of view, and on the other side, seeking to make sure that the harmonized variables are available for most datasets. Thus, *operational comparability* allows researchers to analyze variables across several datasets, i.e. making comparability operational.

The outcome of *operational comparability* is best shown with examples. LIS' concept of *disposable household income* excludes the value of *imputed rent* (market rent that would be paid by homeowners and subsidized or rent-free tenants for a dwelling similar to the occupied one) from its definition, although its inclusion is widely advised in various scientific guidelines. Due to scarce availability and comparability limitations of the concept of *imputed rent* in the source data, a conceptual inclusion of imputed rent would lower *operational comparability*. Thus, users of the LIS databases would only have a limited set of datasets at their disposal.

That said, after careful selection of datasets where *imputed rent* is available, researchers can still add it to the definition of disposable household income. The conceptual framework for *disposable household income* is summarized in Figure 15.2, where it can be seen that *social transfers in kind* (STIK) are also excluded due to similar reasoning.

Likewise, the conceptual treatment of pension assets when defining the measure of *disposable net worth* is also guided by *operational comparability*. While there is scientific consensus that *total assets* should include the totality of pension assets, the current value of all pension assets (individual, occupational, and public) is rarely collected or calculated by the data providers. In order to address the gap between scientific concepts and the availability of data, LIS provides several alternative *net worth* measures that gradually include additional parts of pension assets, hence, allowing researchers to compare

	Cash	Non cash
Labor income	Wages, salaries, bonuses	In-kind earnings
	Profits and losses from self-employment	Own consumption
+ Capital income	Interest and dividends	—
	Rental income	Imputed rent
+ Pensions	Universal and assistance	—
	Contributory public insurance	—
	Occupational and individual	—
+ Public benefits	Family benefits	STIK
	Unemployment benefits	
	Sick pay/work injury/disability benefits	
	Housing/heating benefits	
+ Private transfers	Scholarships, charity	In-kind assistance
	Alimony, remittances	Gifts
- Deductions	Income taxes	
	Social security contributions	
= Disposable household income		

Figure 15.2 Conceptual framework for *disposable household income*.

Nonfinancial assets
+ Financial assets
- Total liabilities
= Disposable net worth
+ Life insurance and voluntary individual pensions
= Adjusted disposable net worth
+ Other pensions
= Total net worth

Figure 15.3 Conceptual framework for *net worth*.

the different wealth measures available in the source data. Figure 15.3 summarizes the various net worth definitions.

Alongside striving for *operational comparability*, LIS harmonization work aims as much as possible to reach full *standardization*. *Standardization* refers to the unified meaning of value codes and labels in the variables across countries and thus enables an easy understanding of the harmonized data. In the LIS databases, many categorical variables include *standardized* categories to create a high degree of *operational comparability* in the harmonized data files. Thus, LIS systematically groups the national information into more broadly defined categories, ensuring more widely available and comparable country samples – another facet of *operational comparability*.

While working on the data, the LIS team follows a set of Stata programs, the *harmonization template*. The dataset-specific programs include three elements: the documentation of variables in the source data, the coding of the final LIS variables, and any relevant discussion about country-specific decisions. Thus, once familiar with this process, each person at LIS can easily follow the harmonization practices carried out by co-workers in other datasets.

Once the LIS team has performed carefully data management, programming, and documentation to reflect the survey-specific situations, various internal routines in the *harmonization template* allow an automatized creation of partial or complete LIS files. These routines automatically create some of the LIS variables, where information can be derived from other variables. The automatized routines simplify, especially the construction of hierarchical sums in the income or balance sheet classifications, and ensure consistency across countries and over time. Furthermore, these procedures enable a consistent treatment of person and item-unit nonresponse patterns in the LIS and LWS datasets.

After presenting the harmonization methods in a rather general and abstract way, some practical examples of the harmonization work are presented below. Figure 15.4 illustrates two points, the complete *standardization* of most LIS categorical variables and the *nesting* of subcategories within the more aggregated levels. For example, the variable *own* (owned/rented housing) contains at the highest level of aggregation the codes *owned* (100) and *not owned* (200); those are widely available across countries. The idea of *nesting* refers to the subgroups within the major groups. These codes are sublevels of *owned* (100) and *not owned* (200); all information in the source data is recoded to any of the 12 specified categories and is thus fully *standardized* and *operationally* comparable. However, depending on the information

own : owned/rented housing

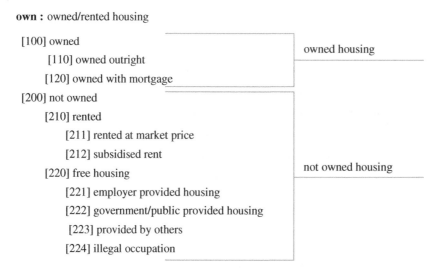

Figure 15.4 Aggregation of subcategories into overall categories.

available in the source data, either subgroups or major categories are constructed. A more detailed breakdown is available for a smaller set of countries. It is essential that users of the harmonized data consult the *Metadata Information System* (METIS) and tabulate the variables so that they can conclude whether in all their selected datasets sufficient level of information is available, e.g. that all datasets allow for analyzing the two subgroups *owned outright* versus *owned with mortgage*.

On the other hand, for variables where the national contents prove impractical to be harmonized, LIS provides dataset-specific categories and labels; technically, LIS variables ending with _c indicate country-specific variables. These variables derive from three specific considerations. First, geographic information follows national groupings, which cannot be further standardized into a common structure. Second, for various classifications, such as highest education level achieved and industry/occupation in the main job, LIS provides more national detail next to the highly grouped information in the standardized variables following international classifications. These country-specific variables open up the possibility of in-depth country case studies embedded in the otherwise harmonized cross-country context. Third, variables about financial literacy, type of business, or subjective health status contain information that is collected differently across countries. Hence, a *standardization* of codes reflecting all various national measures is not possible.

Figure 15.5 exemplifies the harmonization for the *highest level of education achieved*. National education systems differ across countries; thus, researchers are in need to study differences in educational levels when comparing educational qualifications across countries. This is where the harmonization efforts of the LIS team come into play. By closely following the *International Standard Classification of Education* (ISCED) *mappings* by the OECD, Eurostat, and the UNESCO Institute for Statistics (2015), the LIS databases provide a standardized variable. However, this standardized variable cannot simply follow an ideal classification, as it needs to accommodate for cross-national differences in data collection. Thus, for example, code 310 (BA, MA or equivalent, and short cycle tertiary) was nested in the variable codes

Country-specific variable (_c)
educ_c: educational attainment

code	label
31	less than 1st grade
32	1st,2nd,3rd,or 4th grade
33	5th or 6th grade
34	7th and 8th grade
35	9th grade
36	10th grade
37	11th grade
38	12th grade no diploma
39	high school graduate - high school diploma or equivalent
40	some college but no degree
41	associate degree in college occupation/vocation program
42	associate degree in college academic program
43	bachelor's degree (e.g. BA,AB,BS)
44	master's degree (e.g. MA,MS,MENG,MED,MSW,MBA)
45	professional school degree (e.g. MD,DDS,DVM,LLB,JD)
46	doctorate degree (e.g. PHD,EDD)

Standardised variable
educlev: highest completed education level

code	label
100	low, less than upper secondary
110	less than primary
113	never attended
120	primary
130	lower secondary
200	medium upper secondary and post-secondary non-tertiary
210	upper secondary
220	post-secondary non-tertiary
300	high, tertiary
310	BA, MA or equivalent, short-cycle tertiary
311	short-cycle tertiary
312	bachelor or equivalent
313	master or equivalent
320	doctorate or equivalent

Figure 15.5 Country-specific variable and *standardized* variable for educational qualification.

for datasets where educational qualifications were collected with less detail (e.g. the *first stage of tertiary education*). This way, without the need for further documentation, it gets relatively clear, which level of detail is available in a specific dataset. At the same time, by making available the country-specific detail in variable *educ_c*, LIS provides both the documentation of the national categories collected in the source data and the reclassification to the standardized (typically less detailed) comparable categories.

Altogether, the various elements described above exemplify that the *ex post* harmonization process requires a multifaceted skill set to harmonize microdata for the LIS income and wealth databases properly. For a high-quality *ex post* harmonization, persons working on the preparation of the data need expertise on scientific concepts and need to be able to adapt these concepts to the national context. At the same time, they need to understand what it means to provide comparable data. LIS operates under the assumption that a solid understanding of cross-national differences is more relevant than the national expertise to ensure maximum consistency for comparative research. Therefore, *internal* harmonization rather than outsourcing to country experts is the primary mode of working at LIS. The microdata experts pursue the whole harmonization chain and gradually build up invaluable expertise to reflect carefully on country differences and the challenges of harmonizing cross-national data. On the other hand, LIS has established a network of country experts who are regularly consulted while carrying out the harmonization. The broad knowledge and the consultation of national experts ensure a good reflection of the national context.

15.3 Documentation and Quality Assurance

15.3.1 Quality Assurance

Data quality at LIS is ensured at three levels: (i) the careful selection of source datasets for the LIS databases, (ii) the structured harmonization process, and (iii) the careful validation of the produced data before they are included in the LIS databases. Next to this, there is a highly automated infrastructure, making the microdata available in the remote access system and providing the metadata documentation in an entirely standardized way. Figure 15.6 summarizes the three main quality assurance stages. The choice of a Venn diagram mirrors the interdependency of each stage with the other two.

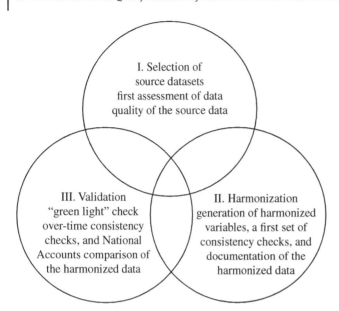

Figure 15.6 The three main stages of quality assurance at LIS.

Selection of Source Datasets

As shown in detail in Section 15.2, the selection and acquisition of source data is a main element of *ex post* harmonization. Therefore, through a careful assessment of the minimum entry criteria described in Section 15.2 for inclusion in its databases, LIS makes sure to select only datasets, which are of suitable quality for harmonization with the LIS conceptual framework, foremost with respect to completeness and representativeness of income or wealth information in the source data.

Harmonization

The structured harmonization process is the crucial element of LIS' quality assurance: the series of applied harmonization methods (described in Section 15.2) ensure that harmonization is carried out with solid expertise, foremost through the data team's fluency with income and wealth concepts, established labor market and education classifications, and multifaceted reflection of comparability problems in *ex post* harmonized microdata.

Validation – "Green Light" Check

A significant element of the harmonization work at LIS is the *over-time consistency check*. To save time for the LIS team to compute over-time comparisons, LIS established a set of flexible *checking tools*, which the team can apply to several datasets simultaneously. These *tools* enable selecting specific variables or sections of variables, allowing filter conditions, and creating a final overview of the harmonized data for the final validation. A first consistency assessment is part of the individual data expert's tasks during harmonization, but also, a central role of data quality coordination for all harmonized datasets generally contributes to the validation stage of the harmonization. In this respect, data quality coordination refers to again going through all consistency files and picking up remaining comparability issues, which are discussed and solved in the team's joint effort.

The primary purpose of checks for categorical variables is to analyze frequency patterns of categories and missing value patterns over time. Automatized Excel sheet overviews for income, expenditure, and balance sheet variables enable a comprehensive view concerning the mean values, percentage of missing values, non-missing and nonzero cases, and percentage of non-missing and nonzero cases in income and wealth items. For mean and median statistics, an automatized reading of consumer price indices and purchasing power parity (PPP) conversion rates from the World Development Indicators (WDIs), allows for exploring data in nominal values in local currency and in real values in International Dollars. All elements are critically assessed for comparability with other datasets. This process foresees going through frequency tables and the preparation of documentation of the harmonized variables over time. Once all intermediate data modules of the harmonized subsections are thoroughly checked for consistency, the final person and household-level files are constructed.

A second major component in the validation stage is comparing various components of income with corresponding numbers from the National Accounts. This enables LIS to understand better the quality of the source and harmonized data with respect to representativeness of the information provided in the data on the country level (see Endeweld and Alkemade 2014). This is a crucial element for any data analysis aimed at evidence-based conclusions about identified patterns in the data. Figure 15.7 groups the harmonized datasets from 2013 to 2016. These groupings present, from left to right, the relevance of administrative data in the data collection phase. Thus, in countries where the microdata is based on administrative data, the inflated numbers from the microdata best resemble the actual numbers in the corresponding National

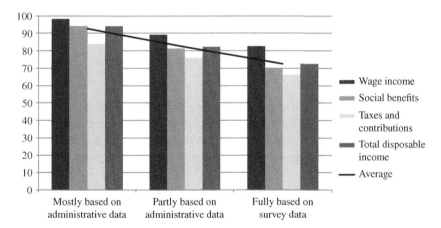

Figure 15.7 LIS to National Accounts ratios in percent – around 2016 or latest year available. Notes: Category "mostly based on administrative data" includes Austria, Denmark, Estonia, Finland, Iceland, Netherlands, Norway, Spain, and Sweden; category "partly based on administrative data" includes Canada, Ireland, Lithuania, and Switzerland; category "fully based on survey data" includes Australia, Brazil, Colombia, Czechia, Germany, Greece, Israel, Italy, Japan, South Korea, Luxembourg, Russia, Slovakia, United Kingdom, United States, and South Africa. Source: Microdata come from the Luxembourg Income Study (LIS) Database, National Accounts figures come from the Annual Detailed Non-financial Sector Accounts from the OECD Stats, June 2021.

Account categories. On the contrary, where microdata solely rely on survey data, the ratios between inflated microdata against official National Accounts are typically the lowest. Additional insights in representativeness of survey data against tax data have recently been proposed by Yonzan et al. (2020), whose approach puts a specific focus on the accuracy of measuring top incomes. However, such a procedure is currently not carried out at LIS yet, awaiting further advances in scientific testing of such methodologies.

In addition to National Accounts comparisons, a specific *data quality protocol* that is run at the end of the harmonization allows assessing the harmonized datasets with respect to: (i) sensitivity to extreme values (i.e. the impact of bottom and top coding procedures for income inequality measures) and (ii) internal consistency of overtime series (to make researchers aware of breaks in series due to changing methodology in the data collection).

15.3.2 Documentation

Documentation at LIS fulfills a central role in the harmonization process. All dataset-specific documentation is provided in the METIS, an interactive selection tool for datasets and contents. METIS provides: (i) detailed documentation of the source data entering the harmonized databases, including information on sampling, collection period, modes of collection and instruments, various data quality aspects (nonsampling error, item nonresponse/imputation, and weighting), and information on the collection of income, wealth, and labor force status; (ii) descriptive statistics for all LIS variables in each dataset; (iii) information on the exact contents of the income, consumption, expenditure, and balance sheet items, possibly with the indication of the country specificities (such as the name of the programs of the social security benefits included in the income variables); and (iv) a series of *notes* to warn researchers about situations where the contents of a variable deviate from the ideal variable definition in the conceptual framework, hence, giving rise to limitations in comparability.

15.4 Challenges to Harmonization

The *ex post* aspect of harmonization is what makes it most challenging. The variety of differences in the national data collection not only creates the need for a substantial reworking of the source data during the harmonization process but also requires clear and accessible documentation. In this section, we describe key challenges of the *ex post* harmonization work and illustrate some examples of limitations of *ex post* harmonized data.

Harmonization of income and wealth data is especially challenging. Checchi et al. (2021) summarize various specific challenges that arise from survey data collection in middle-income countries. In this chapter, we concentrate on presenting two major points. The final measures need to be both exhaustive and without double counting. Regarding exhaustiveness, the main challenge is to decide, which subcomponents to include to construct the best comparable total household income or net worth measure. For example, it is not always clear which in-kind incomes should be included in the harmonized variables. The LIS conceptual framework prescribes leaving out all *nonmonetary universal transfers from government*, whereas in-kind income received within the scope of social assistance (such as food and clothes distributed to precarious households) should be included. It is the distinction between the two that requires a thorough understanding of the country's social system, as well as the availability of detailed information and documentation of the source data.

The risk of double counting arises when information is collected in more than one part of the questionnaire. This situation appears most often in surveys that collect both income and consumption information so that some amounts can be recorded both as in-kind incomes in the income section of the questionnaire and as nonmonetary consumption in the consumption part of the questionnaire. Similarly, the same subcomponents of wealth can be collected at both individual and household levels, creating a danger of double counting as both head and spouse can claim the ownership of the same assets/debts. Persons working on the harmonization need to assess which amounts should be included in which section of the questionnaire.

Although LIS has put a strong effort in setting up a solid conceptual framework and practices for quality assessment and documentation, *ex post* harmonization bears a major challenge. The various survey and administrative data from different countries do not necessarily collect perfectly comparable information. Here we pick up again the concept of *operational comparability* that we introduced in Section 15.2, while clarifying harmonization methods. The principle is guiding LIS of what is possible to harmonize – thus defining the conceptual framework for *ex post* harmonization, including the generic definitions for each harmonized variable. Thus, over the years, LIS refined the variable list to better adopt the harmonization work to refinements in the conceptual standards. Particularly, the enlargement of the scope of *ex post* harmonization to middle-income countries brought a further focus on nonmonetary and in-kind benefits on the agenda of the harmonization work. However, the generic definitions assume availability of all income items in every national data source, whereas not necessarily all information is collected in such a way to fit in these variables perfectly – a major challenge to *operational comparability*.

Figure 15.8a,b illustrate two examples. Example (a) shows differences in the way self-employment income is collected. In country A, all self-employment income, consisting of gains and losses from

	Country A		Country B	
	Personal level variables	Household level variables	Personal level variables	Household level variables
Wage income	✓	Aggregated from personal level	✓	Aggregated from personal level
Self-employment income	✓	Aggregated from personal level	(✓) (excluding farm income)	Aggregated from personal level + farm income
of which: farm income	✓	Aggregated from personal level	✗	✓

(a)

	Country A		Country B	
	Personal level variables	Household level variables	Personal level variables	Household level variables
Private pensions	✓	Aggregated from personal level	✓	Aggregated from personal level
of which: occupational pensions	✗	✗	✓	Aggregated from personal level
of which: personal pensions	✗	✗	✓	Aggregated from personal level

(b)

Figure 15.8 (a) Example: availability of information person versus household level. (b) Example: (non-)availability of most-detailed level.

activity as employer, own-account work, and contributing work in family or farming businesses, is collected at the individual level. In country B, only gains and losses from activity as employer or own-account work are collected at individual level, whereas income from farming activity is collected from a section of household activity without separating out the individual's involvement. Example (b) points to the varying level of detail collected in the data. From a conceptual viewpoint, occupational (second pillar) and individual (third pillar) pensions typically exist side by side in many countries, but in country A, the level of detail is not available in the source data, and hence, the *ex post* harmonized cannot provide the detail either. Both examples highlight the need for comprehensive documentation of the harmonized data.

Operational comparability is particularly difficult to achieve in the classification of social security benefits. Not only does the right conceptual placement require extensive knowledge of the national social security system, but also LIS needs to follow acceptable standards for social benefits cutting across various types of benefits provision, or various needs being relieved. For example, social assistance programs in emerging economies frequently cut across different providers, the state or nongovernmental institutions. At the same time, these programs are set up to relieve individuals of the burden of different risks or needs, such as low income, inadequate education of children, or insufficient nutrition, so that there is no unique placement where they belong to in the conceptual framework (at the same time, general social assistance, family benefits, education, or food benefits). Still, LIS tries to place them in the most appropriate placement for international comparison not to lose the detail for cross-country studies (in this example, general social assistance) and documents these decisions so that researchers better understand, which benefits were included in LIS variables.

Besides the nonavailability of information, other general limitations in comparability of income components arise; source data do not necessarily collect all income items with the same reference period, even within the same survey. In the case of actual income surveys (that collect the regular amount received at the last payment), *annualization* may imply a multiplication by the number of periods in the year if the income is received regularly (such as a weekly wage), but may need to be multiplied by a smaller factor in case the income is not regularly received throughout the year.

More generally, on the one hand, *ex post* harmonization suffers from data providers not following precisely international standards and guidelines for definitions and collection. For instance, for wealth data, definitions of what constitutes business wealth and valuation criteria for assets are still different across countries. On the other hand, following international standards, e.g. ILO employment criteria (ILO 2018), may lead to situations – especially in emerging economies – where a vast majority of the population is classified as mainly employed, but where very few people actually have a regular job, hence, creating a bias when comparing emerging economies to affluent countries. As a result, even though there is not a direct challenge to the harmonization in this example, it points rather to research that needs to interpret the *ex post* harmonized data from different regions in the world.

Last but not least, different cultures and languages cause varying ways of asking questions with differences in wording and understanding of one and the same concept. This is particularly relevant in trying to achieve a harmonized concept of disability. Besides the interpretation, while translating documentation, national institutions may differ. Thus, in country A, disability might be linked to the eligibility for disability benefits, whereas in country B, a question about general limitations in daily activities, might be still influenced by the existing norms and social benefits for disability or invalidity. In country C, where such norms are possibly less common, the same question about limitations in daily activities might more generally lead to a rather subjective than objective answer. Due to these cross-national differences, LIS decided to always add clarifying remarks on the national contents for variable *disability* in METIS.

To sum up, despite the efforts carried out during a careful *ex post* harmonization, not all of the challenges posed by the source data can be solved. This gets clear, for example, when the harmonized data are evaluated against the existence of extreme values in the distribution. Neugschwender (2020) illustrates how extreme values at the top and bottom of the distribution can influence inequality measures such as the Gini and Atkinson indices. It is important to stress that researchers working with data sampled with different procedures need to be evaluated against the possible research objective. Only researchers can implement strategies to overcome the limitations of cross-national limitations in *ex post* harmonization. However, it is imperative that persons working on harmonization recognize limitations and that clear comparability warnings are passed on to the final data users through a sound documentation system. Likewise, it is crucial that data users bear in mind these limitations of *ex post* harmonized data.

15.5 Software Tools

As shown in Section 15.3, LIS puts a strong focus on standardization practices to ensure high-quality data and documentation. Thus, all datasets acquired by LIS to be released in the *LIS* and LWS Databases are entered directly into the METIS. METIS is a user interface consisting of a *frontend* and *backend* application that has been specifically built for the needs of LIS to provide general information on variable definitions, and dataset availability, as well as dataset-specific information on the source data, variable content, their availability, comparability warnings, and basic statistics. The entries in the LIS internal *backend* interface allow a clear overview and easy navigation through the rich information stored in the various underlying databases. Various designated fields allow the LIS team to easily update the databases. Once synchronized with the *frontend* user interface, the information is then accessible to the public, where scholars can explore variables and documentation interactively.

Managing the data and their documentation with an internal *backend* at LIS has the advantage that datasets can be easily created, revised, selected, and grouped by their current work status. Likewise, the *frontend* allows scholars to explore selected information across countries, which can then be easily exported.

For the latest harmonization *template*, LIS currently uses Stata as harmonization software. The various harmonization programs enable an automated run of all LIS/LWS Database variables. The specific country folders contain a standardized *.do*-file structure for variable construction. An internal Stata *.ado*-file allows execution of various pieces of the programs; the syntax recalls the standard folder structure for programs, intermediate data and final data, which enables the possibility of a reoccurring update of the harmonized files once the original microdata is fed again in the system. During the harmonization, with the same Stata *.ado*-file, the various consistency checks (described in Section 15.3) can be carried out, which allows a tailor-made comparison of individual variables from datasets in progress and currently online.

For continuous variables in the income, consumption expenditure, and balance sheet, specific Excel tables were built that report for the selected dataset's mean and median values, percentage of missing values, non-missing and nonzero cases, and percentage of non-missing and nonzero cases over time.

Once the harmonized data are created, the data and documentation are made available for external use. This stage includes a series of actions using various software (e.g. R, SAS, and SQL), including conversion of the Stata microdata to other formats compatible with R, SAS, and SPSS so that the data can be accessed in the remote access system by different software packages.

15.6 Conclusion

As shown in this chapter, *ex post* harmonization of data is not a trivial task, but it is necessary for comparative research that intends to use data from different sources and in different formats. When undertaking the task of *ex post* harmonizing preexisting data, it is imperative that all differences in the source data are carefully reviewed so that appropriate treatment can be applied for the best comparability between harmonized datasets. We also highlighted that quality assurance procedures (including proper infrastructure for the data management) are necessary for harmonization that aims to construct data to conduct high-quality research.

As a pioneer in cross-national *ex post* microdata harmonization, LIS has vast experience conducting the assignment of constructing datasets that enable comparative research on household income, consumption, net worth, portfolio composition, and income and wealth distributions. Through nearly 40 years of experience in harmonization and data dissemination for researchers, LIS has learned that to provide the best quality harmonized data is a combination of efforts that involve many actors. So far, we have not talked about the users of LIS data, who must be able to conduct state-of-the-art research with the cross-national databases. Thanks to the researchers, LIS can maintain these databases since their analyses provide feedback by pointing out some obstacles in the harmonized data and helping with the development of conceptual frameworks needed for the latest research advancements. The LIS users' numerous *LIS/LWS Working Papers* and peer-reviewed publications that use LIS databases can be viewed as proxies for the high quality of the harmonized data. It indicates that the researchers from various social science disciplines trust the harmonized LIS/LWS databases.

The establishment of networks with the researchers, as well as mutual advice, is an additional component of a broadly defined harmonization process that needs to be highlighted here. Thus, LIS also provides the annual week-long workshops in Luxembourg, which provide new and existing users with solid knowledge of LIS databases and the latest methodological advancements in welfare economics. LIS can provide such a service since the team regularly collaborates and interacts with prominent scholars in the field, and the LIS team conducts research with LIS/LWS data.

The regular exchange through the extensive day-to-day user support with many LIS database users leads us to provide crucial general recommendations while working with the harmonized data. Foremost, it is vital to understand differences in data context and consult each dataset's documentation. This recommendation relates to the fact that, despite the efforts carried out during a thoughtful harmonization, not all of the challenges posed by the very different input data can be solved, so that the final harmonized data could still have some comparability issues. For example, Pfeffer and Waitkus (2021) point to the underrepresentation of wealth at the top end of the distribution in wealth surveys. Likewise, nonresponse errors are a serious concern for cross-national comparisons (Ravallion 2015). It is in the sphere of national data providers to implement appropriate methodology during the data collection and editing processes to limit possible biases and misperceptions from cross-national analyses. *Ex post* adjustments can only be seen as a second-best option, and due to their assumptions, such adjustments remain among the researcher's methodological triangulation. But we need to acknowledge that some harmonized variables might differ from the conceptual definitions set up as the standard for harmonization (due to the data collection); therefore, users of the *ex post* harmonized databases are advised to consult the metadata documentation carefully. The remaining comparability issues might have implications for how the analysis should be carried out as well as how the results should be interpreted.

What else have we learned through all these years as an institution that harmonizes and disseminates data for scientific inquiry? In order to provide reliable data for the cross-national interdisciplinary research, the processes of conducting LIS activities are not only related to the good source data, well-implemented harmonization procedures, quality assurance, documentation, infrastructure, and hiring highly skilled labor force. The success of the LIS databases also involves a network with other institutions and researchers to exchange knowledge and expertise that must be implemented into the overall activities of harmonizing institutions. With this exchange, LIS goes beyond the *ex post* harmonization procedures conducted in-house, having the objective to improve standardized practices for collecting income and wealth data. First, LIS provides international guidelines for data producers in the form of collaborative publications and through the direct contact with data providers during harmonization. These are essential factors that help to improve not only the survey data but also the design of questionnaires. Second, through the established network and close ties to experts in the field, the LIS team easily learns about and adopts latest survey developments, advises on potential data improvements that can help comparative research, and provides feedback on quality and consistency of the source data. All these elements are the ingredients to make the efforts of our work thrive and continue to improve the outcome of *ex post* harmonized data.

References

Checchi, D., Cupak, A., and Munzi, T. (2021). Empirical challenges comparing inequality across countries: the case of middle-income countries from the LIS database. In: *Inequality in the Developing World* (ed. C. Gradín, M. Leibbrandt, and F. Tarp), 74–105. Oxford University Press https://doi.org/10.1093/oso/9780198863960.001.0001.

Endeweld, M. and Alkemade, P. (2014). LIS micro-data and national accounts macro data comparison: findings from wave I – wave VIII. LIS Technical Working Paper Series, No.7. LIS: Cross-National Data Center in Luxembourg.

ILO (2018). Statistics on work relationships. 20th International Conference of Labour Statisticians (Geneva, 10–19 October 2018). Geneva, ILO: Department of Statistics. https://www.ilo.org/wcmsp5/groups/public/---dgreports/---stat/documents/publication/wcms:644596.pdf

Neugschwender, J. (2020). Top and bottom coding at LIS. LIS Technical Working Paper Series, No.9. LIS: Cross-National Data Center in Luxembourg.

OECD (2013a). *OECD Guidelines for Micro Statistics on Household Wealth*. Paris: OECD Publishing https://doi.org/10.1787/9789264194878-en.

OECD (2013b). *Framework for Statistics on the Distribution of Household Income, Consumption and Wealth*. Paris: OECD Publishing https://doi.org/10.1787/9789264194830-1-en.

OECD, Eurostat, and UNESCO Institute for Statistics (2015). *ISCED 2011 Operational Manual: Guidelines for Classifying National Education Programmes and Related Qualifications.* OECD Publishing https://doi.org/10.1787/9789264228368-en.

Pfeffer, F.T. and Waitkus, N. (2021). The wealth inequality of nations. *American Sociological Review* 86 (4): 567–602. https://doi.org/10.1177/00031224211027800.

Ravallion, M. (2015). The Luxembourg income study. *Journal of Economic Inequality* 13 (4): 527–547.

Smeeding, T., Schmaus, G., Allegrezza, S. (1985). An introduction to LIS. LIS Working Paper Series, No.1.

The Canberra Group (2001). *Expert Group on Household Income Statistics: Final Report and Recommendations.* Ottawa.

UNECE (2011). *Canberra Group Handbook on Household Income Statistics*, 2e. New York: UN http://digitallibrary.un.org/record/719970.

Yonzan, N., Milanovic, B., Morelli, S., and Gornick, J.C. (2020). Drawing a line: comparing the estimation of top incomes between tax data and household survey data. Stone Center Working Paper Series, No. 27. https://doi.org/10.31235/osf.io/e3cbs.

16

Ex-Post Harmonization of Time Use Data: Current Practices and Challenges in the Field

Ewa Jarosz[1], Sarah Flood[2], and Margarita Vega-Rapun[3]

[1]*University of Warsaw, Faculty of Economic Sciences, Warsaw, Poland*
[2]*IPUMS Center for Data Integration, Minneapolis, USA*
[3]*University College London, Centre for Time Use Research and European Commission Joint Research Centre, London, UK*

16.1 Introduction

Time use surveys provide very rich and versatile data that can be used to study a variety of dimensions of an individual's daily experiences including activities, their context, and related effect. Time use surveys collect information on all activities individuals engage in over 24 hours. The primary instrument for collecting such data is a time diary, which allows for recording daily events as they happen, preserving information about their order, duration, location, and other activity characteristics (see Figure 16.1, for example, the information collected by a time diary). In many recent time use surveys, these instruments have often been expanded to include information about individuals' subjective assessment of their daily experiences, including the associated level of enjoyment, stress, or pain. Topics studied using time use data have also shifted over time to inform knowledge on topics of contemporary public debate and social interests. At the beginning of the twenty-first century,[1] the major issues that have been investigated using time use data have been: gender inequality and domestic division of labor (Sullivan 2000; Sayer 2005), individual and public health (Cawley and Liu 2012; Jarosz 2018), parental time with children (Craig and Mullan 2011; Musick et al. 2016), and the specificity of different life stages (Moen and Flood 2013; Caparrós Ruiz 2017).

The major benefit of time use surveys is that they provide very universal data, which may be used in different fields of research. The applicability of time use studies extends to a very broad set of topics that can be addressed using information on daily behavior. For example, time use methodology was used to investigate changes in people's behaviors due to the coronavirus pandemic (Gershuny et al. 2020a). However, this broad applicability of time use data may generate challenges for survey

---

1 Rudolf Andorka gives an overview of the topics that gained popularity prior to 2000 (Andorka 1987).

Survey Data Harmonization in the Social Sciences, First Edition.
Edited by Irina Tomescu-Dubrow, Christof Wolf, Kazimierz M. Slomczynski, and J. Craig Jenkins.
© 2024 John Wiley & Sons Inc. Published 2024 by John Wiley & Sons Inc.

Time	Main activity	Additional activity	Location	Who was present (1)	Who was present (2)	Who was present (3)	Did you travel	ICT* Use
4:00-4:10	Sleep	–	Home	NA	NA	NA	No	No
4:10-4:20	Sleep	–	Home	NA	NA	NA	No	No
(...)								
7:20-7:30	Self-care	–	Home	–	–	–	No	No
7:30-7:40	Dressing up	–	Home	–	–	–	No	No
7:40-7:50	Dressing children	Preparing breakfast	Home	Children	–	–	No	No
7:50-8:00	Eating	Listening to the radio	Home	Children	Husband	–	No	No
8:00-8:10	Drinking coffee	Checking email	Home	Children	Husband	–	No	Yes
8:10-8:20	Driving to school	–	Own car	Children	–	–	Yes	No
8:20-8:30	Driving to work	Listening to the radio	Own car	–	–	–	Yes	No
8:30-8:40	Working	Checking email	Work	–	–	–	No	Yes
(...)								
10:00-10:10	Coffee brak	Chatting	Work	–	–	Co-worker	No	No

Figure 16.1 An example of a contemporary time use diary: the 2014–2015 United Kingdom Time Use Survey collected by NatCen.

harmonization – because it is impossible to exactly picture the end-user of the harmonized dataset. In fact, it can be anyone, an epidemiologist, an economist, an environmental scientist, or a cultural anthropologist; and across disciplines, investigators may prioritize different aspects of the data in their work (e.g. precise locations and geographic detail versus contact with others).

Time use accounts should be separated from the so-called stylized questions of time use in which a respondent is asked to provide an estimated total duration of a given activity (Kan and Pudney 2008) within a specified time frame, for example, a day or a week. Stylized measures of time use are based on questions such as "How much time do you spend on exercise weekly?" and rely on the respondent's own estimation, which is often less accurate (Kan and Pudney 2008). Compared to stylized questions, the data provided by time use diaries are far more detailed and less prone to bias. In the case of the stylized questions, respondents make approximations using different heuristics and many people are simply not aware of how much time they spend on different activities. These factors lead to estimates that reflect perceptions of time use rather than actual time allocation. It has been argued that time diaries and related time-based data collection techniques are the only reliable method to obtain valid time use estimates (Robinson 1985; Gershuny et al. 2020b). This chapter focuses on the *ex post* harmonization of time use data that are collected via time use diaries.

Time diary data collection became increasingly common during the second half of the twentieth century (as evidenced by federal government's collected time diary data), but time use studies have a much longer history. The earliest time use accounts date back to the beginning of the twentieth century. The first diaries were collected in the United Kingdom by Reeves (1913), and, almost simultaneously, in the United States by Bevans (1913). These were closely followed by Stanislav Strumilin's study carried out in Russia shortly after the Bolshevik revolution (Strumilin 1925). These first diaries captured the daily lives of the working class and peasants and were mostly focused on different forms of labor and production. In the central European context the predecessor of modern time use research was the classic study of daily lives in an Austrian town of Marienthal, which was struck by unemployment during the Great Depression (Jahoda et al. 1971). The monograph provided an account of how time was used by those who had lost their jobs and how it affected their families and the entire community.

With time the scope of time use studies broadened to include not just selected behaviors or categories of respondents but all types of daily activities collected for large and often national samples. The geographic coverage of time use studies increased as well as more scholars and governments acknowledged the usefulness of the methodology, for example, to propose alternative measures of societal well-being or ways to structure social policy (Kahneman et al. 2004; Cornwell et al. 2019). These developments resulted in increased interest in comparative time use studies. The first major effort to produce a cross-national harmonized dataset was the 12 nation comparative time-budget study by Szalai (1966, 1972). Szalai's study was carried out in 1965 and 1966 in the USSR, Hungary, Eastern and Western Germany, Czechoslovakia, Poland, Bulgaria, Yugoslavia, France, United States, and Peru. Besides the diary data, the surveys also collected respondent's background information. Szalai's pioneering work is the first major example of ex ante time use harmonization; it also paved the way for modern cross-national time use data studies. Today many countries across the world have collected at least one time use survey. Europe, North America, and Asia have a good coverage. To date the US, Canada, Australia, Russia and most of the European countries have collected 4 or more surveys per country. At least one time use survey is available for most of the South American countries and for a substantial share of Asian countries including China, India, South Korea or Mongolia. In Africa the coverage is more limited but the number of countries with at least one time use survey is rapidly increasing, and some countries, like South Africa, have collected two or more surveys to date.

The expansive temporal and geographic coverage of time diary data creates excellent opportunities for rich comparative analyses, but there are numerous challenges researchers must overcome to compare the data between countries and across time. Until very recently, with few exceptions such as Szalai's study, time use data collection was done independently across countries, which resulted in severe comparability challenges. In 1994, the Statistical Program Committee recommended a harmonized coordinated time use data collection within the European Community. The first round of the Harmonized European Time Use Survey (HETUS[2]) data collection based on unified data collection guidelines was in 1998–2006; the second round was in 2008–2015. A third round of data collection is currently underway. The guidelines for data collection have been regularly updated with the newest version being published in 2019 (Eurostat 2019).

---

2 https://ec.europa.eu/eurostat/web/time-use-surveys.

The need for comparative time use studies is reflected in the fact that not only HETUS, but many of the currently conducted surveys follow a particular design that allows for comparing them with time use surveys coming from other countries but using a similar data collection framework. Other examples of such *ex ante* harmonization are the American Time Use Survey (ATUS[3]) – collected annually in the US, or the International Classification of Time Use Activities (ICATUS), proposed by the United Nations and applied primarily to data collection in developing countries. While unified frameworks facilitate comparisons within the framework, comparisons of countries across frameworks and combining those surveys based on a standardized framework with those not using a standardized framework require *ex post* harmonization. This is also true for data collected in different world regions or across different decades.

To facilitate comparative time use research including datasets from different countries and decades, in mid-1980s Jonathan Gershuny started the Multinational Time Use Study (MTUS,[4] Gershuny et al. 2020b). In 2020, the database consisted of 105 surveys, the earliest of which is the 1961 UK Time Use Survey (Gauthier et al. 2006). MTUS represents an effort to harmonize existing time use studies from countries across the world. Our experience with time use data harmonization is based on our work on surveys included in MTUS and we use this database as an example of the *ex post* harmonization of time use surveys.

MTUS harmonizes both the diary data and background variables (i.e. household and individual sociodemographic characteristics) to enable comparing the social patterning of time use across time and between nations. Most of the datasets included in MTUS are provided by national institutes of statistics of each country, yet there are cases where time use surveys were collected and provided by other institutions such as ministries or international organizations. The data provider may restrict the access to the survey data, even though they are fully anonymized. For some countries, access to the data may only be allowed after some conditions have been met.

Currently the MTUS dataset include three types of datasets: (i) those that are freely available and can be downloaded by individual researchers after they have agreed to terms and conditions of use; (ii) those that are provided only with authorization by the national statistical office; and (iii) the last category contains surveys that are used internally within the Centre for Time Use Research (CTUR) and cannot be disseminated to third parties due to restrictions imposed by data providers.

The first category of freely available surveys include the sets for the following countries and years: Austria (1992, 2008), Belgium (1966), Bulgaria (1965, 2001, and 2010), Canada (1971, 1981, 1986, 1992, 1998, 2005, 2010, 2015), Czech Republic (1965), Denmark (1964, 1987, 2001), Finland (1979, 1987, 1999, 2009), France (1966, 1974, 1985, 2009), Hungary (1965, 1976, 1986, 1993, 1999, 2009), Norway (1971, 1980, 1990, 2000), South Korea (1999, 2004, 2009), United Kingdom (1961, 1971, 1974, 1983, 1987, 1995, 2000, 2005, 2014), and the United States (1965, 1966, 1975, 1985, 1993, 1995, 1998, and then annually from 2003 until 2018). Surveys available after authorization by the country's statistical institute that collected the data include Sweden (1984, 1991, 2001, 2010). Lastly, surveys available only for CTUR researchers include: Australia (1974, 1987, 1992, 1997, 2006), Germany (1965, 2001, 2012), Italy (2013), and Poland (1965, 2003, 2013).

The majority of the surveys included in the MTUS are randomly sampled nationally representative datasets. Exceptions include those that come from the Szalai study that represents a much smaller

---

3 https://www.bls.gov/tus/data.htm.
4 https://www.timeuse.org/mtus.

sample, usually drawn from a single location that was considered "representative" of the given country. MTUS datasets include a special variable generated in the harmonization process. This variable is used to differentiate between nationally representative surveys and other types of surveys that might, for example, include a non-representative sample, have an atypical diary design, or represent a region (e.g. Basque country).

This chapter is dedicated to harmonizing time use surveys, which is different from harmonizing other types of surveys. Its specificity will be discussed in detail in Section 16.2, but in principle it is the variety of time use data collection techniques that present harmonization challenges unique to this methodology. Time use diaries come in many shapes and formats, representing different decades, conceptual approaches, cultural specificity, and, especially today, they may account also for respondents' subjective assessment of their time use (Sullivan et al. 2020). Comprehensive documentation and relevant quality checks are essential to assure the data is understandable to the end-user.

16.2 Applied Harmonization Methods

The major objective of *ex post* harmonization is to render the data comparable while preserving as much of the original detail as possible. When harmonizing time use diaries we need to take into account the following: (i) the structural integrity of the data – such as whether the activity durations calculated using start and end times match the amount of time reported in the activity; (ii) activity detail and codes which may vary widely from place to place and over time; (iii) amount of additional detail in the time diary – who is the activity for, whether there is any secondary activity report, with whom activity took place – those vary across surveys and may be recoded in different ways.

Harmonization is usually constrained by the dataset with the least amount of detail available, which means that some detail in the original data is lost to make it comparable across time and place. For example, the United States federal time diary data collection – the ATUS – catalogs daily life in extensive detail with more than 400 detailed activity codes about what people do over the course of the day. By contrast, most European countries typically have about 200 activities in the original data. By necessity, some of the granularity of the ATUS data must be collapsed to enable comparisons with the European surveys.

The process of time diary data harmonization may be divided into two major stages. The first step involves standardizing the format of the data, and the second pertains to harmonizing the information collected in the time diary. While these stages are independent of one another, they are closely linked and should be done in a fixed order, as outlined below.

16.2.1 Harmonizing the Matrix of the Diary

Time diary data are typically collected in slots or episodes. In a slot-based data collection, time slots in which respondents may report their activities are predefined; they are usually of 10 minutes duration each. In an episode-based data collection, respondents may report any activity length of one minute or more, which is denoted as an episode of a given activity. For example, a 480-minute night sleep episode in a slot-format diary would consist of 48–10 minute slots for an individual whereas in ATUS it would be one 480 minute episode. Relatedly, activities shorter than 10 minutes in duration would not get reported

		Data collection format	
		Slot	Episode
Data arrangement	Wide	Wide slot (e.g. HETUS)	Wide episodic
	Long	Long slot (e.g. HETUS)	Long episodic (e.g. ATUS)

Figure 16.2 Main data formats in time use diaries.

in a slot-based diary, but are eligible to be reported in an episode-based diary where there is no constraint on the minimum length of an activity.

Figure 16.1 shows an example of a slot-based time diary data collection where respondents report their activities in 10-minute durations. Harmonizing the matrix of a time use diary essentially means rendering the matrices of all datasets into same shape: using episodes or slots as units of observation, and either stacking them one on top of another in the matrix (long format), or arranging all observations from the same diary in a single row (wide format). Overall, with the two criteria, there are four possible combinations that determine the matrix structure (Figure 16.2).

Harmonizing data formats is required because the original diaries come in different forms and shapes that cannot be directly compared. Harmonizing data formats is necessary if a researcher wishes to conduct any analyses using the sequential data. The sequential nature of time diary data is unique to this type of time use data collection. The sequential data indicate the order, timing, and duration of activities as opposed to simply reporting how much time is spent on a particular activity. In addition to being more accurate than stylized measures of time use, time diary data can be used to analyze the sequential and temporal rhythms of daily life. Uses of sequential data might entail, for example, counting episodes of a given activity, examining patterns in the timing and order of activities, or how they are intertwined in a sequence (e.g. Jarosz 2016).

In time use surveys conducted within the HETUS framework the diary day (24 hours) is divided into 144 time slots of equal duration (10 minutes each). That is, all activities are recorded within 10-minute slots. Conversely, in many other frameworks, such as the ATUS, there is no predefined time slot, so episodes reflect the actual duration of an activity, which may be as short as one minute. In other words, there is no fixed duration of an episode across time diary data collections, and the number of all episodes within the sequences varies across diaries. These are the two major examples of how activities may be recorded, but there is great variability in the format of time diary data collected across the world. Other examples of time diaries are those that are structured using predefined slots of various duration (e.g. a mix of 15- and 30-minute long slots), or those in which several activities are recorded in the same slot while not specifying their duration.

All harmonized datasets must use the same unit of time and must be organized in the same way. Typically, when both episodic and slot data need to be converted into a single format, the data needs to be converted into episodic format because episodes cannot be rendered into slots (they would all need to be divisible by 10, which is not the case). For example, the 7 : 30 to 7 : 40 and 7 : 40 to 7 : 50 slots in Figure 16.1 would be combined into a single episode in the harmonized dataset because all of the information reported is identical between the two records. By contrast, the 7 : 10 to 7 : 20 and 7 : 20 to 7 : 30 slots would not be combined into a single episode because the secondary activity reported in column 3, for example, is different across the two slots. Collapsing slot data by converting identical consecutive slots into a single episode saves space and avoids repeated records without losing information.

The choice of data matrix arrangement (wide or long) is optional, as long as it is the same for all harmonized datasets. Wide format data are all located on a single record per individual. Long format data lists each episode or slot on a separate record, and an individual contributes many records. In the case of the MTUS, all data are converted into a stacked (long) format. Specifically, activity records for an individual are collapsed into episodes (if the original format is in slots) and stacked one under another. Some data already comes in long format (e.g. ATUS) while other data need to be transposed. The standardization of matrix arrangement requires that the data be transformed. This may be done using standard transposition commands available in statistical packages[5] or with custom programming. No information is lost during this phase of data transformation.

As a result of these procedures (changing the unit of observation and standardizing the matrix format when needed), all datasets within the harmonized database use the same time units and present the same matrix arrangement. No change is made to the actual data and the original format may be restored by transposing the matrix back to its original format and dividing the collapsed episodes back into the original slots.

16.2.2 Variable Harmonization

Variable harmonization, as opposed to standardizing the format of the data, typically occurs in the second stage. Variables common to most time diaries include characteristics of activities, locations where activities occurred, and who else was present during the activity. In HETUS additional information is collected about the mode of data collection including when the diary was filled in (on an ongoing basis, in the evening of the same day, or on the next day). Such information is not available for ATUS, which is based on recalled information from the day before. This information is typically not harmonized though it may be used for data quality checks. Most diaries record two activities: primary activity and one accompanying (i.e. secondary) activity. However, some diaries (e.g. United Kingdom Time Use Survey 2015) provide up to five accompanying activities. While it is rare for respondents to record more than two activities happening at the same time, it occasionally occurs. As most diaries do not collect more than two activities, and some, such as the ATUS, collect only one, the datasets included in the harmonized MTUS database provide the record for the main and secondary activity only. Overall, harmonizing more activities than that (third, fourth, etc.) is not practiced currently as there are very few surveys that provide such data and so it cannot be compared.

The activities that must be harmonized are the primary (main) and any accompanying (secondary) activity. Overall, the major variable harmonization task is to impose consistency in main and secondary activities. Assuming that they have already been rendered in the same format (as outlined in Section 2.1 above), the original activity codes need to be replaced with the harmonized ones. This is typically done by matching the original code to a list of harmonized activity codes. Matching means that the original code is replaced by the one from the harmonized list that is either the equivalent of the original (e.g. replacing "having a meal" with "eating") or designates a broader category of codes within which the

---

5 Many statistics packages such as R, Stata, and SPSS refer to this operation as "reshaping."

original would fit (e.g. replacing "mending the roof" with "work related to house maintenance"). The list of original activity codes is often much more detailed (the HETUS type of surveys usually includes around 200 unique activity codes, but it may be much more than that), whereas the harmonized activity codebook is less detailed (e.g. in MTUS there are 69 activity codes). As a result, some details about what people were doing are lost during the harmonization process in MTUS.

In theory, the harmonization could be done through hierarchical coding, which would allow preserving more information about the nature of the activity in the original data. In practice, that is not done at the moment because it would make the already lengthy harmonization process even more time-consuming, and could also make it more difficult for the end users to use the data. Unlike for the variables such as ISCO codes, there is no clear hierarchy as to which activities should be higher in order – this is highly dependent on the country context; for example, foraging could be a leisure activity in one country and an activity done to supplement household income in another.

Hierarchical coding does preserve the complexity of the original activity, but it also requires a coder to make decision with regard to how to rank the new activity codes as this is not a simple case of going from a broader (e.g. housework) to narrower (e.g. hoovering) activity. If this procedure is done for every activity it would greatly complicate coders' work and not necessarily simplify the use of the data for the end users as they would need to review codes for each type of activity with hierarchical codes. Taking into account the fact that the complexity of the data is already a barrier to use, simplifying, rather than complicating its structure would be more beneficial.

However, there are also variables, which could be ranked in a clear and self-explanatory way, for example – the copresence (who else was present) variable. Different countries collect the data with varying levels of detail with regard to the reporting of copresence. For example, some countries differentiate between mother, father, grandmother, and grandfather, while others use only a broad category of "parents" or even "other family members." In such cases hierarchical coding is an excellent solution as a single hierarchy can be proposed and maintained across all datasets. It could follow ISCO coding scheme with each additional number (1-digit, 2-digit, 3-digit, etc.) adding more detail to the code, for example, 5 would denote "other family member," 51 "mother," 511 "maternal grandmother," 512 "maternal grandfather," and so on. This is not currently being done, but it is a good solution to the problem of loss of detail for some variables.

Activity harmonization is generally a manual process in which a coder follows detailed harmonization guidelines. This is because even surveys within a single framework use slightly different activity codes. The guidelines specify which original activities correspond to particular harmonized codes or are included in the more general harmonized activity categories (see MTUS user guide for information: Gauthier et al. 2006). Coders are instructed to strictly follow the guidelines instead of making their own decisions. This assures that all datasets are harmonized in the same way and there are no cross-country differences that could be attributed to differences in activities recoding. Occasionally, coding decisions are not covered by the guidelines and need to be made (see point 4 of the chapter) on a case-to-case basis. In these instances, it is critical to note this in the documentation (see point 3).

Main and secondary activities are typically coded the same way in the original data, and therefore, harmonized following the same guidelines. Time use surveys usually come with very few missing values in the main activity. In the case of the MTUS harmonization, some of those missing values may be imputed based on other diary variables. For example, in the case of missing main activity, a secondary activity may be substituted for the main activity if it is available. Alternatively, information regarding activity location (e.g. using a means of transportation) may also be used to provide information on the nature of the main activity (e.g. traveling). Lastly, information may be imputed based on the preceding and following activities. This strategy is used in MTUS. For example, missing values during the night-time hours – before waking up and after falling asleep – are coded as "imputed sleep." Any imputed values are flagged as such so that the end user is aware that the activity was missing from the original data. Information on that is also added to dataset documentation.

Secondary activities are typically highly incomplete, which is not an error but reflects the specificity of individual's time episodes. In all surveys missing secondary activities are just left blank by the provider. In the course of harmonization, they may either be left blank or coded as missing (regular missing value). Based on our experience with harmonizing national time use surveys, usually, around 80% of secondary activities are missing. However, very little additional effort is needed to harmonize the secondary activity, as it uses exactly the same activity codes as the main activity.

Besides the main and secondary activity, time use diaries typically also record information on the activity setting, that is where (location) and with whom (copresence) it took place. These variables are harmonized using a very similar procedure as is used for harmonization of the activity codes. The original location or copresence codes are recoded into a set of harmonized codes. However, unlike the activity categories, sometimes the original location and copresence codes are less detailed than the harmonized codes. If the original code list is less exhaustive than the harmonized one, some harmonized categories are simply not used. When the original codes include more categories, again – just as in the case of activity categories – some detail is lost. For example, the MTUS does not record the copresence of grandparents or coworkers. Instead, the code "other relatives" or "person not living in the household" is used. As a result, the original information is only partially preserved in the harmonized database. Hierarchical coding could be a solution to the problem in some cases. None of the harmonized datasets (HETUS or MTUS) provides this option at the moment.

As in the case of missing activities, occasionally, other information in the data can be used to fill in missing data in the location or copresence field. Sometimes the activity setting allows for recovering missing information; for example, foraging, a visit to the cinema, or a doctor's appointment, all indicate places where these activities happened. With some more effort, in some cases, the missing presence of others may also be retrieved based on the diaries of other household members. While this is rarely done because of the high amount of work and time it potentially requires (diaries of all household members should be checked), it is feasible.

16.2.3 Other Variables

A time use diary is the major but not the only component of a time use survey. Typically, besides the diary, most national time use surveys also collect information on: (i) respondent's sociodemographic characteristics, (ii) household characteristics, and (iii) specificity of each day for which the diary was filled in, for example, how hurried the day was. In some cases (e.g. United Kingdom 2014/2015 TUS available from the UK data services: https://beta.ukdataservice.ac.uk/datacatalogue/studies/study?id=8128) respondent's work schedules are also available. Selected variables from each dataset are subject to harmonization in case of MTUS, including the basic individual's social and demographic characteristics, household information, and essential information about the nature of the day (e.g. calendar date, month, or which day of the week it was or whether it was a holiday).

While not all available variables are harmonized (which would be very cumbersome and time-consuming), the harmonized dataset includes a selected set of commonly used sociodemographic (e.g. age, gender, education, occupation) variables. As background variables are not time use variables, this chapter will not elaborate on them. They are harmonized in line with the established and written-down guidelines proposed by the harmonization team and in this case, the process of harmonization resembles the procedures of data harmonization outlined elsewhere in this book (e.g. Chapters 9–15 in this volume).

16.2.4 Other Types of Time Use Data

Aside from the national time use surveys, which are the focus of this chapter, there are also smaller datasets that collect similar time use information but which are designed to address particular social issues. Some data might be recorded by an observer, instead of a respondent, and such is the case in anthropological studies using observation as a method of data collection – those studies have been referred to as time allocation studies and the sample is usually very small (Gross 1984).

Overall, surveys that are targeted at specific populations may deviate from the standard national-level time use surveys in terms of the data collection technique, study design, type of information collected, or activity codes. Examples of small-scale time use surveys include time diary-based surveys conducted among parents of children with disabilities (Thomas et al. 2011) or children experiencing school lockdown during the pandemic (Asanov et al. 2021). As these studies are unique and very group or time-specific, their value in cross-national comparisons is limited. In such cases a method commonly used as an alternative to the standard time use diary is experience sampling method (ESM), which focuses on fluctuating emotional states and helps to investigate the relationship between behavior or activity setting and affect (Larson and Csikszentmihalyi 2014). ESM collects randomly sampled activity episodes, which means it does not provide a full sequence of activities, but gives much information on subjective aspects of daily lives. The technique has been used on specific, often epidemiological, populations such as dementia patients (van Knippenberg et al. 2017), older adults (Jarosz 2021), or obese children (Salvy et al. 2008). Again, it is unlikely to find two comparable studies using this method as they are primarily meant to explore the situation of a very specific targeted audience.

Ex post harmonization typically does not apply to these types of datasets. They often have very small numbers of observations (in particular for ESM that is under 100 individuals), use subject-specific questions, and are generally unique for a given location or group. They could potentially be harmonized at aggregated level (that is aggregated measures as opposed to detailed time use records), but we are not aware of any example of *ex post* harmonization using ESM or similar data.

16.3 Documentation and Quality Assurance

16.3.1 Documentation

Producing harmonized data and accompanying documentation starts with reviewing the original documentation provided by the data collector. This documentation includes information about how the data were collected, number of observations, sampling method, period when the fieldwork was done, and detailed information on the variables (usually in the form of the questionnaire added to the documentation). The mode of data collection: PAPI or CAWI is also generally provided in the original documentation.

Documentation produced in the process of harmonization should describe all changes made to the original data. That means the process of harmonization should be outlined in a detailed way, starting by providing characteristics of the original datasets and ending with describing the quality checks that were performed on the harmonized dataset. Providing information on the original survey is important in the context of comparing datasets across countries and time. In particular, the target population, sampling frame, sample size, and fieldwork dates should be specified. The documentation should also provide technical information on the original files, such as the number of the original survey files provided by the organization that collected the data, response rate, number of individuals and diary days, survey period, type of diary used for data collection, and the time interval used in the original diary. In addition, the documentation should include information on data quality such as the number of complete diaries per diary day (first or second), missing values, as well as any peculiarities related to the data. A researcher might also like to include information about the type of classification of activities that were used in the source data, but the data user needs to be mindful of the fact that even among the countries, which theoretically should be using the same classification of activities (e.g. HETUS in Europe) there is great variability in the actual codes used, and therefore, the list of original codes (which must be provided in the documentation) should always be consulted.

The second part of the documentation should focus on the process of harmonization itself. The conversion of each variable should be documented. Deviation from specific guidelines about variable harmonization should be documented. For example, if the researcher needs to make a decision with regard to how a given original variable should be converted to a harmonized one, this decision should be documented along with any other peculiarities of the data that the person who is carrying out the harmonization process has found. Of primary importance is that it is clear, which original activity codes correspond to each of the harmonized activity categories. It is helpful if this documentation is adequately presented: tables are commonly used to structure the information on which harmonized code was assigned to which original code. That document needs to contain labels, ideally in English as well as in the original language (if other than English).

The third set of documentation is the syntax which is the code used in statistical software that was employed in data conversion. This should also be made available to end users of the data to allow replicability and promote open science. Occasionally, users may also find errors in the data that were not identified through initial data quality checks. Retaining the harmonization syntax makes it possible to correct such mistakes. The harmonization syntax should be complete including steps taken during every stage of the harmonization process and annotation so that it is understandable also to someone who did not perform the harmonization.

The MTUS user has access to three types of documents: (i) The user guide gives an overview of the MTUS datasets, the variables included on the MTUS dataset, and some suggestions on how to use the dataset, (ii) an overview of the coding procedure that explains step by step how the process of harmonization was done – if someone wants to replicate the harmonization, they should be able to do it following the instructions included in the coding procedure, and (iii) a so-called "readme" file that includes basic information about the survey and the process of harmonization of each survey. This is just an example of documentation that can be provided, but each team should come up with their own best practices.

16.3.2 Quality Checks

According to EUROSTAT recommendations (Eurostat 2019), there should be a specific set of quality indicators for time use surveys that would correspond to the specifics of this type of survey. This includes general recommendations related to reporting about survey characteristics as well as detailed guidelines regarding quality assurance.

These guidelines provide information on possible quality checks for harmonized time use data and conditions that need to be met for the dataset to be deemed of good quality. These conditions must be the same for all datasets, regardless of the specificity of the harmonization process. However, a certain level of flexibility is needed as we cannot expect the same quality data (especially in terms of the level of detail) from the 1960s compared to the most recent surveys.

First, the total duration of all (main) activities recorded in every diary day included in the dataset should be no longer than 1440 minutes, which is the number of minutes each of us has in one day. Even when diarists report on more than a single day, the sum of every single day's episodes should not exceed 1440 minutes. Ideally, all of them should be equal to this number, but we know that occasionally diaries miss some minutes. In that sense, some limited imputations can be done, provided it is documented. Other universal conditions include that there should be no negative or zero values in variables describing duration of each activity (as there is no negative time count). Relatedly, duration should have the minimum value of 1, regardless of the type of the matrix.

While most diaries are complete or almost complete (do not have missing values for main activity sequence), some may be missing a substantial share of activities and in such a case they may be flagged in the dataset. In the case of MTUS, a gap of 90 minutes of missing time between episodes would result in flagging a diary as "low quality," but this is up to the converting team to set their own standards. Overall, it is not recommendable to delete any diary from the sample even if its quality is considered poor. Instead, the user needs to be able to identify the poor-quality diaries and decide if they want to include them in their analysis. The total number of bad-quality diaries could be provided in the documentation to give the user an idea of the quality of the overall dataset. In the case of the MTUS, the "low quality" diaries are also zero-weighted in the weights provided by the MTUS conversion team. That means they are not included in cross-country or across-time analyses if the MTUS weights, which are intended for comparing across the harmonized datasets, are used. A harmonized dataset should also preserve the original survey weights. It is also possible to combine country weights with weights enabling comparisons across the datasets. These decisions are made by the conversion team and each should agree on their own standards that would be applied and documented across all datasets.

Quality checks include also checking for data completeness. For example, a researcher would check for any instances where location is missing but the activity might give a clue to location – or vice versa. Coding consistency should also be checked across the dataset. For example, travel as an activity should not be performed at an indoor location (such as home or school). Simple cross-tabulations reveal such inconsistencies. Errors occurring during the harmonization process should be corrected. Errors existing in the original dataset may be fixed whenever possible but in this case, it should be properly documented. They can also be flagged for the user or left unchanged if we decide to harmonize the original data without corrections. Either way, it is useful to provide a way for the end user to trace the erroneous observations.

Many checks are based on cross-tabulations of activity data with specific background variables. For example, the hours of paid work should be much lower for the weekend diaries, and formal schooling is not expected to occur on Sundays. Because quality checks are run on the harmonized data, the process allows a substantial degree of programming to facilitate the work. As opposed to the data harmonization procedure, very few things are done manually, and this can usually be done only if the algorithm for quality checks flagged issues with the data.

Some more specific (as opposed to systematic) checks may be run while the data are used for analyses. Feedback from researchers working with particular datasets is always valuable as it helps to identify peculiar errors that can be corrected. The end users of the data should therefore be encouraged to report any issues to the agency that harmonized the data.

Lastly, there is also an issue of intercoder reliability. During the checks, it is typically addressed using the "readme" files mentioned above. These files list all of the original codes and the harmonized codes they were assigned to in the harmonization process. That applies to all variables in the time use diary that were subject to harmonization, such as the main and secondary activity, copresence, and location. Though all coders must use the same harmonization guidelines, in some cases they may need to decide on a case-by-case basis, which harmonized activity code would best represent an original code in a given country context. Once harmonization is completed and readme file is produced, another coder should review the document and evaluate the harmonization original coder's harmonization decisions especially regarding the activities. Usually there are few disagreements with regard to how activities should be coded and in such cases consulting additional sources might be useful. Knowing specific country context and local habits is very helpful in solving these issues.

16.4 Challenges to Harmonization

Harmonization of time use data comes with particular challenges that, to add to the complexity of the process, are often specific to a given dataset. In many cases, they are linked to the fact that a researcher needs to make multiple compromises while converting original data into a harmonized format. It is challenging to keep the data as close to the original as possible, and, at the same time, make it comparable across very different countries.

Second, the data need to be detailed but also they should not overwhelm the end user. As time use sequences are generally one of the most complex types of data encountered in the social sciences, harmonization should not add to this complexity but rather, whenever possible, make the data easier to analyze. In the cases where there are major country differences in time diary variables and when the responses could be ranked (such as in the example of copresence provided earlier in this chapter) using hierarchical coding can be useful. It would not complicate the data too much for the user since he or she can simply use the most general code (e.g. 1-digit) and ignore additional detail available if those are not needed in the analyses.

However, we would not recommend hierarchical coding for activities, because of the already large number of codes, the great variety of activities, the differences in meaning of the same activity across different cultures, and the difficulty of ranking them in a meaningful way that would allow establishing a straightforward hierarchy. The problem with coding some activities is not about linking them with a certain more general category but rather in the fact that they may straddle different broad categories and the coder would still need to make the decision where to assign a given activity. The probability that the end user would review all detailed codes to identify a certain three-digit activity is low and we consider this approach not worth time and effort.

Conversion of time use data raises both technical and conceptual challenges. While the quality of the data has generally improved greatly since the first time used surveys, countries with limited experience with these types of data sometimes devise atypical ways of recording the data, which may later pose problems in the conversion process. There are often inconsistencies within the dataset itself, and the data are coded in ways that need to be well understood before they are harmonized. Original documentation makes it easier to understand a dataset and, as mentioned earlier, reviewing these materials is always the first step in harmonization. In some cases, a conversion process reveals coding errors within the original dataset, such as recording more than 1440 minutes of activities for a single person per day, repeated activity episodes, or inconsistencies between information provided for a single episode (e.g. contradictions between location and activity type). Some of these errors can be corrected, but many are carried further into the harmonized dataset with a flag in the dataset or a relevant note in the data documentation. Theoretically, an alternative to preserving some errors that cannot be fixed would be to delete a given record. However, in the harmonization process, this is not an option, as that would mean removing an observation that might, under some circumstances, be meaningful to the end user. Instead, it makes sense to let the user decide whether he or she wants to use the record – and provide them with the information about existing problems.

Regardless of their complexity, technical issues and coding errors which are occasionally present in the original data do not pose conceptual challenges. Such challenges arise when activities are harmonized across countries. Each diary reflects cultural intricacies regarding how the days usually unfold, and what particular activities mean in a given social context. Original activities usually provide more information than the harmonized ones, as they are more detailed and better reflect particular cultural specificities. The researcher harmonizing the data should preserve as much of the original information as possible; in practice, it often comes down to a choice regarding how the activities should be recoded. For example, handcrafts, sewing, or knitting in developing countries may denote activities related to mending clothes, but may also imply work done to supplement household income. In affluent societies the same activities are more likely to be hobby or leisure time activities. In a similar way, foraging may be seen as a productive physical activity, a leisure activity, and an activity intended to boost household income (if the products are later sold on a local market), but in the harmonized version it can only be coded as one of those. Should these activities be coded in the same way regardless of where they are carried out, or should the code be chosen, while taking into account the country context? Either choice entails consequences for the end user. First of all, depending on the research topic, a person might be interested in very different aspects of a given activity (e.g. metabolic expenditure versus unpaid work) – and not all potential needs may be satisfied. Secondly, in the context of cross-national comparisons, such differences in the meaning (or multiple meanings) of an activity may be important. This particular challenge of harmonizing time use data is related to the otherwise major advantage of such data which is that they may be used to examine a very broad range of topics. A harmonized dataset may be used to address topics from an individual's daily physical activity (Harms et al. 2019) to household energy consumption patterns (Archer et al. 2013). As shown in the examples provided above, activities straddling several dimensions need to be simplified and recoded into a single category. Depending on the particular topic for which the data are later used, it may or may not matter for the study.

Clear guidelines are provided in the MTUS (Gauthier et al. 2006) to ensure that all activities are recoded in the same way for all countries. This is a conceptual choice, but it does not mean there is no other way to do it. For example, activities can be harmonized based on their major role in the given cultural context (returning to the foraging example – this may be a type of work to supplement household income or a hobby). However, that would require an extensive knowledge of each cultural context and detailed information about the nature of the activity in the data, the latter of which is usually missing. A possible solution might also be for the original variable to be kept in string format, if the user wanted to consult it. That makes the data slightly more complicated but allows for more detailed analyses.

A more systemic challenge, but still related to the area of activity harmonization, lies in the fact that any harmonized activity list would always be better fitted to reflect a particular cultural context. For example, as the MTUS was first designed to harmonize the data coming primarily from Western Europe (HETUS) and the United States (ATUS), its list of activities is less relevant for developing countries which typically use the ICATUS categorization in the original data collection tools. As a result, the "costs" of harmonizing the data is loss of detail or reinterpretation of the meaning of activities for a given culture is higher for countries with activity categories that are the least similar to those found in the HETUS and the ATUS. The categorization used in MTUS is conceptually and culturally more distant from ICATUS than it is from HETUS or ATUS. Perhaps that calls for a new categorization system that more effectively bridges the differences between activities common in developing and developed nations. That has not been done yet.

Further challenges with harmonizing time use data surround the handling of additional information collected alongside the time diary. As outlined earlier in this chapter, there is a considerable variation in the scope and content of additional information collected. For example, some countries go into far greater detail, such as the United States, which captures distinctions between household children and between types of non-household members such as friends, or coworkers. Not all these details are preserved in the harmonized dataset because most countries do not provide such information. When selecting the variables for harmonization MTUS has chosen to focus on the variables that are available in the original data for most countries.

Furthermore, in some cases, there are subtle differences in how certain variables are operationalized in different datasets. For example, some countries report a copresence of a child aged up to 18 years old, while others report copresence of a child aged up to seven years old. Overall, the variation with regard to the age of the child reported in the copresence column is quite broad. Across the original datasets, this variable is therefore not fully comparable. A researcher harmonizing the data needs to use the original information. They may not be using the variable at all, but that would mean losing potentially important data.

What is commonly done at the moment is to use approximations (such as collapsing categories of similar age, for example, children below 5 and below 7) and add flag variables as well as information in the data documentation. As a result, in some cases the variables in the harmonized dataset also represent slightly different definitions which may be compared in certain circumstances, but not unconditionally. Importantly, any existing differences in the variables are described in detail in documentation files for a specific dataset and they need to be taken into account by the end users.

As mentioned earlier, the alternative to the current practice in the field might be using hierarchical coding schemes where end users are provided with meaningfully ranked codes representing different possible responses that were used in the original data (e.g. 11 – with a child, 111 – with a child aged 0–7, 112 – with a child aged 0–9, and 113 – with a child aged under 18 years old).

16.5 Software Tools

The harmonization of MTUS is largely a hand-crafted process. While there is a standard set of guidelines for harmonizing data into MTUS, the source data are so variable that it is difficult to develop one-size fits all tools. Rather, harmonization is done in statistical software[6] with a series of recode and conditional statements. Each file is processed separately with a set of syntax files restructuring the data as needed and harmonizing it according to MTUS guidelines. After the majority of the harmonization is completed, an additional set of steps are completed to construct a standardized set of weights and to make any final adjustments to the data. At this point in the process, a final set of data quality checks are also performed to ensure data integrity.

The CTUR, the expert organization in time diary data harmonization and main developer of MTUS, has partnered with IPUMS at the University of Minnesota to produce IPUMS Time Use. IPUMS Time Use is a web-based data access system that makes harmonized MTUS data available to the research community (https://timeuse.ipums.org). IPUMS Time Use leverages IPUMS specialized software for simplifying and streamlining researcher access to harmonized time diary data. The IPUMS Time Use system has been modified specifically to accommodate the complex structure of time diary data. Figure 16.3 shows the IPUMS MTUS variables selection page where researchers can select person or activity variables and where they can select predefined time use variables or construct their own using harmonized time diary attributes such as activity, secondary activity, location, and with whom (copresence).

MTUS data harmonized by CTUR are placed into the IPUMS system for delivery via the data access system (https://ipums.org/projects/ipums-time-use). Each variable code and label is converted to formally structured data for a dynamically generated display via IPUMS Time Use. The conversion of static

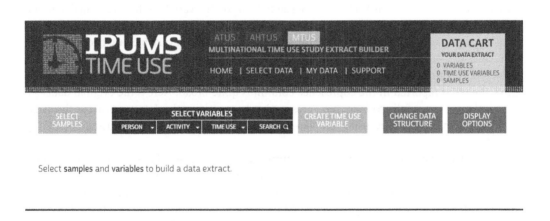

Figure 16.3 Screenshot of IPUMS MTUS variable selection page.

6 The specific software used has varied over time. Files are currently processed using Stata.

Word or PDF documents into formally structured metadata is required by the IPUMS system to populate the website. The metadata-driven approach is a permanent investment in data infrastructure since the metadata can be read and displayed by any software thus allowing data access system upgrades without having to develop new web pages by hand. The approach also makes documentation available at the variable level so that users can see documentation about a variable in a single place and comparability issues across datasets at a glance rather than having to open separate documentation files for each dataset of interest.

Using IPUMS, researchers can select which of the harmonized datasets they would like to download along with the background and time diary variables they would like to include in their customized data file. Data formats are customizable allowing researchers to choose between datasets formatted for particular statistical packages and data structured in different formats. Selected datasets are pooled for researchers, meaning that they download a single file with the variables they want and in the format of their choice instead of multiple data files with all variables included. IPUMS stores the exact specifications of each requested data extract. This functionality allows researchers to easily add additional variables or datasets to previously requested data extracts and to generate new data extracts with the additional data included.

A key feature of IPUMS Time Use is the ability for researchers to create custom time use variables, which indicate the number of minutes per day spent in various combinations of activities, locations, and copresence specifications. This feature significantly lowers the barriers to access by eliminating the need for researchers to be able to manipulate the episode-level data and instead aggregate minutes spent in specific types of episodes. For example, a researcher can specify that they want the number of minutes per day spent engaging in paid work at home while someone else was present or the number of minutes per day spent outside with someone else. These aggregations can be complicated for the novice user of time diary data but are dramatically simplified by using the time use variable creation component of IPUMS Time Use.

16.6 Recommendations

Harmonizing time use surveys may be challenging, especially for those with less experience in handling and analyzing time use data. For this reason, the available harmonized time use databases, such as the MTUS, and, in particular, the IPUMS Time Use data extract builders were generated. These resources provide much simpler access to time use surveys from across the world.

As outlined in this chapter, occasionally even the harmonized database may include variables that differ across countries in some regards. They may also carry uncorrected errors from the original diaries if corrections were not possible or would require substantial changes to the dataset. For this reason, the essential initial step for anyone intending to use a harmonized dataset is to read the user guide and specific documentation for every dataset to be included in the analyses. The documentation includes information on any differences within the harmonized variables, informs which variables should be treated with caution – e.g. due to their low quality in the original dataset, and gives an overview of any other problems found in a given dataset. Sometimes researchers are interested only in the aggregated measures of time such as the total daily duration of a given activity. For those scholars, the MTUS and IPUMS MTUS provide an aggregated file in which a single row represents a single day for a given individual and provides summary of time for all activities.

As emphasized throughout this chapter, the process of harmonization of a time use dataset is usually linked with some degree of detail being lost. In some cases, this does not affect the analyses, in others – in particular studies focusing on a narrow range of behaviors or addressing a specific topic in greater detail (e.g. food-related behaviors, cultural consumption, household energy use, etc.), – the generic harmonized dataset may not be the optimal choice. For such studies, we recommend that the researchers produce their own harmonized dataset using the original data files.

To produce your own harmonized dataset, we advise that you follow the procedure proposed in this chapter, that is, first, convert all datasets into the same format – without changing their actual content. Next, the diaries need to be checked for errors or missing values, and missing values may be imputed where possible. Harmonization of activities – when intended for a particular project – should not pose problems, which stands in contrast to harmonization done for the database available for broader use, such as the MTUS. This is because the researcher knows exactly what aspects of activities are of interest to them and may therefore focus on preserving information addressing these particular issues. The source data, all conversion syntax files, documentation, and the final datasets should all be preserved during the harmonization process.

Throughout the harmonization process, it is advisable to preserve as much detail as possible (potentially even by keeping the source variables in the harmonized data, though this currently is not a common practice), and flag the diaries that were found to have technical issues, such as a long-missing overall time in the sequence. This is because during later computations it is easy to forget about these issues, but they may hamper the analyses.

Lastly, any changes should be documented in a separate file – for the record but also to inform any other users of the dataset. We also recommend keeping an easy-to-read syntax file and retaining variables linking the harmonized file with the original one (that would usually be household identifier, respondent identifier, and diary identifier). In this way, it will be possible to get back to the original data in case any new variables need to be generated.

References

Andorka, R. (1987). Time budgets and their uses. *Annual Review of Sociology* 13: 149–164.

Archer, E., Shook, R.P., Thomas, D.M. et al. (2013). 45-year trends in women's use of time and household management energy expenditure. *PLoS One* 8: e56620. https://doi.org/10.1371/journal.pone.0056620.

Asanov, I., Flores, F., McKenzie, D. et al. (2021). Remote-learning, time-use, and mental health of Ecuadorian high-school students during the COVID-19 quarantine. *World Development* 138: 105225. https://doi.org/10.1016/j.worlddev.2020.105225.

Bevans, G.E. (1913). *How Workingmen Spend their Spare Time, 2018*. Franklin Classics Trade Press.

Caparrós Ruiz, A. (2017). Adolescents' time use in Spain: does the parental human capital matter? *Child Indicators Research* 10: 81–99. https://doi.org/10.1007/s12187-016-9368-0.

Cawley, J. and Liu, F. (2012). Maternal employment and childhood obesity: a search for mechanisms in time use data. *Economics and Human Biology* 10: 352–364. https://doi.org/10.1016/j.ehb.2012.04.009.

Cornwell, B., Gershuny, J., and Sullivan, O. (2019). The social structure of time: emerging trends and new directions. *Annual Review of Sociology* 45 (1): 301–320.

Craig, L. and Mullan, K. (2011). How mothers and fathers share childcare. *American Sociological Review* 76: 834–861. https://doi.org/10.1177/0003122411427673.

Eurostat (2019). Harmonised European time use surveys (HETUS) 2018 guidelines. Luxembourg. https://doi .org/10.2785/926903. https://ec.europa.eu/eurostat/documents/3859598/9710775/KS-GQ-19-003-EN-N .pdf/ee48c0bd-7287-411a-86b6-fb0f6d5068cc?t=1554468617000.

Gauthier, A.H., Gershuny, J., and Fisher, K. (2006). Multinational Time Use Study User's Guide and Documentation. https://www.timeuse.org/sites/ctur/files/public/ctur_report/5715/mtus-user-guide-r5.pdf.

Gershuny, J., Sullivan, O., Sevilla, A. et al. (2020a). A new perspective from time use research on the effects of lockdown on Covid-19 behavioral infection risk. *PLoS One* https://doi.org/10.1371/journal .pone.0245551.

Gershuny, J., Harms, T., Doherty, A. et al. (2020b). Testing self-report time-use diaries against objective instruments in real time. *Sociological Methodology* 50: 318–349. https://doi.org/10.1177/0081175019884591.

Gross, D.R. (1984). Time allocation: a tool for the study of cultural behavior. *Annual Review of Anthropology* 13: 519–558. https://doi.org/10.1146/annurev.an.13.100184.002511.

Harms, T., Berrigan, D., and Gershuny, J. (2019). Daily metabolic expenditures: estimates from US, UK and polish time-use data. *BMC Public Health* 19: 453. https://doi.org/10.1186/s12889-019-6762-9.

Jahoda, M., Lazarsfeld, P., and Zeisel, H. (1971). *Marienthal: The Sociography of an Unemployed Community*, 1e. New York: Routledge.

Jarosz, E. (2016). The duration and dynamics of leisure among the working population in Poland. A time-use approach. *World Leisure Journal* 58: 44–59. https://doi.org/10.1080/16078055.2015.1088467.

Jarosz, E. (2018). Lifestyle behaviors or socioeconomic characteristics? Gender differences in covariates of BMI in Hungary. *Obesity Science and Practice* 4: 591–599. https://doi.org/10.1002/osp4.316.

Jarosz, E. (2021). What makes life enjoyable at an older age? Experiential wellbeing, daily activities, and satisfaction with life in general. *Aging & Mental Health* 1–11. https://doi.org/10.1080/13607863.2021 .1916879.

Kahneman, D., Krueger, A.B., Schkade, D. et al. (2004). Toward national well-being accounts. *The American Economic Review* 94: 429–434. https://doi.org/10.1257/0002828041301713.

Kan, M.Y. and Pudney, S. (2008). 2. Measurement error in stylized and diary data on time use. *Sociological Methodology* 38: 101–132. https://doi.org/10.1111/j.1467-9531.2008.00197.x.

van Knippenberg, R.J.M., de Vugt, M.E., Ponds, R.W. et al. (2017). Dealing with daily challenges in dementia (deal-id study): an experience sampling study to assess caregiver functioning in the flow of daily life. *International Journal of Geriatric Psychiatry* 32: 949–958. https://doi.org/10.1002/gps.4552.

Larson, R. and Csikszentmihalyi, M. (2014). The experience sampling method. In: *Flow and the Foundations of Positive Psychology*, 21–34. Netherlands, Dordrecht: Springer.

Moen, P. and Flood, S. (2013). Limited engagements? Women's and Men's work/volunteer time in the encore life course stage. *Social Problems* 60: 206–233. https://doi.org/10.1525/sp.2013.60.2.206.

Musick, K., Meier, A., and Flood, S. (2016). How parents fare. *American Sociological Review* 81: 1069–1095. https://doi.org/10.1177/0003122416663917.

Reeves, M.P. (1913). *Round about a Pound a Week*. London: G Bell and Sons Ltd. https://www.gutenberg.org/ files/58691/58691-h/58691-h.htm.

Robinson, J. (1985). The validity and reliability of diaries versus alternative time use measures. In: *Time, Goods, and Well-Being* (ed. F.T. Juster and F. Stafford), 33–62. Ann Arbor, MI: Survey Research Center Institute for Social Research University.

Salvy, S.-J., Bowker, J.W., Roemmich, J.N. et al. (2008). Peer influence on children's physical activity: an experience sampling study. *Journal of Pediatric Psychology* 33: 39–49. https://doi.org/10.1093/jpepsy/jsm039.

Sayer, L.C. (2005). Gender, time and inequality: trends in women's and men's paid work, unpaid work and free time. *Social Forces* 84: 285–303. https://doi.org/10.1353/sof.2005.0126.

Strumilin, S. (1925). Byudzhet vremeni rabochikh v 1923–1924 g [Time Budgets of Russian Workers in 1923–1924]. Plan khozyaistvo 7:NA.

Sullivan, O. (2000). The division of domestic labor: twenty years of change? *Sociology* 34: 437–456. https://doi.org/10.1177/S0038038500000286.

Sullivan, O., Gershuny, J., Sevilla, A. et al. (2020). Time use diary design for our times – an overview, presenting a click-and-drag diary instrument (CaDDI) for online application. *Journal of Time Use Research* 1–17. https://doi.org/10.32797/jtur-2020-1.

Szalai, A. (1966). The multinational comparative time budget research project: a venture in international research cooperation. *The American Behavioral Scientist* 10: 1–4.

Szalai, A. (1972). *The Use of Time. Daily Activities of Urban and Suburban Populations in Twelve Countries.* The Hague: Mouton.

Thomas, M., Hunt, A., Hurley, M. et al. (2011). Time-use diaries are acceptable to parents with a disabled preschool child and are helpful in understanding families' daily lives. *Child: Care, Health and Development* 37: 168–174. https://doi.org/10.1111/j.1365-2214.2010.01156.x.

Part IV

Further Issues: Dealing with Methodological Issues in Harmonized Survey Data

17

Assessing and Improving the Comparability of Latent Construct Measurements in *Ex-Post* Harmonization

Ranjit K. Singh[1] and Markus Quandt[2]

[1]*Survey Design and Methodology, GESIS – Leibniz Institute for the Social Sciences, Mannheim, Germany*
[2]*Survey Data Curation, GESIS – Leibniz Institute for the Social Sciences, Cologne, Germany*

17.1 Introduction

The chapter addresses a challenge that most *ex post* harmonization projects face: how to reconcile differences in the measurement instruments, that were used in the various source data collections, to create a harmonized dataset? Specifically, we focus on survey measurement instruments, i.e. survey items, that measure latent constructs. Latent constructs are concepts that cannot be directly observed but only inferred indirectly. Examples are attitudes, values, emotions, personality traits, motivations, aptitude, or intentions. The social science literature on *ex post* harmonization of latent constructs is still sparse. At the same time, there is great potential for improving the quality of *ex post* harmonized data if latent constructs are harmonized with state-of-the-art methods.

We first address a core issue of assessing and improving the comparability of different instruments measuring the same latent construct: the relationship between measurement and reality. Next, we explore the aspects of comparability that this chapter focuses on comparability of constructs, comparability of reliability, and comparability of units of measurement. For each of these, we will discuss the central idea, which problems arise from a lack of comparability, and how to assess and improve comparability. Finally, we discuss the added challenges of harmonizing international or cross-cultural data.

17.2 Measurement and Reality

What would successfully *ex post* harmonized data look like? Ideally, we should be able to interpret all harmonized data points as if they had come from a single source survey. Respondents with the score "3" on the harmonized target variable for, say, political interest should be very similar in their true interest in politics, regardless of which source survey they participated in, and of which instrument was used there.

This simple notion is complicated by the nature of latent constructs. In measurement theory, latent constructs are the underlying, unobservable causes for observed phenomena (Bollen 2002). For example, if a person voluntarily prefers the consumption of political news over other media content (observed phenomenon), we might infer that that person is strongly interested in politics (not directly observable cause). In other words, we can only infer a respondent's latent construct intensity indirectly. Thus, establishing if an instrument measures the intended construct and ideally only the intended construct is crucial. A second distinction between the latent construct and its measurement is that the measurements are always subject to random error.

Another point to consider is that latent constructs can be continuous, such as attitudes, or discrete, such as identities. In this chapter, we will focus on continuous constructs, meaning that we can order respondents by their construct intensity, with infinitesimally small differences. However, the observable measures of these constructs, such as Likert-scales, are usually not continuous, but ordinal. The numerically coded instrument responses or scores are ordinal representations of a continuous latent construct. Each score can be thought to represent a segment of that continuous dimension. According to their observed scores, respondents from the population are then assigned to the corresponding segments of the latent dimension. Consider a five-point scale for political interest. The 13% of Germans with the lowest political interest might choose a "1." The next 24% of Germans with more political interest may choose a "2," and so on. Thus, each measurement instrument for the same construct can differ in the way it maps the continuous construct onto the ordinal measurement scale.

These complexities of latent constructs must be overcome to harmonize them. In this chapter, we distinguish three areas of comparability. First, **validity** or "construct match," which means we have to ensure that different source instruments measure the same construct. Second, **reliability differences** because different source instruments may incur different levels of random measurement error. Third, **units of measurement**, which addresses the problem that different instruments may map the same true construct intensity onto different numerical values. In these areas, we make a distinction between psychometric multi-item question batteries, such as personality questionnaires or aptitude tests, but also attitude scales in social surveys and single-item instruments, i.e. instruments with only one question. This distinction is important because multi-item and single-item instruments often require different approaches to *ex post* harmonization. Multi-item instruments have many desirable statistical features that help us capture latent constructs. However, the constraints of survey research mean that single-question instruments are far more common in the social sciences.

17.3 Construct Match

Before combining different source instruments, we must assess whether these instruments measure the same construct. This is closely tied to the idea of validity: do instruments measure the construct we intend to measure (Moosbrugger and Kelava 2012)? Comparability across instruments then extends this requirement such that all different instruments must validly measure the same single construct.

17.3.1 Consequences of a Mismatch

If instruments do not measure the same construct, then combining measurements into a single variable leads to a conceptually heterogeneous variable. Values derived from different source instruments would have different substantive implications and analyses across the whole variable cannot be meaningfully interpreted. However, expecting a perfect construct match is unrealistic in the social sciences. How strictly the construct match requirement is interpreted, and which deviations are permissible depends on the construct in question, the conventions of the respective discipline, and the research questions that the *ex post* harmonized data should answer.

17.3.2 Assessment

In some cases, we may forego formal analyses and instead rely on "face validity": "How interested would you say you are in politics" (European Social Survey 2018) and "How interested would you say you personally are in politics?" (ISSP; GESIS 2016) will most likely tap into the same construct. However, sometimes deceptively minor item wording differences across instruments can mean that different constructs are captured. For example, apparent antonyms often do not evoke the same construct dimension. Asking about trust or mistrust captures different meanings (Saunders et al. 2014).

17.3.2.1 Qualitative Research Methods

There are many established qualitative approaches to assessing construct validity. These approaches may also help us decide if two instruments should be combined. One common approach, often discussed under the term **content validity** (Moosbrugger and Kelava 2012), is to systematically interview survey methods, linguistic, or subject domain experts about possible construct mismatches. Expert judgment also comes into play when we assess the comparability of two instruments through the lens of the theoretical background of the *ex post* harmonization project.

Apart from expert approaches, instruments can also be assessed by interviewing members of the target population or people similar to the target population in important attributes (such as sociodemographic). *Ex post* harmonization practitioners may draw inspiration from **qualitative pretests** on how to conduct such interviews (Lenzner and Neuert 2017). And, while face-to-face cognitive interviews are costly, web-probing may provide a more viable alternative (Lenzner and Neuert 2017). It may also help to look up pretesting reports for related instruments in **pretest databases**, such as the GESIS Pretest database (GESIS 2020). Even if the exact instruments are not covered, pretest reports may still inform about ways to (mis) understand important constructs in the same subject domain.

17.3.2.2 Construct and Criterion Validity

In the realm of quantitative methods, validity is usually assessed through (inter) correlations with other substantive variables. Assessing **construct validity** focuses on the relationship of the measure and other constructs where we have a good theoretical understanding of the expected associations between our construct and the validation constructs (Sherman et al. 2011). **Convergent validity**, for example, examines if the instrument is closely correlated with other instruments measuring very similar constructs, while **divergent validity** inverts this logic, by affirming that the instrument is not perfectly correlated with measures of other neighboring, but distinct constructs (Sherman et al. 2011). If the construct is embedded in an elaborate network of theoretical relationships, we can employ tests of **nomological validity** (Preckel and Brunner 2017); for example, via structural equation modeling (SEM). Here, the full pattern of correlations with other constructs is tested in regard to how far it meets theoretical expectations. Lastly, **criterion validity** focuses on the relationship of the construct of interest and relevant outcomes (Price 2017). A latent measure of a respondent's health should be predictive of health-related outcomes, for example.

For the purpose of assessing comparability prior to harmonizing the different instruments, all these forms of validity testing can in principle be applied to each separate instrument in their respective source dataset. If comparable patterns emerge, construct comparability is likely. However, there typically is a lack of viable criterion variables in *ex post* harmonization settings, because obviously, these criterion variables would need to be comparable across the source datasets for themselves. Otherwise, differences in the correlation patterns across the source datasets could be attributable both to incomparability between the harmonization candidate instruments and to incomparability of the criterion variables, not allowing firm conclusions on either. Furthermore, the variables available for validation and those included in the substantive research efforts based on the harmonized data overlap to a large degree. This can save redundant effort but also runs the risk of introducing a circular logic into our validation. We assume that constructs are comparable, because our substantive analyses have the expected results, and we trust these expected results because we assume that constructs are comparable. To sidestep this, at least some degree of external validation might be desirable, e.g. with variables or data sources outside the substantive project.

17.3.2.3 Techniques for Multi-Item Instruments

For multi-item instruments, the most popular approach to testing construct validity (and thus construct comparability) is **factor analysis**. Factor analysis is geared toward continuous latent variables and conventionally at least ordinal-scaled manifest indicators (although approaches for nominal data exist; e.g. Revuelta et al. 2020). For the sake of brevity, we do not explicitly address equivalent analysis techniques for analyzing categorical manifest indicators (e.g. Item Response Theory and Salzberger and Sinkovics 2006) or categorical latent variables (e.g. Latent Class Analysis and Kankaraš and Moors 2009), although those allow similar approaches.

If we have several multi-item instruments we can examine the factor structure of each instrument in a confirmatory factor analysis (CFA). A CFA can demonstrate that all instruments measure the same number of factors, and how they relate to each other. If we collect data on both instruments from the same respondents, we can test the assumption that both instruments measure the same construct directly. This is also possible if one instrument has several items, and one has only one. Furthermore, if the instruments' data come from separate samples, the CFA approach can be naturally combined with the construct validities discussed above. Multigroup confirmatory factor analysis (MG-CFA) was specifically designed to statistically test for the (non) differences of correlational patterns of the "same" model across different samples (cf. Chapter 18). Since CFAs are nothing else than precisely specified SEMs, such models can easily include tests for the divergent and convergent validation of particular constructs, and in the MG-CFA setup, would also directly reveal where relevant validity differences between samples exist.

17.3.2.4 Improving Construct Comparability

There are some strategies for mitigating construct mismatch. However, not all types of mismatches can be overcome *ex post*. There are two basic types of construct incomparability. First, a construct **mismatch with no overlap**, meaning that the instruments capture completely different constructs. In this case, there is little to be done. If the instruments simply refer to different constructs, they cannot be combined. Second, a **construct mismatch with overlap**, in the sense that both instruments capture variance of the same construct, but one or both also capture variance of other constructs. In that case, it is imaginable to coax out only the variance components relating to the common construct by controlling for the variance introduced by competing constructs. One way to approach this is with bifactor CFA models where indicators load on two factors simultaneously (Brown 2015). Such models allow us to separate what is common across all indicators and what is unique about subsets of indicators. Similarly, several variables in a source survey can be used as indicators to construct synthetic measures via **SEM** *in lieu* of a suitable raw measure. For example, if the construct of interest (e.g. socioeconomic status) was not measured, but many of its component causes were (e.g. income, education, and profession), then we can estimate what is called a formative indicator factor model to estimate the construct (Brown 2015).

17.4   Reliability Differences

Reliability, as a common conceptualization of measurement accuracy, is our second aspect of comparability. Even if two instruments relate to the same construct, they may differ in their accuracy, meaning that they measure the same construct with different levels of random error. As in all research, we would like to obtain measurements with as little random error as possible (Moosbrugger and Kelava 2012). However, in *ex post* harmonization, we also must be mindful of different levels of random errors when combining data from different sources.

17.4.1   Consequences of Reliability Differences

Reliability difference means that different instruments incur different levels of random measurement error. If data from those instruments are combined into a harmonized dataset, then segments of this joint variable differ systematically in their reliability. This is relevant because reliability attenuates substantive correlations with other variables (Raykov and Marcoulides 2011): even if two constructs were perfectly correlated in reality, the correlation of the instruments measuring these two constructs would always be underestimated due to random error.

Attenuation occurs in single-source data as well. In harmonized data, the issue is complicated by the fact that attenuation varies across segments of the same harmonized target variable. This may cause methodological artifacts. Imagine that we wanted to predict political participation with a political interest in two countries. While political participation was measured with the same reliability in both countries, political interest was measured with different instruments that were not equally reliable. Even if at construct level, the true correlation of political interest with political participation were the same in both countries, our model would show a stronger effect in the country with the more reliable instrument, because there is less attenuation occurring.

17.4.2 Assessment

Consequently, we want to assess the reliability of data measured with different source instruments. Reliability can be conveniently estimated for **multi-item instruments**. So-called measures of internal consistency treat each item relating to a construct as a separate (and thus repeated) measurement of the same construct. The most well-known measure, Cronbach's Alpha, relies on very restrictive assumptions and thus tends to underestimate reliability. Modern CFA-based measures of **congeneric reliability** (e.g. Omega) circumvent such restrictions and are preferable (Raykov and Marcoulides 2011). For **single-item instruments**, estimating reliability is harder or even unfeasible. We require multiple instances of measurement with the same instrument for each person to estimate a **test–retest reliability** (Tourangeau 2020). Within a typical *ex post*-harmonization setup using secondary data, this is only possible within panel surveys repeatedly measuring the same construct for each respondent. More elaborate approaches that can be applied to multi- as well as single-item instruments include the **multitrait–multimethod approach** (MTMM; Tourangeau 2020). However, such approaches mandate the collection of additional data.

17.4.3 Improving Reliability Comparability

If a difference in reliability has been detected it should be declared in the documentation of the *ex post* harmonized dataset. Knowing the individual reliabilities, data users can employ a **correction for attenuation due to measurement error** (CAME; Charles 2005) in their analyses. CAME attempts to estimate the correlation that would have been obtained if the instruments were measured with perfect reliability. However, CAME is not a panacea. It corrects the estimated correlation parameters but also widens their confidence intervals. Otherwise, instruments with low reliability would be unfairly favored (Hunter and Schmidt 2004).

17.5 Units of Measurement

The **units of measurement** are the third aspect of comparability. The units of measurement describe the relationship between measured numerical scores of an instrument with the true construct intensity they represent. In single-item instruments capturing continuous latent constructs, this usually means that each response score of an instrument represents a specific segment of the continuum of possible construct intensities (Price 2017).

Thus, in *ex post* harmonization different instruments usually have different units of measurement. A "3" in one instrument for political interest likely represents a different range of political interest than a "3" in another instrument. This is readily apparent in instruments with a different number of response categories (i.e. scale points). However, all other design features of the instruments can play a role too.

For example, the question wording might determine how easy a statement is to agree with, i.e. item difficulty (Moosbrugger and Kelava 2012). Even with the same response scale, we might observe higher scores when asking if respondents are "interested" in a topic than when asking if they are "passionate" about it. Similarly, the response labels, the visual layout, and many other aspects can influence responses (Moosbrugger and Kelava 2012). Thus, response scores measured by different constructs are rarely directly comparable, even when their measured constructs matched perfectly.

17.5.1 Consequences of Unit Differences

Combining data from instruments with different units creates heterogeneous target variables which can bias analyses performed with the harmonized dataset. Imagine a harmonized dataset where data from different populations was measured with different instruments. If we now find an empirical difference between those two populations, such as a mean difference, this could either be a true population difference, a **measurement difference**, or a mix of the two. Neither adding source instruments as control variables to the model nor applying a z-standardization solves this, because these approaches cannot disentangle measurement differences and true differences. Unit differences can also **bias correlations** we estimate. This is most easily understood in (but not limited to) the case where two constructs are uncorrelated. Imagine data from two surveys A and B sampling the same population. In each survey, we would observe a circular cloud of dots as bivariate scatterplots for the two constructs. However, if we combine the data of surveys A and B in a joint dataset without harmonizing measurement scores first, the two-point clouds will probably not align. Even if measurement units differ only by an additive constant, the clouds will be shifted in relation to each other. The statistical result would be a spurious correlation between the two variables for the joint data, even though there is no correlation in each separate survey. In essence, we now observe Simpson's paradox (Kievit et al. 2013) in our data: a different correlation across the combined data may occur in each source survey, even if the source surveys capture the same population. Now of course, in our simple example with data from the same population, we can remove the bias easily by adding the source survey A or B as a control variable. However, the same correlation bias can occur if surveys A and B sample different populations and also for true correlations that are not zero. In such cases, population differences and instrument differences become entangled again and need to be tackled with suitable harmonization procedures.

17.5.2 Improving Unit Comparability

To make measurement units comparable, we have to transform their original scores such that the "new" scores for the harmonized instrument point to the same range of true scores, regardless of what the source instrument was. Only then respondents who, for example, are similarly interested in politics are represented by similar numerical values in the harmonized dataset, regardless of the source survey they participated in. Note that the only requirement here is that after harmonization, the same score should imply the same construct intensity regardless of the instrument used. This means that units of measurements need to be made comparable, but how we scale the units in our harmonized data is arbitrary. In this chapter, we assume that responses are harmonized toward a chosen reference instrument. That means the scores can be interpreted in terms of that instrument. Interpreting raw scores, that is in terms of their operationalization, is common practice in the social sciences. However, you can of course choose other scaling schemes, such as norming scores to a chosen population at a specific point in time (Kolen and Brennan 2014; Price 2017).

There are different strategies for harmonizing measurement units, which we organize into three categories: controlling for specific instrument characteristics, harmonizing units based on repeated measurements, and harmonizing units based on measurements obtained from the same population.

17.5.3 Controlling for Instrument Characteristics

Strategies controlling for specific instrument characteristics are usually either semantic or based on instrument design aspects. **Semantic harmonization techniques** focus on the word meaning, especially on response options. This may be as simple as domain experts deciding which response options are roughly the same across instruments. Quantitative approaches meanwhile attempt to empirically scale the meaning of specific response options and quantifiers (e.g. "very much") along an intensity continuum (e.g. de Jonge et al. 2017). The idea is that respondents whose true score is near one of these positions along the intensity continuum will tend to choose the corresponding response category. This technique reached some popularity in research on constructs asked with very similar question wording but very different response scales, such as happiness.

Other techniques control specific instrument design aspects. A basic technique here is **linear stretching**. Instruments with different numbers of scale points are harmonized by setting the instruments' maximum scores and minimum scores equal and scaling all options in between with equal distances (de Jonge et al. 2017). Harmonizing a 5-point scale to match a 7-point scale would result in the transformed scores 1, 2.5, 4, 5.5, and 7. It is also possible to scale instruments to a different range; often to a range from 0 to 100 as POMP (percent of maximum possible) scores. Sometimes, the harmonized values are also scaled to decrease variance differences (Slomczynski and Skóra 2020). However, linear stretching cannot control for any other instrument differences, such as differences in question or response category wording, or visual layout.

One way of overcoming these shortcomings is to control several instrument design aspects at once. This becomes possible if information about aspects of the instrument's design (e.g. number of scale points, response scale polarity, or positive versus negative wording) is systematically added to the harmonized dataset via specially constructed methodological variables (Słomczyński et al. 2016; Slomczynski and Tomescu-Dubrow 2018). Such instrument metadata can then be used as **control variables** in substantive analyses (Kołczyńska and Slomczynski 2018; Slomczynski et al. 2021). This approach achieves a high level of transparency and flexibility for the end-users of the harmonized data. Due to the nested nature of *ex post* harmonized data (survey program – country – year – respondent) **multilevel modeling** approaches are an obvious choice to explicitly model these methodological controls (see Chapter 20 in this volume). Furthermore, because we have different constructs measured with similar instruments and the same constructs measured with different instruments, MTMM approaches may be helpful as well (Raykov and Marcoulides 2011). However, including every detail of every instrument in a model might not be feasible or even advisable. Furthermore, if some instruments are perfectly correlated to substantive variables such as the country or time, we might not be able to disentangle instrument differences from true effects.

17.5.4 Harmonizing Units Based on Repeated Measurements

Repeated measurement strategies usually present respondents with several items targeting the same construct, for example, psychometric **multi-item instruments**. With such an instrument, we obtain several measurements of the same true score within each respondent, and we obtain a distribution of true scores across all respondents. Psychometry then uses this interplay of within- and between-respondent information to derive estimations of each respondent's true score, for example, in factor analyses or item response theory models (Raykov and Marcoulides 2011). Crucially for harmonization, we also obtain information on how each item relates to the true scores and thus the underlying construct (Brown 2015). In factor analysis, this is represented by the factor loadings (regression coefficients of the item onto the true score) and the intercepts (constants scaling the measurement units of each item).

In *ex post* harmonization, we can make direct use of this if we have two multi-item instruments which happen to share some items. Many psychometric tests intentionally include such shared "anchor" items that facilitate harmonization. Based on such anchor items, the two instruments can be easily harmonized using **equating** under the **NEAT design** (nonequivalent anchor test design; Kolen and Brennan 2014). As the term already implies, we can harmonize multi-item instruments with shared items even across different ("nonequivalent") populations.

A second application of repeated measurements is **multiple imputation**. If we have data where the same respondents answered both instruments intended for harmonization, then multiple imputation can align the instruments' units of measurement (Siddique et al. 2015). This can occur in panels, but also in cases where different surveys are administered together (e.g. ALLBUS and ISSP Germany). If the instruments were not used together in the source surveys you are interested in, imputation can still be performed if you add another dataset where other respondents answered both instruments (Siddique et al. 2015). With this so-called external calibration data, multiple imputation can easily transform instrument scores into a common instrument format. One hurdle for *ex post* harmonization projects is that multiple imputation does not impute singular values into the harmonized dataset. Instead, the process generates several imputed versions of the dataset, and all subsequent analyses are performed once for each dataset and then aggregated (Siddique et al. 2015). This may be considered a prohibitive effort by some data users.

17.5.5 Harmonizing Units Based on Measurements Obtained from the Same Population

The third strategy is based on repeated measurements, which would most likely have to come from fresh random samples drawn from (approximately) the **same population** for calibration purposes. Since the samples stem from the same population, the distribution of the underlying construct (the true score distribution) should be similar in each sample. But again, the distributions of observed scores obtained from the different instruments may differ from each other. As we know that each sample and instrument describe a similar population, we can conclude that the observed distribution differences are due to the different instruments. If we then align these distributions by their centers, shape, and range, we thereby effectively match the scores of different instruments so that each (transformed) numerical score points to the same segment of the population distribution. Hence, a "3" in the chosen reference instrument and a "3" in another transformed instrument now refer to similar segments of the construct intensity distribution. This is the basic idea behind **observed score equating** using a **random group design** (Kolen and Brennan 2014; Singh 2020a).

A crucial point here is that observed score equating harmonizes the instruments and not just the specific data. Equating two instruments generates equivalents for each instrument score of one instrument in the format of the other instrument. These equivalents can then be applied to other data using the same instruments as well. And since observed score equating uses only the observed distributions, it is equally applicable to multi-item and single-item instruments.

Lastly, the random group design necessary for observed score equating can be achieved in different ways. First, with method experiments. Online experiments are certainly possible (Singh 2020b), but the sample should be diverse enough that we can collect sufficiently many observations for all possible instrument responses. Second, equating can also be done with data from probabilistic surveys which sample the same population in the same year (Singh 2020b). Third, even if the two instruments were never used in the same year in the same population, we may still link them if we find a relay instrument. Via chained equating, we might link the first instrument to the relay instrument in one year and then link the relay instrument to the second instrument in a completely different year. And finding such direct and relay links can be easily automated.

However, sometimes data for both instruments drawn from the same population cannot be acquired. Then you can attempt to use data from different populations and mitigate population differences by taking other variables into account. In equating this is called the **nonequivalent groups with covariates (NEC) design**. It should be noted, however, that the NEC design is only a second-best approach that should only be used if there are neither shared anchor items for the NEAT design nor data from the same population for the random group design (Wiberg and Bränberg 2015). One issue is that the NEC design can only mitigate population differences that are represented by the available covariates. It also assumes that the relationship between covariates and the instruments we want to harmonize is stable across instruments and populations (Wiberg and Bränberg 2015). Despite these limitations, the NEC design is a flexible new approach that may not be able to fully harmonize instruments, but it might mitigate comparability issues in cases where no other approach is applicable.

17.6 Cross-Cultural Comparability

Until now, all issues raised apply to national as well as international research. However, instrument comparability in cross-cultural contexts adds further issues to the common challenges detailed above. Within our framework, cross-cultural comparability creates additional challenges on the construct level and the units of measurement level. To simplify our discussion somewhat, we use the term "cross-cultural" to encompass harmonization across countries, cultures, and languages. Below, we first discuss cross-cultural comparability for measurement instruments with only one item, additional options opened up by multi-item instruments are briefly addressed after that.

17.6.1 Construct Match

The cross-cultural perspective adds a level of complexity to the idea of different instruments measuring the same construct because by design, the different data sources do not relate to the same population.

Broadly speaking, we can distinguish two problem complexes. First, does the construct exist at all in a similar form in all the populations? Second, can the construct be measured with similar instruments across populations? In practice, those two issues are often closely related, of course.[1] In the following, we will discuss these issues alongside different strategies to assess cross-cultural comparability. Please note that unlike reliability and scaling issues, substantial comparability violations can often not be mitigated *ex post*. In such cases combining instruments might not be advisable at all. There is for example a long-standing debate about how much the very common political left–right ideological dimension is transferable between different polities, where the terms "left" and "right" describe quite different "policy contents" across cultures, which is reflected in how citizens from those cultures understand them (Zuell and Scholz 2019).

17.6.1.1 Translation and Cognitive Probing

Cross-cultural comparability often relies on ex-ante harmonization efforts. Many international survey programs employ different **translation** procedures to align the meaning of instrument wording across cultures as closely as possible. This ranges from blinded back-translation strategies (Cheung and Rensvold 1998) to elaborate strategies such as the TRAPD (Translation, Review, Adjudication, Pretest, Documentation) method (Behr and Shishido 2016).

Such systematic translation efforts certainly go a long way in ensuring comparability, but they do not guarantee it. For example, translation approaches cannot easily solve the problem that idiomatic expressions exist across cultures but may come with very different associations and connotations. Assessing such problems may be helped by probing the understanding of instrument wordings through **qualitative cognitive interviews**. Similar to pretesting (Lenzner and Neuert 2017), we could ask respondents from different cultures about the meaning of the terms (e.g. Zuell and Scholz 2019). Generally, the more evidence there is that careful ex-ante harmonization was already conducted, the less *ex post* harmonization researchers need to worry about assessing such sources of potential incomparability on their own.

17.6.2 Reliability

The approaches for assessing reliability and the consequences of a reliability divergence are much the same in cross-cultural research as they are in single-culture research settings. However, it is important to note that reliability always also depends on the population in which an instrument was applied (Price 2017). Thus, if we assess reliability in one country, we cannot automatically assume that the instrument will perform equally reliably in other countries. Careful translation of instruments can mitigate this problem somewhat, but cross-cultural differences may well remain. For example, imagine asking about attitudes on a political topic that is part of the current public discourse in one country, but largely ignored in another. Even if the instrument is well translated, we would still expect lower reliability in the country where many respondents think about the topic for the first time during the survey.

---

1 Note that such comparability issues are not exclusive to cross-cultural research. Similar issues can occur between different groups within countries as well (e.g. highly and less educated respondents), however, the potential heterogeneity in how such groups respond to instruments is often less obvious.

17.6.3 Units of Measurement

Comparable units of measurement for instruments with only one item may be an even larger challenge for cross-cultural research than for other settings. Even if instruments are carefully translated, there is no guarantee that the same score refers to the same construct intensity (true score) across cultures. In *ex post* harmonization, we usually face two different cases. First, the case that (nominally) the same instrument was used across cultures in different language versions, as done in international surveys. Second, that different instruments were used in different cultures when we need to harmonize independently conducted surveys.

17.6.3.1 Harmonizing Units of Localized Versions of the Same Instrument

By translating instruments into the local language(s), cross-cultural surveys create different versions of the same instrument. Usually, the same response format is used, and much care is invested into translating the question and response labels. However, this cannot guarantee that the instrument versions have the same units of measurement. With multi-item instruments, such issues are tackled with formal tests for measurement invariance described later. However, for single-item instruments such statistical procedures cannot be applied. This means that the same numerical score might imply different construct intensities across different countries.

17.6.3.2 Harmonizing Units Across Cultures and Instruments

If we want to harmonize two instruments for the same construct, instrument A in country A and instrument B in country B, we can employ different strategies. First, we can research if an established measurement invariant multi-item instrument for the same construct exists in another survey with comparable survey characteristics that were fielded in the countries we are interested in. With this, we can get an estimate of the true country differences and this can help us correct cross-cultural bias in the instruments we are interested in.

We can also use **observed score equating** in some instances (Kolen and Brennan 2014). Specifically, we can augment international survey programs with data from national surveys with probability samples. Equating is used within each country to harmonize the national survey data with the respective national sample of the international program. The comparability across nations is then established by the ex-ante harmonization of the international survey program. However, since construct distributions change over time, the approach requires at least one year in which the national and international survey programs overlap.

17.6.4 Cross-Cultural Comparability of Multi-Item Instruments

With multi-item instruments, cross-cultural comparability is easier to establish. Especially in the case of cross-cultural comparability of national versions of the same instrument, we have powerful tools at our disposal. Often, factor analytical models are used to establish cross-cultural **measurement invariance**. This is often done with **MGCFA**. As we discussed earlier, CFAs allow us to examine the factorial structure of an instrument, and how the items of the instrument relate to the underlying factor. MGCFAs do the same, but in different groups (here: different cultures) in parallel. As we employ stricter measurement invariance assumptions step by step, we gain insight into all three aspects of comparability:

construct match, reliability divergence, and units of measurement. As the use of MGCFA for assessing cross-cultural comparability is described at length in Chapter 18, we only provide an intuition here. For the construct match, MGCFAs test if the basic dimensional structure of the instrument holds across cultures. This so-called configural measurement invariance assesses if the same items build the same factors across cultures. This step may reveal items, which are understood very differently in different cultures, for example. The next step is metric measurement invariance, which tests if factor loadings are similar across cultures. Since measures of congeneric reliability are derived from the factor loadings, metric invariance also tests for equal reliability across cultures. Lastly, for comparable units of measurement the even stricter scalar measurement invariance tests if the instrument is numerically comparable across cultures (Cieciuch et al. 2019). Lastly, it should be noted that factor analysis is not the only approach to establish cross-cultural measurement invariance. Other approaches include **IRT-Models** or **Latent Class Analysis** (Davidov et al. 2018).

However, in *ex post* harmonization, we often also want to harmonize different multi-item instruments for the same construct and not just localized versions of the same instrument. Here, we have to differentiate two cases. Firstly, the case that the two instruments share at least some items. We can then try to establish measurement invariance via those shared items. This is, in essence, the **NEAT research design** we already discussed for equating (Kolen and Brennan 2014). The second case is that two different multi-item instruments were used with no shared items. Here, we cannot employ MGCFAs or the NEAT design. Cross-cultural comparability then can only be established the same way as for single-item instruments. However, even then multi-item instruments retain some advantages. For example, reliability can be estimated more easily for multi-item instruments. And via tests of their dimensional structure, we already have gained some information about the construct match.

17.7 Discussion and Outlook

We hope to have sensitized the reader for the multi-facetted issue of instrument comparability. We hope to have made the challenges of comparability and harmonization more transparent with our three-step comparability heuristic. Firstly, the construct match should establish that different instruments measure the same construct at all. Secondly, the assessment of reliability divergence, asks if measurements obtained from different instruments or in different populations have similar accuracy. And thirdly, matching the units of measurement, to achieve that the same score in the dataset always represents the same true construct intensity. For each step, we deliberately refrained from giving strict guidance. Instead, we tried to give a taste of the range of possible approaches. *Ex post* harmonization is an issue that touches many disciplines, and both data sources and substantive topics are often interdisciplinary. Hence, it is important to tailor approaches to the conventions of the respective disciplines and research fields. However, we also feel that as *ex post* harmonization matures as a methodology, the conventions might change as well. For example, throughout the chapter, we mentioned approaches that require the collection of additional data. Currently, it is less common for *ex post* harmonization projects in the social sciences to budget such methodological experiments into their grant proposals. However, collecting harmonization samples is already common in integrative medical and epidemiological research (Siddique et al. 2015) and we hope that the social sciences will follow suit in due time.

It is also important to note that this chapter offers only a first foray into the universe of different methods and their finer points. We selected methods that we feel will be most beneficial for common cases of *ex post* harmonization in the social sciences. But obviously, the literature on scaling, linking, and equating knows many more research designs and statistical approaches than we could describe here (e.g. Kohlen and Brennan 2014; Price 2017; Dorans and Puhan 2017).

Lastly, we want to stress that *ex post* harmonization projects are often daunting enterprises and may already push the boundary of what can be realistically expected from researchers. As such we would encourage the reader to take this chapter as a collection of informative ideas and helpful tools. Each issue tackled and each mitigating method applied already allow for an increase in comparability and thus quality of the *ex post* harmonized data. But given the limits of available data, very often even important issues simply cannot be solved satisfactorily. In those cases, it will be crucial to make the limits of the achieved harmonization transparent in the documentation of the research or of the harmonized datasets.

References

Behr, D. and Shishido, K. (2016). The translation of measurement instruments for cross-cultural surveys. In: *The SAGE Handbook of Survey Methodology* (ed. C. Wolf, D. Joye, T.W. Smith, and Y.-c. Fu), 269–287. London, UK: SAGE Publications.

Bollen, K.A. (2002). Latent variables in psychology and the social sciences. *Annual Review of Psychology* 53 (1): 605–634. https://doi.org/10.1146/annurev.psych.53.100901.135239.

Brown, T.A. (2015). *Confirmatory Factor Analysis for Applied Research*, 2e. New York: The Guilford Press.

Charles, E.P. (2005). The correction for attenuation due to measurement error: clarifying concepts and creating confidence sets. *Psychological Methods* 10 (2): 206–226. https://doi.org/10.1037/1082-989X.10 .2.206.

Cheung, G.W. and Rensvold, R.B. (1998). Cross-cultural comparisons using non-invariant measurement items. *Applied Behavioral Science Review* 6 (1): 93–110. https://doi.org/10.1016/S1068-8595(99)80006-3.

Cieciuch, J., Davidov, E., Schmidt, P., and Algesheimer, R. (2019). How to obtain comparable measures for cross-national comparisons. *KZfSS Kölner Zeitschrift Für Soziologie und Sozialpsychologie* 71 (1): 157–186. https://doi.org/10.1007/s11577-019-00598-7.

Davidov, E., Schmidt, P., Billiet, J., and Meuleman, B. (ed.) (2018). *Cross-Cultural Analysis: Methods and Applications*, 2e. New York: Routledge, Taylor and Francis Group.

Dorans, N.J. and Puhan, G. (2017). Contributions to score linking theory and practice. In: *Advancing Human Assessment: The Methodological, Psychological and Policy Contributions of ETS* (ed. R.E. Bennett and M. von Davier), 79–132. Cham: Springer International Publishing https://doi.org/10.1007/978-3-319-58689-2_4.

European Social Survey (2018). *ESS 1–8, European Social Survey Cumulative File, Study Description*. Bergen: NSD – Norwegian Centre for Research Data for ESS ERIC.

GESIS (2016). ISSP 2016 – Citizenship II, Variable Report: Documentation release 2016/08/24, related to the international dataset Archive-Study-No. ZA6670 Version 2.0.0. No. 2016, 08; Variable Reports.

GESIS (2020). GESIS Pretest Datenbank. https://pretest.gesis.org/Pretest.

Hunter, J.E. and Schmidt, F.L. (2004). *Methods of Meta-Analysis: Correcting Error and Bias in Research Findings*, 2e. London: Sage.

de Jonge, T., Veenhoven, R., and Kalmijn, W. (2017). Diversity in survey items and the comparability problem. In: *Diversity in Survey Questions on the Same Topic: Techniques for Improving Comparability* (ed. T. de Jonge, R. Veenhoven, and W. Kalmijn), 3–16. Basel: Springer International Publishing https://doi.org/10.1007/978-3-319-53261-5_1.

Kankaraš, M. and Moors, G. (2009). Measurement equivalence in solidarity attitudes in Europe: insights from a multiple-group latent-class factor approach. *International Sociology* 24 (4): 557–579. https://doi.org/10.1177/0268580909334502.

Kievit, R.A., Frankenhuis, W.E., Waldorp, L.J., and Borsboom, D. (2013). Simpson's paradox in psychological science: a practical guide. *Frontiers in Psychology* 4: https://doi.org/10.3389/fpsyg.2013.00513.

Kołczyńska, M. and Slomczynski, K.M. (2018). Item metadata as controls for ex post harmonization of international survey projects. In: *Advances in Comparative Survey Methods* (ed. T.P. Johnson, B.-E. Pennell, I.A.L. Stoop, and B. Dorer), 1011–1033. Hoboken, NJ: Wiley https://doi.org/10.1002/9781118884997.ch46.

Kolen, M.J. and Brennan, R.L. (2014). *Test Equating, Scaling, and Linking*, 3e. New York: Springer https://doi.org/10.1007/978-1-4939-0317-7.

Lenzner, T. and Neuert, C.E. (2017). Pretesting survey questions via web probing – does it produce similar results to face-to-face cognitive interviewing? *Survey Practice* 10 (4): 1–11. https://doi.org/10.29115/SP-2017-0020.

Moosbrugger, H. and Kelava, A. (2012). *Testtheorie und Fragebogen-konstruktion*, 2e. New York: Springer.

Preckel, F. and Brunner, M. (2017). Nomological nets. In: *Encyclopedia of Personality and Individual Differences* (ed. V. Zeigler-Hill and T. Shackelford). Cham (CH): Springer https://doi.org/10.1007/978-3-319-28099-8_1334-1.

Price, L.R. (2017). *Psychometric Methods Theory into Practice*. New York: Guilford Press.

Raykov, T. and Marcoulides, G.A. (2011). *Introduction to Psychometric Theory*. New York: Routledge.

Revuelta, J., Maydeu-Olivares, A., and Ximénez, C. (2020). Factor analysis for nominal (first choice) data. *Structural Equation Modeling: A Multidisciplinary Journal* 27 (5): 781–797. https://doi.org/10.1080/10705511.2019.1668276.

Salzberger, T. and Sinkovics, R.R. (2006). Reconsidering the problem of data equivalence in international marketing research: contrasting approaches based on CFA and the Rasch model for measurement. *International Marketing Review* 23 (4): 390–417. https://doi.org/10.1108/02651330610678976.

Saunders, M.N., Dietz, G., and Thornhill, A. (2014). Trust and distrust: polar opposites, or independent but co-existing? *Human Relations* 67 (6): 639–665. https://doi.org/10.1177/0018726713500831.

Sherman, E.M., Brooks, B.L., Iverson, G.L. et al. (2011). Reliability and validity in neuropsychology. In: *The Little Black Book of Neuropsychology* (ed. M.R. Schoenberg and J.G. Scott), 873–892. Boston, MA: Springer.

Siddique, J., Reiter, J.P., Brincks, A. et al. (2015). Multiple imputation for harmonizing longitudinal non-commensurate measures in individual participant data meta-analysis. *Statistics in Medicine* 34 (26): 3399–3414. https://doi.org/10.1002/sim.6562.

Singh, R.K. (2020a). Harmonizing instruments with equating. *Harmonization: Newsletter on Survey Data Harmonization in the Social Sciences* 6 (1): 11–18.

Singh, R.K. (2022). Harmonizing single-question instruments for latent constructs with equating using political interest as an example. *Survey Research Methods* 16 (3): 353–369. https://doi.org/10.18148/srm/2022.v16i3.7916.

Slomczynski, K.M. and Skóra, Z. (2020). Rating scales in inter-survey harmonization: what should be controlled? And how? *Harmonization Newsletter on Survey Data Harmonization in the Social Sciences* 6 (2): 16–28. Warsaw: Cross-national Studies: Interdisciplinary Research and Training Program.

Slomczynski, K.M. and Tomescu-Dubrow, I. (2018). Basic principles of survey data recycling. In: *Advances in Comparative Survey Methods* (ed. T.P. Johnson, B.-E. Pennell, I.A.L. Stoop, and B. Dorer), 937–962. Hoboken, NJ: Wiley https://doi.org/10.1002/9781118884997.ch43.

Słomczyński, K.M., Tomescu-Dubrow, I., and Jenkins, J.C. (2016). *Democratic Values and Protest Behavior: Harmonization of Data from International Survey Projects.* IFiS Publishers.

Slomczynski, K.M., Tomescu-Dubrow, I., and Wysmulek, I. (2021). Survey data quality in analyzing harmonized indicators of protest behavior: a survey data recycling approach. *American Behavioral Scientist* 66 (4): 412–433. https://doi.org/10.1177/00027642211021623.

Tourangeau, R. (2020). Survey reliability: models, methods, and findings. *Journal of Survey Statistics and Methodology* 9 (5): 961–991. smaa021. https://doi.org/10.1093/jssam/smaa021.

Wiberg, M. and Bränberg, K. (2015). Kernel equating under the non-equivalent groups with covariates design. *Applied Psychological Measurement* 39 (5): 349–361. https://doi.org/10.1177/0146621614567939.

Zuell, C. and Scholz, E. (2019). Construct equivalence of left-right scale placement in a cross-national perspective. *International Journal of Sociology* 49 (1): 77–95. https://doi.org/10.1080/00207659.2018.1560982.

18

Comparability and Measurement Invariance

Artur Pokropek

Institute of Philosophy and Sociology, Polish Academy of Sciences, Warsaw, Poland

Since the 1980s, comparative sociology saw a rapid growth in the number of international cross-country surveys and their importance for sociology, which even expanded in the twenty-first century (Davidov et al. 2014). Together with this expansion, also the interest to measure constructs that are not directly observed, such as personality traits, attitudes, values, worldviews, and norms, significantly increased. Such constructs are usually considered as continuous latent variables and the observed responses to questions are treated as manifestations of those latent traits (Brown 2015; Jöreskog 1971). For example, the level of religiosity is usually measured by several questions on general religiousness and religious practices (e.g. Lemos et al. 2019), generalized political trust is measured by several questions referring to trust in political institutions like the parliament; politicians and political parties, while the level of political participation is routinely identified by asking whether respondents had taken part in various forms of political activities (e.g. Kostelka 2014).

While much progress has been made to ensure conceptual comparability of attitudinal indicators and multiple item indices, until a few years ago problems of possible nonequivalence of these measures were rarely considered by sociologists when doing group comparisons. Instead, sociologists often assumed that if questionnaire items had been carefully developed, tested, and well translated, their meaning was approximately equivalent in heterogeneous populations, i.e., cross-culturally, overtime, between genders, age groups, or ethnicities. This pervasive methodological oversight has led to potential flaws in the conclusions of some influential research (Davidov 2009; Ippel et al. 2014).

A good example comes from Billiet (2003). The European Social Survey (ESS) study constructed a scale measuring the level of religiosity. The scale consists of several questions about general religiousness and religious practices. Turkey turned out to be the country with the lowest average value of this indicator. A particularly low rate was observed among women. One of the questions concerned the frequency of visits to the church/mosque. For cultural reasons in Turkey, women do not go to mosques regularly. Another example was provided by Pokropek et al. (2017). In their study they showed that the scale to measure socioeconomic status in Programme for International Student Assessment (PISA), which was

Survey Data Harmonization in the Social Sciences, First Edition.
Edited by Irina Tomescu-Dubrow, Christof Wolf, Kazimierz M. Slomczynski, and J. Craig Jenkins.
© 2024 John Wiley & Sons Inc. Published 2024 by John Wiley & Sons Inc.

based on questions referring to ownership of different goods was not fully comparable. Authors have noticed and confirmed by empirical investigations that some indicators of socioeconomic status are not comparable across all countries. For example, having a car may not indicate socioeconomic status in the same way in the United States as in Japan. While in the United States car ownership is virtually universal (because distances between locations are large and the costs associated with maintaining a car are low), in Japan car ownership is less common even in relatively wealthy families (as distances are shorter between locations, public transportation is widespread and efficient, and the cost of maintaining a car is high). The same logic applies to within-country changes over time in the prevalence of certain possessions. For instance, between 2000 and 2012 there was a twofold increase in the availability of mobile phones in developed countries, which significantly changed the meaning of this item in the context of a wealth indicator.

The problems described above have led to formulating the problem of comparability in a more formal way and developing statistical devices, which allow to formally test data from many cultures in terms of comparability. Such formulation came to fruition in the concept of measurement invariance. Horn and McArdle (1992, p. 117) define measurement invariance as a situation where "under different conditions of observing and studying phenomena, measurement operations yield measures of the same attribute." And Mellenbergh (1989) proposed to formulate this assumption in terms of conditional independence:

$$f(U \mid \theta, G = g) = f(U \mid \theta) \tag{18.1}$$

Measurement invariance holds when the probability of given response U using response function f given the latent trait θ is conditionally independent of group membership G for given group g. It is worth noting that in the context of cross-country comparisons violation of the conditional independence assumption is mostly called measurement non-invariance, while in psychometrics the term differential item functioning is used (Differential Item Functioning (DIF); Holland and Wainer 2012; Millsap 2011).

It is worth noting that the definition of measurement invariance does not impose lexical equivalence or even content equivalence. Items are only expected to be related to measurement traits in the same way among investigated groups. This makes the concept of measurement equivalence and the tools designed to test and adjust it useful for harmonization of different studies because we assume that differently formulated questions in different groups may measure in a sufficiently similar way to the construct under study. The problem comes down to a pure empirical issue, testable by statistical models and possible for some statistical adjustment.

18.1 Latent Variable Framework for Testing and Accounting for Measurement Non-Invariance

The most popular approach to address problems of group comparisons and potential compatibility problems in cross-country settings is the latent variable framework (Davidov et al. 2014). In this chapter, the investigation is restricted to the case where indicators are continuous; however, the approaches that are discussed here are easily applicable to binary (Lord 1980; van der Linden and Hambleton 2013) and polytomous indicators (van der Linden and Hambleton 2013).

Let y_{pgi} be a continuous indicator (response to the item) for a latent factor F_g. According to the Multiple-Group Confirmatory Factor Analysis (MG-CFA) modeling, the relation between indicator y_{pgi} and factor g is given by

$$y_{pgi} = \tau_{ig} + \lambda_{ig} F_{pg} + u_{pgi} \qquad (18.2)$$

where i denotes the item index, p the person index, and g the group index, τ_{ig} and λ_{ig} are the item parameters, threshold (in some contexts named intercepts), and loading, respectively, F_{pg} is a factor reflecting the latent trait that is assumed to be normally distributed in each group and u_{pgi} is an error term. The graphical representation of two groups is presented in Figure 18.1.

u_{11} Unless some restrictions on parameters are imposed, the MG-CFA model is not identified because not all item loadings and item thresholds can be simultaneously estimated along with group means $\alpha = (\alpha_g)$ and standard deviations $\psi = (\psi_g)$. The standard approach in multiple-group settings is to constrain all item parameters to be the same in all groups and fixing the mean together with variance of one of the groups to 0 and 1, respectively. Alternative identification strategies are possible but this is the most common in multiple-group settings (see Svetina et al. 2020). This model is often referred to as *a scalar* model and assumes full-invariance, i.e. equivalence, of item parameters (both loadings and thresholds). Assuming conditional independence of the responses, the models could be estimated using various techniques including maximum marginal likelihood (Bock and Zimowski 1997), weighted least squares estimations and their extensions (Beaducel and Herzberg 2006; Li 2016) as well as Bayesian estimation (Fox 2010). This is the model that is the starting point for most of the measurement invariance analysis.

18.2 Approaches to Empirical Assessment of Measurement Equivalence

The MG-CFA approach is illustrated in Figure 18.1. This is an example with two groups, where one latent factor is measured by five indicators. The invariance analysis effectively boils down to checking whether corresponding item parameters (item thresholds and loadings) in each group are equal. If all the parameters are exactly the same, we have the full invariance of the measured items and can assume comparability of the latent factors F_{g1} and F_{g2}. If corresponding parameters differ, the difference suggests possible comparability problems.

Various MG-CFA models and their extensions that are used for comparability analysis might be classified using two dichotomous criteria. The first is based on the question of whether the adopted approach

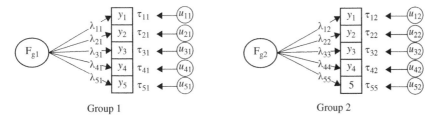

Group 1　　　　　　　　　　　　　　　Group 2

Figure 18.1　Graphical representation of MG-CFA model.

Table 18.1 Four approaches to assessing measurement invariance (the relevant statistical models listed in parentheses).

		Does the approach allow for small "natural" differences in all item parameters between groups?	
		NO	**YES**
Does the approach allow for large item parameter differences between groups for some items?	NO	1. Classical invariance analysis (Scalar MG-CFA)	3. Approximate invariance (BSEM)
	YES	2. Classical partial invariance (Partial MG-CFA)	4. Approximate partial invariance (MG-CFA with alignment, partial BSEM)

Note: MG-CFA – Multi-Group Confirmatory Factor Analysis; BSEM – Bayesian Structural Equation Modeling.

allows for some measured items to have large parameter differences[1] between groups. Such differences might be caused by serious translation flaws or systematic differences in respondents' understanding of questions in different groups. The second criterion relates to allowing for small "natural" differences between groups in all item parameters. Natural differences here mean small differences that capture the sampling variation and also reflect the assumption that small differences in the meaning of questions are inevitable and should be reflected in the measurement model. In contrast, older approaches viewed invariance dichotomously and assumed either perfect invariance or non-invariance of item parameters. The four approaches, which result from the application of the two criteria outlined above, are in sketched Table 18.1.

18.2.1 Classical Invariance Analysis (MG-CFA)

The Classical Invariance Analysis is based on MG-CFA models and assumes that several levels of comparability exist to be tested using specific model constraints (Table 18.1, Cell 1). The first level denotes *configural invariance,* that is, the situation where the structure of the measurement is the same in all groups. In other words, the same set of items load on some latent construct in each group.

Configural invariance is a condition necessary, but not sufficient for valid cross-group comparisons. Such comparisons are valid only in the presence of *metric equivalence* (or *weak factorial invariance*), i.e., when the loadings of indicators on latent factors are equal across groups. *Metric invariance* indicates that a change in the mean value of the underlying construct is equally reflected in individual items across different groups. Therefore, the MG-CFA fit test for the model with strong restrictions for factor loadings $\lambda_{ig} = \lambda_{ig'}$ is effectively the test for *metric equivalence*. In it, the factor loadings for indicator i are restricted to be the same across all groups.

Neither configural nor metric invariance justifies assuming full comparability. This can only be achieved in the presence of *scalar invariance* (or *strong factorial invariance*). The scalar invariance requires that, in addition to factor loadings, the intercepts of each item should be the same ($\tau_{ig} = \tau_{ig'}$ and

---

1 The differences here refer to "true" item parameters, not their estimates.

$\lambda_{ig} = \lambda_{ig'}$ across all groups). *Scalar invariance* allows for valid cross-group comparisons of not only regression coefficients but also the means of latent variables.[2]

18.2.2 Partial Invariance (MG-CFA)

Full equivalence (i.e., invariance of the intercepts and slopes for all items) in real data is rarely achievable. Therefore, in practice, many researchers are satisfied with *partial equivalence,* where some item parameters are equal across countries, while others are set to be estimated freely by partially relaxing the requirements of the classical invariance MG-CFA analysis constraints (Table 18.1, Cell 2).

The approach allows for identifying comparable measures across groups, however, there is no consensus on how many comparable items are required to achieve unbiased estimates of latent traits. Some researchers claim that the partial invariance model requires that at least two-factor loadings and intercepts should be equal in each country and that at least one item per factor loading and intercept (the so-called anchor item) should be equal across all countries (Byrne et al. 1989), which was confirmed in simulation studies (Pokropek et al. 2019). Clearly, the greater the number of items – whether factor loadings or intercepts – that satisfy the invariance criteria detailed above, the more reliable comparisons across different groups. The partial invariance approach, therefore, justifies treating indicators of latent concepts as being to some extent comparable, providing a clear signal that the less comparable items are, the less valid the conclusion.

The problem with the partial invariance approach is that there are several methods for detecting non-invariance, however, not all of them are suitable for situations with more than two groups and most of the methods, even those with the word "multiple-groups" in their name, were designed and tested for 3 or 4 groups only (Fidalgo and Scalon 2010; Woods et al. 2013).

Other methods of detecting non-invariant parameters, like iterative pairwise procedure proposed by Asparouhov and Muthén (2014) or methods based on so-called modification indices criteria (MI), also referred to as the univariate *Lagrange multiplier* (Sarris et al. 1987; Sorbom 1989) proved to have limited efficiency (Finch 2016; Lin 2020). However, works on more robust methods were undertaken with some promising results (e.g. Robitzsch and Lüdtke 2018; Köhler et al. 2021; Pokropek and Pokropek 2022).

18.2.3 Approximate Invariance

The Classical Invariance Analysis sets clear parameter constraints for item parameters by indicating that item-related parameters are fixed across all groups to be equal, i.e., $\tau_{ig} = \tau_{ig'}$ and $\lambda_{ig} = \lambda_{ig'}$. In contrast, in the approximate invariance approach these constraints are relaxed by assuming that item-related parameters are only approximately equal: $\tau_{ig} \approx \tau_{ig'}$ and $\lambda_{ig} \approx \lambda_{ig'}$ constraints (Table 18.1, Cell 3). This is done by allowing for cross-group variation between item parameters, like in MG-CFA, but using zero-mean small-variance informative priors via Bayesian analysis (Asparouhov et al. 2015). Here, the MG-CFA can

---

2 Sometimes an additional level of invariance is defined referred to as residual invariance (strict or invariant uniqueness), invariance of items' residuals or unique variances. Residual invariance is of debatable importance because it is inconsequential to interpretation of latent mean differences (Vandenberg and Lance 2000) and represents a highly constrained model that usually is not plausible to hold. Because of that, many researchers omit to test this type of measurement invariance.

be thought of as a hierarchical or multilevel analysis, in which random parameters depicting factor loadings and means accommodate small cross-group differences in item characteristics, which allow for meaningful comparisons of latent constructs (Muthén and Asparouhov 2013).

Typically to model approximate invariance Multiple-Group Bayesian Structural Equation Modeling (MG-BSEM) is used where non-informative priors are used for all parameters except for the parameters defined for the allowed wiggle room in item measurement parameters. In the most common implementation, the differences between item parameters are expressed in terms of standard normal distribution with a mean of zero and a between-group variance of parameters that needs to be predefined, usually in the range of 0.001 and 0.1. If differences between parameters are the same between all groups, a prior variance of 0.01 in an MG-BSEM model implies a cross-group parameter variance of 0.005, a prior variance of 0.05 implies a cross-group parameter variance of 0.025, etc.

The task of choosing priors might also be described as defining a model that seeks the golden middle between model fit and imposing cross-group measurement equality constraints that allow meaningful comparisons of parameters of interest while still reflecting, to some extent, the true cross-group measurement parameter differences. The biggest problem with this approach is that guidelines for applied researchers with respect to the selection of the priors in MG-BSEM modeling are still largely missing (Seddig and Leitgoeb 2018) and only a few studies were trying to provide some practical recommendations (Pokropek et al. 2020b; Arts et al. 2021).

18.2.4 Approximate Partial Invariance (Alignment, BSEM Alignment, Partial BSEM)

Finally, the approximate partial invariance approach allows for both large differences in some of the item parameters for some of the groups and small differences for all other item parameter constraints (Table 18.1, Cell 4).

The method that perfectly fits this category is alignment optimization. It was proposed by Asparouhov and Muthén (2014) and then further developed by other researchers (Robitzsch 2020; Pokropek et al. 2020a). The method replaces the cross-group equality constraints with a procedure similar to rotation in exploratory factor analysis (EFA). An algorithm estimates a solution that minimizes overall differences between group parameters using the simplicity function, which is optimized for a few large non-invariant parameters and many approximately invariant parameters.

More specifically, first, configural measurement models are estimated for each group, where each model is identified by setting the mean to zero and the standard deviations to one while all item parameters are estimated freely in each of the groups. Because the means and standard deviations of latent variables are functions of item parameters:

$$\lambda_{ig} = \frac{\lambda_{ig,0}}{\psi_g} \text{ and } \tau_{ig} = \tau_{ig,0} - \frac{\lambda_{ig,0}}{\psi_g}\alpha_g,$$

(18.3)

where $\lambda_{ig,0}$ and $\tau_{ig,0}$ are loadings and thresholds from a configural model, respectively, shifting values of means $\alpha = (\alpha_g)$ and standard deviations $\psi = (\psi_g)$, of groups would result in changes to item parameters. This relation is used in the second step of the alignment algorithm where the procedure searches for configuration of means and standard deviations that would minimize the difference between item parameters between groups. To achieve this Muthén and Asparouhov (2014) proposed to use a loss function that would be the same for loadings and thresholds:

$$f = \sqrt[4]{x^2 + \varepsilon} \tag{18.4}$$

where x is the difference between item parameters and ε is a small number such as 0.0001 to make the function differentiable. Alignment procedure penalizes differences in item thresholds and item loadings between groups and hence minimizes the extent of measurement non-invariance according to the loss function. There are various possibilities for specifying the implementation of the alignment optimization. Like the other approaches, this one, too, is not without its drawbacks. The biggest one is that we do not have good measures of model fit here and similar to BSEM approaches the level of regularization (here defined by the loss function) needs to be manually chosen.

Finally, the combinations of the approaches introduced above can also establish the approximate partial invariance. The BSEM might be used jointly with alignment optimization algorithms resulting in models that have a few large nonequivalent parameter differences and many approximately equivalent parameters. Moreover, the BSEM assumption that allows for small nonequivalence margins might be relaxed for some of the item parameters for some chosen groups. This can produce models similar to the partial invariance approach that have a few large nonequivalent parameter differences estimated without Bayesian priors and many approximately equivalent parameters estimated with zero-mean small-variance informative priors.

Paradoxically, the biggest problem with this group of methods is their astonishing flexibility that allows models to fit almost all data making measurement invariance a vague concept that could be freely stretched and therefore leaving comparability uncertain.

18.3 Beyond Multiple Indicators

Methodologists and statisticians provide many solutions for testing and accounting for measurement invariance. This is true for constructs that could be measured by multiple indicators. Now the methodology faces the task of creating tools to determine comparability for constructs measured by a single item. This is not so easy; however, some possibility exists. Constructs measured by a single item can be tested in groups, assuming that there are some higher-order factors that are governing the relations between items. For instance, education, income, and cultural capital could be modeled under the umbrella of general socioeconomic status and those three items could be used to formulate MG-CFA models and their extensions and therefore use the entire range of statistical and methodological devices for multi-indicator constructs.

18.4 Conclusions

Measurement invariance modeling has become an inseparable companion to quantitative cross-cultural, cross-time, and cross-survey studies. In the world of quantitative analysis, there is little doubt that measurement invariance should not be assumed but ought to be tested. In proper scientific research, some pieces of evidence that measurement invariance assumptions are fulfilled should be provided directly or indirectly (Davidov et al. 2014). This is important because the lack of measurement invariance may preclude drawing valid inferences about cross-country or cross-time differences (Stegmueller 2011; Van de Vijver and Tanzer 2004). Without measurement invariance analysis we cannot be sure to what extent varying regression coefficients or scale means mirror "true" differences, and to what extent the observed differences are due to systematic measurement error and measurement biases (Boer et al. 2018; Cieciuch et al. 2019).

It needs to be stressed that methodologically rigorous development of measurement instruments and in some situations properly performed harmonization is a necessary (but not sufficient) condition for achieving measurement invariance and therefore valid comparisons. The tools described here will not fix serious problems of construct comparability or badly designed items, nor wrongly performed harmonization. The measurement invariance methodology is, however, important because it sets necessary conditions for valid comparisons and allows testing them. Moreover, in some situations where measurement non-invariance is properly diagnosed statistical models could be used to achieve valid comparisons. This could be done by detecting non-invariance for some items in some groups and accounting for it, modeling approximate invariance if full invariance is not fulfilled, or excluding whole groups (usually countries) from comparisons if they pose a problem.

On one hand, there are many measurement invariance methods to use that would fit many different situations. The level of statistical refinement is high and results from simulation studies are very promising (e.g. Pokropek et al. 2019). On the other hand, the measurement invariance modeling is not a completed project. It still waits for new developments, better solutions, and more direct answers to practical problems. This short review was trying to show how much we already know and how sophisticated we are in some parts of research on comparability. It was also trying to show how much we do not know and directions for further studies, which will allow for further development of quantitative comparability studies.

References

Arts, I., Fang, Q., Meitinger, K., and Schoot, R.V.D. (2021). Approximate measurement invariance of willingness to sacrifice for the environment across 30 countries: the importance of prior distributions and their visualization. *Frontiers in Psychology* 2911: 624032.

Asparouhov, T. and Muthén, B. (2014). Multiple-group factor analysis alignment. *Structural Equation Modeling: A Multidisciplinary Journal* 21 (4): 495–508.

Asparouhov, T., Muthén, B., and Morin, A.J. (2015). Bayesian structural equation modeling with cross-loadings and residual covariances: comments on Stromeyer et al. *Journal of Management* 41: 1561–1577. https://doi.org/10.1177/0149206315591075.

Beauducel, A. and Herzberg, P.Y. (2006). On the performance of maximum likelihood versus means and variance adjusted weighted least squares estimation in CFA. *Structural Equation Modeling* 13 (2): 186–203.

Billiet, J. (2003). Cross-cultural equivalence with structural equation modeling. In: *Cross-Cultural Survey Methods*, vol. 16 (ed. J.A. Harkness, F.J.R. van de Vijver, and P.P. Mohler), 247–264. New York: Wiley.

Bock, R.D. and Zimowski, M.F. (1997). Multiple group IRT. In: *Handbook of Modern Item Response Theory*, 433–448. New York, NY: Springer.

Boer, D., Hanke, K., and He, J. (2018). On detecting systematic measurement error in cross-cultural research: a review and critical reflection on equivalence and invariance tests. *Journal of Cross-Cultural Psychology* 49 (5): 713–734.

Brown, T.A. (2015). *Confirmatory Factor Analysis for Applied Research*, 2e. New York: Guilford Press.

Byrne, B.M., Shavelson, R.J., and Muthén, B. (1989). Testing for the equivalence of factor covariance and mean structures: the issue of partial measurement invariance. *Psychological Bulletin* 105: 456.

Cieciuch, J., Davidov, E., Schmidt, P., and Algesheimer, R. (2019). How to obtain comparable measures for cross-national comparisons. *KZfSS Kölner Zeitschrift für Soziologie und Sozialpsychologie* 71 (1): 157–186.

Davidov, E. (2009). Measurement equivalence of nationalism and constructive patriotism in the ISSP: 34 countries in a comparative perspective. *Political Analysis* 17: 64–82.

Davidov, E., Meuleman, B., Cieciuch, J. et al. (2014). Measurement equivalence in cross-national research. In: *Annual Review of Sociology*, vol. 40, 55–75. Annual Reviews Inc.

Fidalgo, Á.M. and Scalon, J.D. (2010). Using generalized Mantel-Haenszel statistics to assess DIF among multiple-groups. *Journal of Psychoeducational Assessment* 28 (1): 60–69.

Finch, W.H. (2016). Detection of differential item functioning for more than two groups: a Monte Carlo comparison of methods. *Applied Measurement in Education* 29 (1): 30–45.

Fox, J.P. (2010). *Bayesian Item Response Modeling: Theory and Applications*. New York: Springer.

Holland, P. and Wainer, H. (2012). *Differential Item Functioning*. Hillsdale, NJ, USA: Lawrence Erlbaum Associates, Inc.

Horn, J.L. and McArdle, J.J. (1992). A practical and theoretical guide to measurement invariance in aging research. *Experimental Aging Research* 18 (3): 117–144.

Ippel, L., Gelissen, J.P.T.M., and Moors, G.B.D. (2014). Investigating longitudinal and cross cultural measurement invariance of Inglehart's short post-materialism scale. *Social Indicators Research* 115: 919–932.

Jöreskog, K.G. (1971). Statistical analysis of sets of congeneric tests. *Psychometrika* 36: 109–133.

Köhler, C., Robitzsch, A., Fährmann, K. et al. (2021). A semiparametric approach for item response function estimation to detect item misfit. *British Journal of Mathematical and Statistical Psychology* 74: 157–175.

Kostelka, F. (2014). The state of political participation in post-communist democracies: low but surprisingly little biased citizen engagement. *Europe-Asia Studies* 66 (6): 945–968.

Lemos, C.M., Gore, R.J., Puga-Gonzalez, I., and Shults, F.L. (2019). Dimensionality and factorial invariance of religiosity among Christians and the religiously unaffiliated: a cross-cultural analysis based on the International Social Survey Programme. *PLoS One* 14 (5): e0216352.

Li, C.H. (2016). Confirmatory factor analysis with ordinal data: comparing robust maximum likelihood and diagonally weighted least squares. *Behavior Research Methods* 48 (3): 936–949.

Lin, L. (2020). *Evaluate measurement invariance across multiple-groups: a comparison between the alignment optimization and the random item effects model* (Doctoral dissertation, University of Pittsburgh).

Lord, F.M. (1980). *Applications of Item Response Theory to Practical Testing Problems*. Hillsdale, NJ: Erlbaum.

Mellenbergh, G.J. (1989). Item bias and item response theory. *International Journal of Educational Research* 13 (2): 127–143.

Millsap, R.E. (2011). *Statistical Approaches to Measurement Invariance*. New York: Taylor and Francis Group.

Muthén, B., & Asparouhov, T. (2013). New methods for the study of measurement invariance with many groups. Retrieved from https:// http://www.statmodel.com/download/PolAn.pdf.

Muthén, B. and Asparouhov, T. (2014). IRT studies of many groups: the alignment method. *Frontiers in Psychology* 5: 978.

Pokropek, A. and Pokropek, E. (2022). Deep neural networks for detecting statistical model misspecifications. The case of measurement invariance. *Structural Equation Modeling: A Multidisciplinary Journal* 1–18.

Pokropek, A., Borgonovi, F., and McCormick, C. (2017). On the cross-country comparability of indicators of socioeconomic resources in PISA. *Applied Measurement in Education* 30 (4): 243–258.

Pokropek, A., Davidov, E., and Schmidt, P. (2019). A Monte Carlo simulation study to assess the appropriateness of traditional and newer approaches to test for measurement invariance. *Structural Equation Modeling: A Multidisciplinary Journal* 26 (5): 724–744.

Pokropek, A., Lüdtke, O., and Robitzsch, A. (2020a). An extension of the invariance alignment method for scale linking. *Psychological Test and Assessment Modeling* 62 (2): 305–334.

Pokropek, A., Schmidt, P., and Davidov, E. (2020b). Choosing priors in Bayesian measurement invariance modeling: a Monte Carlo simulation study. *Structural Equation Modeling: A Multidisciplinary Journal* 27 (5): 750–764.

Robitzsch, A. (2020). LP loss functions in invariance alignment and Haberman linking with few or many groups. *Stats* 3 (3): 246–283.

Robitzsch, A. and Lüdtke, O. (2018). A Regularized Moderated Item Response Model for Assessing Differential Item Functioning. Talk given at the VIII. In *European Congress of Methodology, Jena, Germany.*

Sarris, W.E., Satorra, A., and Sörbom, D. (1987). The detection and correction of specification errors in structural equation models. *Sociological Methodology* 17: 105–129. https://doi.org/10.2307/271030.

Seddig, D. and Leitgoeb, H. (2018). Exact and Bayesian approximate measurement invariance. In: *Cross-Cultural Analysis: Methods and Applications*, 2e (ed. E. Davidov, P. Schmidt, J. Billiet, and B. Meuleman), 553–579. New York, NY: Routledge.

Sorbom, D. (1989). Model modification. *Psychometrika* 54: 371–384.

Stegmueller, D. (2011). Apples and oranges? The problem of equivalence in comparative research. *Political Analysis* 19 (4): 471–487.

Svetina, D., Rutkowski, L., and Rutkowski, D. (2020). Multiple-group invariance with categorical outcomes using updated guidelines: an illustration using Mplus and the lavaan/semtools packages. *Structural Equation Modeling: A Multidisciplinary Journal* 27 (1): 111–130.

Van de Vijver, F. and Tanzer, N.K. (2004). Bias and equivalence in cross-cultural assessment: an overview. *European Review of Applied Psychology* 54 (2): 119–135.

Vandenberg, R.J. and Lance, C.E. (2000). A review and synthesis of the measurement invariance literature: suggestions, practices, and recommendations for organizational research. *Organizational Research Methods* 3 (1): 4–70.

van der Linden, W.J. and Hambleton, R.K. (ed.) (2013). *Handbook of Modern Item Response Theory*. Springer Science & Business Media.

Woods, C.M., Cai, L., and Wang, M. (2013). The Langer-improved Wald test for DIF testing with multiple groups: evaluation and comparison to two-group IRT. *Educational and Psychological Measurement* 73 (3): 532–547.

19

On the Creation, Documentation, and Sensible Use of Weights in the Context of Comparative Surveys*

Dominique Joye[1], Marlène Sapin[2], and Christof Wolf[3]

[1]*Faculty of Social and Political Sciences, University of Lausanne, Lausanne, Switzerland*
[2]*Swiss Centre of Expertise in the Social Sciences FORS, Lausanne, Switzerland*
[3]*GESIS – Leibniz Institute for the Social Sciences and the Faculty of Social Sciences at the University of Mannheim, Mannheim, Germany*

19.1 Introduction

Weights in survey research are a disputed topic: some survey methodologists insist on the use of weights (Lavallée and Beaumont 2015) even if the discussion on the construction and impact of these weights is sometimes not fully explicit (Gelman 2007a). Others fear that if weights are used, they will modify results "too much." The idea of trimming or reducing the range of weights could be seen as an indication of taking this direction, even if the statistical reasoning is different (Potter and Zeng 2015). While the discussion seems rather vivid between statisticians (Gelman 2007a, 2007b and comments in *Statistical Science*, Vol. 22 (2); Little 2012), it seems far less developed from a social sciences perspective and in the context of survey data harmonization.

The fact that different disciplines have different perspectives on weighting, is quite interesting, specifically when a renowned statistician like Kish (1994) insists on the difference between design and analysis – the latter probably being more familiar to users of surveys. Kish (1994) writes,

> Weighting belongs to analysis, and its most important forms have much in common with substantive analysis. It represents a form of control by removing disturbing variables from those variables that should belong to the definitions of the predictor variable. For example, should the mean family income in multinational (or domain) comparisons be adjusted for cost of living, and/or for sizes of families, and/or for urban/rural differences? However, some other aspects of weighting are related to sample design and belong to statistical analysis. For example, some weighting may

---

*An earlier version of this chapter appeared as Dominique Joye, Marlène Sapin, and Christof Wolf 2019: Weights in Comparative Surveys? A Call for Opening the Black Box. Harmonization. Newsletter on Survey Data Harmonization in the Social Sciences 5 (2), 2–16. https://www.asc.ohio-state.edu/dataharmonization/wp-content/uploads/2019/12/Harmonization-Newsletter-v5n2-Fall-2019-FINAL.pdf.

> increase variances, hence may perhaps be foregone for lower mean square errors. Weighting for unequal selection probabilities relates to the sample design (p. 173).

In this chapter, we take the position of social scientists interested in promoting the sensible use of international surveys by opening a debate on weighting – a topic too often confined to informal exchanges between colleagues. If we look at handbooks like Wolf et al. (2016) it is interesting to see a lot of attention dedicated to survey harmonization of questions and measurement instruments but far less on the comparison of field organization and its impact on comparability of data (but see also Joye and Wolf 2022). Weights are a further element, both related to field conditions and adaptation of questions to a specific context, that has to be discussed from a comparative perspective (Zieliński et al. 2019). This is even more important when observing a radical change in the way surveys are organized in different countries: change of modes, change in response rates, and increase in data linking and in the combination of information sources (Joye et al. 2016). Of course, we do not have room to discuss all these elements, but we want to emphasize that weighting is even more complicated in such a complex environment and that it is related to all steps of the survey lifecycle.

Before continuing, let us recall two important elements of survey methodology, as they can influence our perspective. The first one is the idea of Total Survey Error (TSE, see for example Weisberg 2005), which argues that researchers should consider all elements affecting the quality of a survey and take into account the constraints. In this sense, weighting is certainly one parameter that influences the quality of the survey. Some of our colleagues have even proposed making it quality criteria from a comparative perspective (Zieliński et al. 2019), even if they focus on formal aspects rather than on research-oriented definitions of quality. At the same time, weighting is influenced by elements of design and implementation. In this context, the idea that weighting is a tool for nonresponse correction is enough to convince researchers that weighting, either with a single weight or a set of weights, could be seen as crucial when considering surveys. However, they should not be considered in isolation but in the context of the survey's design.

The comparative perspective is the second crucial aspect. An often-asked question is whether the best comparability is obtained by using identical procedures in each component of the comparative survey, most often in each country, or if it is better to adapt the procedure in order to find an optimal strategy in each case (Lynn 2003). This is clearly a challenge when considering weighting.

Books on survey methodology, such as the one by Groves et al. (2004), mention weighting as a standard procedure, with a division between weights linked to the design and those correcting for some events as nonresponse or other post-stratification adjustments.[1] The *Guidelines for Best Practice in Cross-Cultural Surveys* takes the same approach:

---

1 In the literature, the definition of these different steps varies slightly. For simplification, we emphasize the difference between weights linked to the statistical design and those linked to further corrections, from nonresponse to calibration. The scientific literature, for example Zielińskiet al. (2019), sometimes addresses a further element of weighting for nonresponse considering the distribution for specific items and not only a complete survey. We will not discuss this specific aspect here: outside technical complexity, a large part of this is related to the same logic as post-stratification weights.

To help correct for these differences, sampling statisticians create weights to reduce the sampling bias of the estimates and to compensate for noncoverage and unit nonresponse. An overall survey weight for each interviewed element typically contains three adjustments: (1) a base weight to adjust for unequal probabilities of selection; (2) an adjustment for sample nonresponse; and (3) a post-stratification adjustment for the difference between the weighted sample distribution and population distribution on variables that are considered to be related to key outcomes (Survey Research Center 2016, p. 657).

However, the authors of these guidelines underline from the beginning that such procedures have pros and cons. In short, weighting can reduce coverage bias, nonresponse bias, and sampling bias at the country or study level. Yet,

weighting can increase sampling variance and, when forming nonresponse adjustment classes, it is assumed that respondents and nonrespondents in the same adjustment class are similar: this is a relatively strong assumption. If the accuracy of the official statistics used to create post-stratification adjustments differs by country, comparability across countries can be hampered (Gabler and Häder 1997). In addition, if the post-stratification adjustments do not dramatically impact the survey estimates, consider not using the adjustment. (Survey Research Center 2016, pp. 659, 660)

This short summary shows some of the difficulties of using weights in surveys. However, these different points also have to be discussed in more detail. First, we will follow a rather classical line of presentation, discussing design weights and post-stratification weights before proposing general strategies for the analyst.

19.2 Design Weights

There is a broad consensus on the use of design weights for random samples with unequal selection probabilities. In these cases, we can correct the resulting data by using these inclusion probabilities. For example, oversampling a region or using a sample based on households is not a problem, as the inclusion probabilities are known, even if the precision of the global sample could be lower according to such a design. The European Social Survey (ESS), like other surveys, uses the idea of "effective size" and proposes some ways to compute it (Lynn et al. 2007). In some cases, however, inclusion probabilities are difficult to compute:

The use of random route techniques is strongly discouraged. The reasons for this are (a) it is rarely possible to implement such techniques in a way that gives all dwellings even approximately equal selection probabilities; (b) it is not possible to accurately estimate these probabilities and, therefore, to obtain unbiased estimates; and (c) the method is easily manipulated by interviewers to their advantage, and in ways that are hard to detect. Instead, as a last resort if no better method is possible, we permit the use of area sampling with field enumeration. How to do this in a way that is consistent with ESS principles is set out in Section 2.1.3. (European Social Survey Round 9

Sampling Guidelines: Principles and Implementation. The ESS Sampling and Weighting Expert Panel, 26 January 2018, p. 6)[2]

If an address-based sampling procedure is used, the number of eligible respondents per household, i.e. reduced household size, should be included when calculating a design weight. It should be noted that using a variable to compute a design weight does not automatically exclude this variable from being used when constructing post-stratification weights. Gelman (2007a) mentions, for example, that correcting elements of design based on the size of the household often causes a strong correction because of the differences of lifestyle, and then probability of contact, between single or multiple-person households.

We note that design weights have been used by the ESS for many years (Häder and Lynn 2007) before they turned to adding post-stratification weights in the most recent editions.

19.2.1 What to do?

In summary, we propose using as much information from the design as possible, though we acknowledge that it might not always be easy to assess the effect of each aspect of the sampling design on selection probabilities. In the same vein, try to avoid random route and other forms of sampling where inclusion probabilities are difficult or impossible to compute.

Implement design weights whenever necessary, even when considering also using calibration weights in the second step: some authors, such as Haziza and Lesage (2016), argue that a one-step calibration is less robust than a propensity score and then calibration. More research on the links between multistep weighting, from design to post-stratification, is certainly useful for gaining a better understanding of the impact of weighting.

From a comparative perspective, it could be seen as desirable to align sample designs in the different countries as much as possible and thus, end up with the same or very similar design weights. However, there are limits to harmonizing sampling procedures because available sampling frames differ greatly. For example, sampling frames, which allow the selection of a respondent without involving an interviewer are preferable because they generally lead to lower nonresponse bias (Eckman and Koch 2019). If persons are sampled from person registers with equal probability the design weight would be one. However, not all countries have person registers that can be used for sampling. Instead, practitioners often have access to address registers, or they resort to creating their own list of addresses by some random walk process involving a specific selection mechanisms. In both cases, first households are identified and then from these households the respondent is identified. In such cases, design weights are proportional to one over the number of eligible household members.

---

2 A detailed study of what happens in the random route survey in a Western context can be found in Bauer (2014). For a concrete discussion, see Díaz de Rada and Martínez (2014) and Eckman and Koch (2019).

In some countries, parts of the population may be oversampled. For example, in Germany surveys will often select East-Germans with a higher probability than West-Germans in order to be able to compare both groups. Again, this choice of sampling should be reflected in appropriate design weights.

Thus, design weights will typically differ between surveys from different countries depending on the type of available sampling frame and other sampling decisions. We recommend taking these differences seriously and use design weights to adjust for country-specific selection.

19.3 Post-stratification Weights

Another family of weights cover corrections based on information of auxiliary variables. For reason of space and complexity we will not go into technical details (see Kalton and Flores-Cervantes 2003; Lavallée and Beaumont 2016). The use of auxiliary variables is heavily discussed. For example, Särndal and Lundström (2005, pp. 129 et seq.) provide a literature review on this topic and Oh and Scheuren (1983) emphasize a number of crucial issues in the field. Among these are: the tendency to choose weightings for convenience rather than appropriateness; the importance for secondary analysis to provide a set of weights rather than a single one; and, finally, the importance of keeping the impact of nonresponse to a minimum[3] (pp. 180, 181), thereby linking weighting to the survey's full lifecycle. The idea of minimizing the nonresponse rate in order to reduce possible bias, and then apply weighting, is also mentioned by Lynn (1996). The practical result of such a strategy is described in empirical studies, such as the one by van Goor and Stuiver (1998). As our reasoning shows weighting through post-stratification and nonresponse analysis are linked, the former being largely based on the latter and the literature on this being very useful in such a context (see for example Brick 2013).

Traditionally, auxiliary variables and weighting classes were developed based on the availability of variables and the judgment of the statisticians (Oh and Scheuren 1983). Predictors of response, key outcome statistics, and domains are considered in this process. Demographic variables, such as age, sex, race, and geography, were – and still are – frequently chosen, even though they may not be effective in reducing bias (Peytcheva and Groves 2009). Many of these are population-based adjustments that, for the controls, use data from a recent census. Furthermore, when the number of respondents in a cell of the cross-classification of the variables is below the threshold set for the survey, then cells are collapsed to avoid large adjustment factors (Brick 2013).[4]

---

3 In fact, nonresponse bias rather than nonresponse by itself should be minimized (Groves 2006) but this does not change the reasoning presented here.

4 This is frequently the case when considering some minorities underrepresented in the main sample. Some astonishing results in a study about culture in Switzerland were linked to a weight of more than four given to a young foreigner with low education but very highbrow tastes. The same type of weighting effect was mentioned in the polls for the American Presidential Election (Cohn 2016).

In contrast, not using enough categories could also be a problem. For example, in Switzerland, the foreign population is important and is often underrepresented in surveys. However, weighting only with the category "foreigner" will give more importance to the foreigners who have answered the survey. Nevertheless, these foreigners are perhaps the most integrated and, therefore, such a weighting scheme does not correct the data at all for the category of less-well-integrated foreigners (Lagana et al. 2013; Lipps et al. 2011).

Some authors have suggested using paradata, such as the number of contacts, as auxiliary variables in weighting (Biemer et al. 2013). Paradata are interesting to consider because they are linked to the data-production process. However, the expectations associated with this approach were not always fulfilled. Furthermore, in a comparative frame, where the modes and organizations of the field most likely will differ, creating comparable paradata can be a challenge.

In fact, Little and Vartivarian (2005) and Bethlehem et al. (2011) claim that the problem of weighting is the link between response behavior on the one side and the target variable on the other. "The auxiliary variable selection can be summarised" Bethlehem et al. (2011) write, "as the search for variables that replace the unknown relation between target variables and response behaviour" (p. 249). In reference to Särndal and Lundström, they continue by stating, "an ideal auxiliary variable has three features: (i) it explains well the response behavior, (ii) it explains well the survey variables, and (iii) it identifies the most important domain for publication of the survey statistics. By important domains, they (Särndal and Lundström) mean subpopulation that appear in publications of statistics based on survey" (p. 249).

Identifying the right auxiliary variables is, therefore, a difficult problem to solve because the most accessible variables, such as sociodemographic variables, are not necessarily related to the response behavior or to the variables of interest. Furthermore, if we consider different variables of interest, the set of predictors may well vary, implying that the weights should differ from one analysis to another. And, once again, this means that the situation can be different in different countries.

To go further, it may be good to return to the debate between statisticians. For example, in the discussion of a seminal paper by Gelman (2007a), Lohr (2007) writes,

> Gelman's (2007a) paper begins with the statement 'Survey weighting is a mess.' I do not think that survey weighting is a mess, but I do think that many people ask too much of the weights. For any weighting problem, one should begin by defining which of the possibly contradictory goals for the weights are desirable. Social scientists interested primarily in relationships among variables may value optimality under the model above all other features (p. 175).

Three important points that we have to consider from this text are crucial here: (i) the importance of documentation, which is a challenge, as weights are often quite technical in the way they are built; (ii) the distinction between the estimation of a single statistic or the reasoning on models; and (iii) that we cannot ask too much of the weights, meaning that we should not invent information that is not present in the data. In the same line, Little (2012) writes,

> From a CB [calibrated Bayes] viewpoint, it is useful to distinguish the case where the variables defining the sampling weights (e.g., the strata indicators in Example 1 above) are or are not included as predictors in the model. If they are, then design weighting is unnecessary if the model is correctly specified. However, from a CB perspective, a comparison of estimates from the

weighted and unweighted analysis provides an important specification check, since a serious difference between a design-weighted and unweighted estimate is a strong indicator of misspecification of the regression model (p. 317).

Once again, we have in this quote the idea to consider models and not just designs. How to build the models is, of course, crucial. The stability of the result with or without weights could also indicate the quality of the estimation.[5] All this discussion is often highly technical, though the consequences can be rather substantial. However, some authors, such as Brick (2013), emphasize that a better knowledge of the crucial argument is often missing:

> The central problem, in our opinion, is that even after decades of research on nonresponse, we remain woefully ignorant of the causes of nonresponse at a profound level. This may be a harsh critique, given all the progress we have made in many areas. We better understand methods to reduce nonresponse due to noncontact in surveys and have made substantial strides in this area. We also have a much better understanding of correlates of nonresponse. Over time, studies have replicated the correlations between demographic and geographic variables and nonresponse rates (e.g. Groves and Couper 1998; Stoop et al. 2010). These are important developments but have not led to a profound understanding of the causes of nonresponse (Brick 2013, p. 346).

Of course, from a comparative point of view, this is even more difficult as the mechanisms governing survey participation may not only vary by survey but also by country or social group. Let us use an example. In a comparative project, imagine taking education as a post-stratification variable. Is it related to the response process? Probably yes, even if such a relation could be quite different from one country to another. Is it related to the target variables? This is likely true in some cases, but also less probable for other analyses. Therefore, having such a variable as an explicit control rather than as an obscure weight could allow researchers to learn more about the relations. However, this is not the end of the story. Is it possible to measure education internationally? The response is probably yes, according to work around the International Standard Classification of Education (ISCED). Is it possible, however, to have a common measure of reference in different countries, and on which basis? Even if the European Labour Force Survey (LFS), is often used to provide data on education in different European countries and may be considered a valid reference, it is not without criticisms, as the LFS is a survey that is itself adjusted in some ways in the countries. Furthermore, countries outside of Europe may not be running LFS and, therefore, the question of the homogeneity of sources is a problem that Ortmanns and Schneider (2016) describe clearly.

------

5 Speaking about the quality of the weighting procedure, it could happen that negative weights are computed. Valliant et al. (2013, p. 370) mention that these estimators could be unbiased in theory; "However, negative weights could have a serious effect on some domain estimates, and users are generally uncomfortable with weights that are negative. In fact some software packages will not allow negative weights." Bethlehem et al. (2011, p. 265) write, "Negative and large weights are signs of an ill-conceived model for non-response and target variables," or in the same vein, "another reason to have some control over the values of the adjustment weights is that application of linear weighting may produce negative weights. Although theory does not require weights to be positive, negative weights should be avoided, since they are counter-intuitive. Negative weights cause problems in subsequent analysis, and they are an indication that the regression model does not fit the data well" (p. 237).

In summary, in comparative surveys, the question of weighting is complicated not only by the difficulty of having homogeneous and reliable sources for additional information but by the questions of similarities in the response and nonresponse process, as well as in the link between weighting variables and potentially varying variables of interest. Thus, returning to the alternatives proposed by Lynn (2003), adapting the weighting strategy to the local conditions of each country could be more appropriate than choosing an identical procedure for all countries.

19.3.1 What Should be Done?

According to the arguments discussed so far, we should be careful when calculating and using post-stratification or calibration weights, keeping in mind that weights – at least in principle – should be adapted to the research question. When computing point estimates, such as the mean of a variable, a weighting strategy is probably the most appropriate one. However, this is not necessarily the case when developing statistical models. Regardless, detailed documentation of reasons for using weights must be available.

This is clearly the strategy proposed by SHARE when they refer to the paper by Solon et al. (2015[2013]) that mentions,

> In Section II, we distinguished between two types of empirical research: (1) research directed at estimating population descriptive statistics and (2) research directed at estimating causal effects. For the former, weighting is called for when it is needed to make the analysis sample representative of the target population. For the latter, the question of whether and how to weight is more nuanced . . . . Our overarching recommendation therefore is to take seriously the question in our title: What are we weighting for? Be clear about the reason that you are considering weighted estimation, think carefully about whether the reason really applies, and double-check with appropriate diagnostics (p. 20).

In summary, good documentation, double checks, and taking seriously the question "what are we weighting for" are key elements for using post-stratification weights in the context of comparative surveys.

But how can we assure that users note the information on weights and their construction? This is a general concern in secondary data analysis and even more pressing when the data are harmonized and originate from different countries, time points and surveys. In addition to providing detailed written documentation, it could be useful to provide some of the pertinent information directly with the data. This could be done in at least two ways:

- First, such information could be directly included into the central data file, which every analyst will use. For example, we could include a variable in the dataset that describes for each weighting variable, which factors were used in its construction. This variable would have several digits, each indicating if a certain characteristic was used to construct the weight. For example, the first digit could be for gender, the second for age, the third for education, the fourth for place of residence, and the fifth for ethnicity. Then 11010, for example, would indicate that gender, age and place of residence was taken into account when calculating the weights.
- Second, it would be even more reasonable to provide such information in a separate dataset containing information on each sample. Of course, such a dataset would not have to be limited to information on weights but could also include data on sampling characteristics, e.g. number of sampling stages,

response rates etc. This data can then be linked to other datasets of a harmonization project such as the respondent data or, where available, contact data, interviewer data or aggregate country- or year-level data. Conceptually, this second alternative is more stringent, having the information at the right level and being more economical in term of redundancy in the data.

This example shows that, in particular in the case of harmonization of complex datasets, the way to present crucial aspects of documentation – and weighting information is part of this – is depending on traditions, but, at the same time, that there is room for innovations to distribute data and metadata in a coherent frame.

19.4 Population Weights

As we mentioned above, there is no consensus concerning the use and necessity of weights for data users of comparative surveys. In particular, the question of how to consider the country-specific characteristics of a survey is still open. This is also true for the question of how to take into account differences of population size between countries. In a recent publication, Kaminska and Lynn (2017) suggest using routine population weights according to the population of the country. This proposition is also interesting to mention because Lynn more readily defended the sole use of design weights in the first edition of the ESS. Going back in the history of statistics, Kish (1999[2003]) discusses six options when considering "multiple population" surveys, which is the case of international projects:

1) Do not combine the data from different countries. This means to exclude comparisons.
2) Harmonize survey measurements in each case, but do not use a common analysis. This just produces statistics by country and compares them.
3) Use an equal weight for each country in the combined analysis.
4) Weight with the sample size, and eventually with the effective size of each sample, knowing that there are more institutional reasons than statistical ones governing the size of each national sample.
5) Population weight should pay careful attention to the reference population (inhabitants, citizens, voters, etc.) in light of the envisaged analysis.
6) Often in multipopulation situations . . . we may well have comparable surveys from several diverse countries of a continent (or the world), but neither all the countries, nor a probability sample of them. (. . .) One may think of constructing "pseudo-strata" from which the available countries would be posed as "representative selections." Someone stratum could have only a single, available, large country. Another stratum could have 2 (or 3) countries . . . . This artificial "pseudo-stratification" procedure may be preferable to simply adding up the available countries into an artificial combination . . . . (Kish 1999, p. 132)

The choice between these alternatives must be made according to substantive reasons. Otherwise, Kish's (1999, p. 295 [in the 2003 reprint]) personal preference is for solutions 4 or 5 rather than solutions 1 or 2, meaning researchers should consider the comparative dimension. He mentions that with some giants like China or India, solution 3 or 5 could be difficult; 5 because of the increase in variance if the sizes are too different. In the case of ISSP, for example, Iceland has 360,000 inhabitants and China

1,386,000,000 nearly 3850 times more than that of Iceland, which means there is probably also much more variance in many of the indicators.

But according to Kish, it is also important to take into account the impact of using population weights on bias. In other words, if the country has no influence on the overall results, then the impact of such weights will be negligible. The last recommendation given by Kish is "that the combination of population surveys into multi-population statistics needs a good deal of research, both empirical and theoretical – and especially together" (Kish 1999[2003], p. 295).

19.4.1 What Should be Done?

In the absence of the research proposed by Kish, we suggest a differentiated strategy:

1) If the goal is to produce a descriptive statistic for all countries combined, then follow Kaminska and Lynn's (2017) work and use population weights for the total sample, for example, an estimate for the European Union.
2) If the idea is to compare different countries on the aggregate level, then there is no need to use population weights, as the figures are produced country by country. This is the case in Welzel and Inglehart's (2016) approach, except when the values are the mean (by country) of a factor analysis of the pooled data set, where the question of weighting according to population could be asked again.
3) If scalar equivalence through multigroup factor analysis is confirmed, measurement in the different countries is identical and may be either directly compared or combined into a joint analysis disregarding country membership, i.e. the variable "country" has no influence in the definition of the structure. Of course, this does not mean that any comparison is impossible if this condition is not fulfilled.
4) Otherwise, a multilevel perspective takes into account the country's effect, and then we do not need population weights either. This does not mean that weighting is not necessary in a multilevel perspective. As Pfeffermann et al. (1998) note, multilevel does not mean "do not weight": "When the sample selection probabilities are related to the response variable even after conditioning on covariates of interest, the conventional estimators of the model parameters may be (asymptotically) biased" (p. 24).

The category "country" plays an ambiguous role in the context of international comparative surveys. On one side, the field is organized most often along this division, which means that we have a lot of effects that are linked to field organization, "survey climate" (Loosveldt and Joye 2016) and, sometimes, modes. On the other side, by having specific institutions, policies, and so on, "country" is an important, aggregated category and has to be taken into account in the debate on individualistic- versus ecological-fallacies (Alker 1969). But, when taking context into consideration, other aggregates, such as regions or even social groups (e. g. social classes or networks), have to be considered in the modelling strategy.

19.5 Conclusion

In concluding, we propose some recommendations for producers and users of comparative, harmonized surveys. For survey producers, the main point is certainly documentation. It is necessary to understand the survey data. However, documentation is also important from an ethical perspective; some authors see weighting as a form of data manipulation. Of course, we know the difficulty of presenting such technical information, even more, when considering the extraordinary amount of information that has to be provided; but the challenge of providing detailed documentation is worth the effort. This challenge is also underlined by DeBell (2018, p. 161) asking about weights: "(i) More accessible guidance; (ii) Increased transparency of methods; (iii) Clearly replicable methods; and (iv) Standardized/comparable methods rather than ad hoc approaches."

We add to this the proposal to consider different sets of weights. Design weights should be distinguished from post-stratification weights and these from population weights. The creation of these weights should follow the same logic in each of the involved countries although the concrete procedure may differ according to sampling design and response process. We encourage all harmonization projects, in particular those applying *ex-ante* harmonization, to develop clear guidelines for the construction and calculation of weights. Though designs may differ and availability of external information for the construction of post-stratification weights may not be identical in all cases such guidelines would ensure that in those cases where the same design feature or the same external data are available, they would be included in the creation of weights in the same way. Furthermore, from a comparative perspective, some thoughts about the processes that produce the need for weighting and a discussion of their similarities or dissimilarities between countries should be included in the documentation to help data users decide if they should include weights in their analyses or not. A table summarizing the variables used for post-stratification in each country could be a useful element of the documentation.

For survey users, more attention should also be given to the question when to use weights. In particular, a challenge is how to consider them when modelling relationships between variables. In this sense, the theoretical perspective on what we are doing in analysis is crucial. In the same line, how to consider the variable "country" is also crucial.

We would advocate for more exchanges between statisticians and social scientists to better understand the underlying conceptualization of each discipline. Furthermore, we sometimes have the impression that survey users are more oriented toward analysis than toward understanding data quality and limitations. The lack of attention given to weights could be related to the perception that it is a problem for survey producers and not survey users. If we are correct, the issue of weighting could also be more incorporated in the regular training and in summer schools.

In any case, we need more analyses on the effect of weighting and the way to take such a factor into account, specifically with regard to the idea of combining theoretical and empirical perspectives. Kish (1994) already made a similar proposal almost 30 years ago and his comments are still relevant today. Referring to them, we hope that our contribution is a call to "open the black box" and experiment further in the field, specifically from a comparative perspective.

As a last practical recommendation, we suggest using design weights whenever they are available – under the assumption that they correctly reflect the inclusion probabilities of respondents. This holds for descriptive as well as for multivariate analyses. Concerning post-stratification weights, we recommend comparing analyses with and without using them – this could be considered a robustness check. Oftentimes the use of post-stratification weights will not make a difference in terms of substantive conclusions drawn from the analyses. Alternatively, for multivariate analyses it could be advisable to use the characteristics that went into calculating the post-stratification weights directly as control variables. This approach gives users a clearer understanding of the influence these factors may have and inform them in how far differing participation rates are relevant for their research question.

References

Alker, H.R. (1969). A typology of ecological fallacies. In: *Quantitative Ecological Analysis in the Social Sciences* (ed. H. Dogan and S. Rokkan), 69–86. London: MIT press.

Bauer, J.J. (2014). Selection errors of random route samples. *Sociological Methods & Research* 43 (3): 519–544.

Bethlehem, J., Cobben, F., and Schouten, B. (2011). *Handbook of Nonresponse in Household Surveys*. Hoboken N.J: Wiley.

Biemer, P.P., Chen, P., and Wang, K. (2013). Using level of effort paradata in non-response adjustments with application to field surveys. *Journal of the Royal Statistical Society Series A: Statistics in Society* 176: 147–168.

Brick, J.M. (2013). Unit nonresponse and weighting adjustments: a critical review. *Journal of Official Statistics* 29: 329–353.

Cohn, N. (2016). How One 19-Year Old Illinois Man is Distorting National Polling Averages. New York Time (12 October 2016). https://www.nytimes.com/2016/10/13/upshot/how-one-19-year-old-illinois-man-is-distorting-national-polling-averages.html (retrieved 24 April 2023).

DeBell, M. (2018). Best practices for creating survey weights. In: *The Palgrave Handbook of Survey Research* (ed. D.L. Vannette and J.A. Krosnick), 159–164. Palgrave.

Díaz de Rada, V. and Martínez Martín, V. (2014). Random route and quota sampling: do they offer any advantage over probably sampling methods? *Open Journal of Statistics* 2014 (4): 391–401.

Eckman, S. and Koch, A. (2019). Interviewer involvement in sample selection shapes the relationship between response rates and data quality. *Public Opinion Quarterly* 83: 313–337.

Gabler, S. and Häder, S. (1997). Deviations from the population and optimal weights. In: *Eurobarometer: Measurement Instruments for Opinions in Europe (ZUMA-Nachrichten Spezial 2)* (ed. W.E. Saris and M. Kaase), 32–44. Mannheim, Germany: ZUMA.

Gelman, A. (2007a). Struggles with survey weighting and regression modeling. *Statistical Science* 22 (2): 153–164.

Gelman, A. (2007b). Rejoinder: struggles with survey weighting and regression modeling. *Statistical Science* 22 (2): 184–188.

van Goor, H. and Stuiver, B. (1998). Can weighting compensate for nonresponse bias in a dependent variable? An evaluation of weighting methods to correct for substantive bias in a mail survey among Dutch municipalities. *Social Science Research* 27: 481–499.

Groves, R.M. (2006). Nonresponse rates and nonresponse bias in household surveys. *Public Opinion Quarterly* 70: 646–675.

Groves, R.M. and Couper, M.P. (1998). *Nonresponse in Household Interview Surveys*. New York: John Wiley & Sons.

Groves, R.M., Floyd, J.F. Jr., Couper, M.P. et al. (2004). *Survey Methodology*. Hoboken N.J.: Wiley.

Häder, S. & Lynn, P. (2007). "How representative can a multi-nation survey be?" in Jowell R. et al., Measuring Attitudes Cross-Nationally, Sage, London, 33–52, https://doi.org/10.4135/9781849209458.

Haziza, D. and Lesage, E. (2016). A discussion of weighting procedures for unit nonresponse. *Journal of Official Statistics* 32: 129–145.

Joye, D. and Wolf, C. (2022). Challenges for comparative surveys in Europe: five theses. In: *Reflections on European Values. Honouring Loek Halman's Contribution to the European Values Study* (ed. R. Luijkx, T. Reeskens, and I. Sieben), 84–98. Tilburg: Tilburg University Press.

Joye, D., Wolf, C., Smith, T.W., and Fu, Y. (2016). Survey methodology, challenges and principles. In: *The Sage Handbook of Survey Methodology* (ed. C. Wolf, D. Joye, T.W. Smith, and Y. Fu), 5–16. Sage, London.

Kalton, G. and Flores-Cervantes, I. (2003). Weighting methods. *Journal of Social Statistics* 19 (2): 81–97.

Kaminska, O. and Lynn, P. (2017). Survey-based cross-country comparisons where countries vary in sample design: issues and solutions. *Journal of Official Statistics* 33: 123–136.

Kish, L. (1994). Multipopulation survey designs: five types with seven shared aspects. *International Statistical Review/Revue Internationale de Statistique* 62: 167–186.

Kish, L. (ed.) (1999). Cumulating/combining population surveys. *Survey Methodology* 15: 129–138. reprinted in Kalton G. and Heeringa S. (eds) (2003). Leslie Kish, Selected Papers. Hoboken: Wiley.

Laganà, F., Elcheroth, G., Penic, S. et al. (2013). National minorities and their representation in social surveys: which practices make a difference? *Quality & Quantity* 47: 1287–1314.

Lavallée, P. and Beaumont, J.F. (2015). Why We Should Put Some Weight on Weights. Survey Insights: Methods from the Field, Weighting: Practical Issues and 'How to' Approach, Invited article. https://surveyinsights.org/?p=6255 (accessed 23 April 2023).

Lavallée, P. and Beaumont, J.F. (2016). Weighting: principles and practicalities. In: *The Sage Handbook of Survey Methodology* (ed. C. Wolf, D. Joye, T.W. Smith, and Y. Fu), 460–476. London: Sage.

Lipps, O., Laganà, F., Pollien, A., and Gianettoni, L. (2011). National minorities and their representation in Swiss surveys (I): providing evidence and analysing causes for their under-representation. In: *Surveying Ethnic Minorities and Immigrant Populations: Methodological Challenges and Research Strategies* (ed. M. Lago and J. Font), 242–269. Amsterdam University Press.

Little, R.J. (2012). Calibrated Bayes: an alternative inferential paradigm for official statistics. *Journal of Official Statistics* 28: 309–372.

Little, R.J. and Vartivarian, S. (2005). Does weighting for nonresponse increase the variance of survey means? *Survey Methodology* 31: 161–168.

Lohr, S.L. (2007). Comment: struggles with survey weighting and regression modelling. *Statistical Science* 22: 175–178.

Loosveldt, G. and Joye, D. (2016). Defining and assessing survey climate. In: *The SAGE Handbook of Survey Methodology* (ed. C. Wolf, D. Joye, T. Smith, and Y. Fu), 67–76. London: Sage.

Lynn, P. (1996). Weighting for nonresponse. In: *Survey and Statistical Computing 1996* (ed. R. Banks), 205–214. Chesham: Association for Survey Computing.

Lynn, P. (2003). Developing quality standards for cross-national survey research: five approaches. *International Journal of Social Research Methodology* 6: 323–336.

Lynn, P., Häder, S., and Gabler, S. (2007). Methods for achieving equivalence of samples in cross-national surveys: the European social survey experience. *Journal of Official Statistics* 23: 107–124.

Oh, H.L. and Scheuren, F.J. (1983). Weighting adjustement for unit nonresponse. In: *Incomplete Data in Sample Surveys*, vol. 2, Part 4 (ed. W.G. Madow, I. Olkin, and D.B. Rubin), 143–184. New York: Academic Press.

Ortmanns, V. and Schneider, S.L. (2016). Can we assess representativeness of cross-national surveys using the education variable? *Survey Research Methods* 10: 189–210.

Peytcheva, E. and Groves, R.M. (2009). Using variation in response rates of demographic subgroups as evidence of nonresponse bias in survey estimates. *Journal of Official Statistics* 25: 193–201.

Pfeffermann, D., Skinner, C.J., Holmes, D.J. et al. (1998). Weighting for unequal selection probabilities in multilevel models. *Journal of the Royal Statistical Society Series B (Statistical Methodology)* 60: 23–40.

Potter, F. and Zeng, Y. (2015). Methods and issues in trimming extreme weights in sample surveys. Presentation in *Proceedings of the Joint Statistical Meetings 2015 Survey Research Methods Section*, Washington. Available at http://www.asasrms.org/Proceedings/y2015/files/234115.pdf (accessed 30 October 2019).

Särndal, C.-E. and Lundstöm, S. (2005). *Estimation in Surveys with Nonresponse*. Hoboken: Wiley.

Solon, G., Haider, S.J., and Wooldridge, J.M. (2015). What are we weighting for? *Journal of Human Resources* 50: 301–316. Also published in 2013 as working paper at the address https://www.nber.org/papers/w18859.pdf, accessed 31 October 2019.

Stoop, I., Billiet, J., Koch, A., and Fitzgerald, R. (2010). *Improving Survey Response. Lessons learned from the European Social Survey*. Chichester: Wiley.

Survey Research Center (2016). *Guidelines for Best Practice in Cross-Cultural Surveys*. Ann Arbor, MI: Survey Research Center, Institute for Social Research, University of Michigan https://ccsg.isr.umich.edu/wp-content/uploads/2019/06/CCSG_Full_Guidelines_2016_Version.pdf (accessed 23 April 2023).

Valliant, R., Dever, J.A., and Kreuter, F. (2013). *Practical Tools for Designing and Weighting Survey Samples*. New York, Heidelberg, Dordrecht & London: Springer.

Weisberg, H.F. (2005). *The Total Survey Error Approach: A Guide to the New Science of Survey Research*. Chicago: University of Chicago Press.

Welzel, C. and Inglehart, R.F. (2016). Misconceptions of measurement equivalence: time for a paradigm shift. *Comparative Political Studies* 49: 10681094.

Wolf, C., Joye, D., Smith, T.W., and Fu, Y. (ed.) (2016). *The Sage Handbook of Survey Methodology*. London: Sage.

Zieliński, M.W., Powałko, P., and Kołczyńska, M. (2019). The past, present, and future of statistical weights in international survey projects: implications for survey data harmonization. In: *Advances in Comparative Survey Methods. Multinational, Multiregional, and Multicultural Contexts (3MC)* (ed. T.P. Johnson, B.-E. Pennel, I. Stoop, and B. Dorer). Hoboken: Wiley.

20

On Using Harmonized Data in Statistical Analysis: Notes of Caution

Claire Durand

Department of Sociology, Université de Montréal, Montréal, Québec, Canada

20.1   Introduction

This chapter aims at examining the main statistical issues that any researcher faces when generating and eventually analyzing a data set harmonized *ex post*. We first present the challenges faced when combining the data. How we deal with these issues will often determine what we can – or cannot – do with the data set afterward. Next, we tackle three issues that stem from analyzing harmonized data sets, that is dealing with time, missing values, and weighting.

We illustrate these questions with concrete examples using a harmonized data set described in Durand et al. (2020) and available at the Scholars Portal Dataverse https://doi.org/10.5683/SP2/TGJV6G. This data set comprises 1327 surveys from 17 international projects conducted in the world from 1995 to 2017. The research aimed at studying trust in institutions with a specific emphasis on countries outside the Consolidated Democracies. Indeed, most research and most publications on this topic focus on only one region, usually Europe or North America (Cole and Cohn 2016), therefore, reducing substantially the variety of contexts in which institutions operate and citizens assess them.

20.2   Challenges in the Combination of Data Sets

Three steps must be taken when combining data sets. First, we must select the appropriate data and measures. Second, we must decide which analytical procedure(s) to use. Finally, once these decisions have been taken, it is necessary to make sure that the analyses performed on the harmonized data sets will not lead to biased results.

Harmonization means putting on the same *scale – a target variable –* various measures of the same concept. The main characteristic of a concept is its unicity. It refers to one idea, one notion. In our own project, we aimed at studying institutional trust. There are many conceptualizations and definitions of trust, including discussions on the lack of a common definition (PytlikZillig and Kimbrough 2016; Cole and

Cohn 2016). A major issue concerns considering trust as a psychological state or as a process (PytlikZillig and Kimbrough 2016). However, debates include interpersonal and social trust, as well as institutional trust.

When we focus on institutional trust, we notice a consensus, at least at the empirical level. There is rare variability in the question wordings used to measure institutional trust itself. A safe and conservative approach consists in sticking to questions that use either of the terms "trust" or "confidence," which pertain to institutions and measure intensity. Cole and Cohn (2016) argue that trust and confidence are two different concepts. However, most languages have only one word for trust or confidence and Kolczynska and Slomczynski (2019) found no significant difference in estimates of trust in parliament in Europe whatever the term used for trust.

There are differences, however, in the answer scales used to measure trust/confidence, in the number of choices offered, in the polarity, and in the direction of the scales (Kolczynska and Slomczynski 2019; Kwak and Slomczynski 2019; Kwak 2020), but they all aim at measuring the level or intensity of trust.

The concept of institution is more equivocal. There is much heterogeneity in the targets for which trust is assessed. The list includes the Church as well as the government and the political parties; some targets refer to an organization (e.g. the Electoral Commission), others to a system (e.g. the justice system), to a ruling body (e.g. the government or, the parliament), even to specific individuals who represent institutions (e.g. the President or, the Prime Minister), or groups of individuals with a similar role (e.g. politicians or, religious leaders). We need to deal with this heterogeneity in deciding which measures to combine, how to combine them, and how to analyze the combined data set.

20.2.1   A First Principle: A No Censorship Inclusive Approach

The first question to deal with is which data sets to select. This requires defining theoretical and empirical criteria. Which concept do we want to measure, and which measures of this concept should we keep? A short answer to this question is to keep as many as possible, provided that all the measures deal with a similar attitude or behavior, for example, the level of satisfaction or agreement with a policy, trust in a specific institution, or the presence of a specific behavior.

With these constraints in mind, we selected all the survey projects that included questions measuring the level of trust in at least one institutional object. We used the English version of the data sets and selected all the questions referring to trust or confidence. In most languages used for interviews, there was only one word for trust. In addition, following a literature review on the relationship between reliability and the number of answering categories in scales (Durand et al. 2021), we selected only the projects that used scales of four or more categories.

In order to combine the different measures, we need to harmonize to a common target scale. The scales used varied from 4-point to 11-point scales, with 4-point scales (65%) and 7-point scales (20%) being the most used. We harmonized to a middle ground 7-point target scale (see Durand et al. 2021). There are different ways to harmonize scales, all with their advantages and disadvantages. Kwak and Slomczynski (2019), for example, harmonized to an 11-point scale (0–10). The transformed scales have the same mean but not the same range, which means that the minimum and maximum of each scale differ. In our case, we aimed at keeping the same mean, minimum, and maximum for all the scales. We thus performed a linear transformation that stretches 4-point scales and shrinks 10 and 11-point scales.

The last decision to make is whether to keep all the different "versions" of the various measures of the concept of interest. The decision is rather easy to take when few and rather similar measures are used but it is not always the case. We may start the combination process with a clear view of the concept of interest and of which measures fit the definition. However, combining data is a "work in progress" and, as we discover and combine new data sets, new issues emerge.

Following classical measurement theory (Sullivan 2009), we conceptualize the different questions as samples of all the questions that could be asked to measure the concept of interest. Therefore, when building the harmonized data set, we keep all the variables that aim at measuring the concept of interest and introduce a new variable each time we run into a new question wording or object (institution in our case). In this way, there is no need to take *a priori* decisions on which variable to keep or drop and the original information stays readily available. Making *a priori* decisions would be risky. We could end up feeling that we "force" measures to fit into a harmonized target measure that is not totally appropriate, or we could discard information that is relevant to our study. The approach suggested here is a *no censorship* approach where decisions about which variable to keep or discard and how to group variables in one or a few target measures are taken *a posteriori*, once all the empirical information becomes available to help the decision-making process.

Proceeding this way requires gathering the relevant information on the original question wordings and scales used. Since each survey project tends to use the same scales and wordings across countries and waves, the information is easily gathered at the project level.

This way of dealing with the combination phase has many advantages. If we keep only the surveys with the same measures of the concept of interest, we deprive ourselves of relevant information and we end up with a dataset that is biased by the selection of measures. We may discard information that pertains to the very regions and countries that are our focus of interest and we may substantially reduce the amount of information on each country and over time that is available for analysis since different projects often complement each other in terms of coverage (Tomescu-Dubrow and Slomczynski 2016).

For example, if we are interested in trust in religious institutions, the question asked in most countries of Judaeo-Christian tradition pertains to trust in the Church. However, in some countries of Latin America and Africa, questions pertain to trust in the Catholic and Protestant Churches, and in countries of Muslim tradition, they pertain to religious leaders. We suggest to keep all the variables related to religious institutions and decide afterward if they may be combined in one unique measure – trust in religious institutions – based on the comparability of the empirical results. If the results show no difference between trust in the Protestant and Catholic churches and if the trends in trust over time are similar for all the religious institutions, we can group these institutions together. As a robustness check, we can add a variable that controls for the different question wordings. We can also run separate analyses for each wording. However, these wordings are used in a specific context, and therefore it is difficult to determine whether differences may be attributed to the wording/object or to the context.

20.2.2 A Second Principle: Using Multilevel Analysis and Introducing a Measurement Level

The next step requires deciding how to analyze the different measures of a concept. In the combined data file, each line of data corresponds to one respondent and answers to all the questions that were asked from that respondent. If only one question is asked from each respondent, it is easy to compute a target harmonized variable. However, when respondents are asked a varying number of questions pertaining to the concept of interest, combining the measures requires an extra step. Trust in institutions, for example, may be viewed as one generic concept that applies to institutions that vary between projects, between countries, and over time.

In the harmonized data set referred to above, keeping all the different institutions surveyed as they came up when combining new data sets led to 133 variables assessing trust in 133 institutions. A minimum of three and a maximum of 23 institutions were assessed in each survey, for an average of 12.5 institutions per survey, and respondents were asked 11.6 questions on average.

In order to analyze trust in a manageable number of institutional categories, we take advantage of the properties of multilevel analysis (Snijders and Bosker 2012; Hox 2010; Raudenbush and Bryk 2002; Raudenbush et al. 2016) and introduce a within-respondent level. Snijders and Bosker (2012) point out that analyzing multiple dependent variables simultaneously is the way to go for four main reasons: (i) it is possible to estimate the correlation between measures; (ii) "the additional power may be. . . considerable if the dependent variables are strongly correlated while at the same time, the data are very incomplete, that is, the average number of measurements available per individual is considerably less than m (the total number of measures)" (p. 283); (iii) it is possible to compare the effect of one independent variable on the different dependent variables; and (iv) finally, it is possible to perform a single test for the effect of one variable on the whole set of dependent variables.

Introducing a measurement level means moving the focus of analysis to answers instead of respondents and therefore, on within-respondent variance. To do this, we restructure the data to have one line per question answered by each respondent. In addition to the information on the characteristics of the respondent and the source of data, each line now includes, in our case, the answer to one trust question and a variable that informs on the institution that was assessed. In the process, we drop the respondents who did not answer any trust questions and the questions that were not answered by a respondent. There is no unit or item nonresponse left. The procedure transforms the 1,829,218 respondent-lines with 133 trust variables into 21,209,889 answer-respondent-lines with *one trust variable* and one variable with 133 categories informing on the institution that was assessed.

Following this procedure, it is easy to group institutions into larger categories through a single recoding. The 133 institutions are recoded into 16 larger categories belonging to 4 thematic groupings: the political system, the public administration, the civil society, and the economic system. Durand et al. (2021) present the criteria used to group the institutions. The grouping may not be perfect but, since the data include the original information, it is easy to perform robustness checks to assess whether any change in the groupings has an impact on the results. It is also possible to select individual institutions for analysis. In terms of analysis, computing dummy variables allows for estimating the difference in trust per institutional grouping.

Using multilevel analysis on these data has the unique advantage of allowing for an estimation of the distribution of variance within, as well as between, respondents. In our case, two-thirds of the variance in trust could be attributed to the institutions assessed, while around one-quarter was attributable to differences between respondents (Durand et al. 2021). Since respondents vary their answers according to the institution assessed, it confirms that institutional trust is a measure of how much people rely on specific institutions in a given context (Mattes and Moreno 2017; Durand et al. 2021) rather than a "psychological state of mind," as proposed by some authors (PytlikZillig and Kimbrough 2016).

If we had restricted ourselves to analyzing only the questions asked in the same way, we would have ended up analyzing political trust – the most studied dimension of trust – and either trust in parliament – the most asked question – or an average of trust in three to five institutions (Durand et al. 2021). We would have deprived ourselves– and researchers – of the major part of the collected information on institutional trust in the world. Using only one variable or an average of a few variables also deprives us of information on the variation in the assessment of different institutions by each respondent.

20.2.3 A Third Principle: Assessing the Equivalence of Survey Projects

If survey projects produce different estimates of the phenomenon studied, the differential presence of survey projects at different places and times will bias the results. The detailed information on specific methodological characteristics of the projects is rarely available (Zielinski et al. 2017). On their part, Kwak and Slomczynski (2019) assess survey quality and conclude that it explains an extra 5% of the variance in three trust measures. However, each survey project features a "cluster of methods," and it is difficult to partial out the specific elements that may explain differences between sources of data.

The main issue is one of collinearity. Not only does each project use specific methodological features but projects are also conducted in specific areas and periods. It, therefore, becomes difficult to assess whether differences in estimates are due to differences between regions or countries or to differences in methods. We illustrate these questions with an example from our data.

One obvious methodological feature of the projects is the original format of the scale used to measure the concept of interest. The scale length is highly related to the polarity and the direction of the scale – all the scales with seven points or more are unipolar and ascending (Durand et al. 2021) – leading to high collinearity between the different scale characteristics. Consequently, we decided to introduce the scale lengths as dummy variables for each length compared with 7-point scales.

Another issue stems from the fact that different survey projects are conducted in different countries and years. Many of them – the Barometers, the European Social Survey (ESS), the Latin America Public Opinion Project (LAPOP), etc. – are regional. Which part of the differences between survey projects may be attributed to the source of data – including the scale used – and which part to the national or regional context?

A first examination of the data shows that all else equal, LAPOP's estimates are on average higher than the estimates produced by the other projects conducted in the same region – Latin America – at the same time. LAPOP uses a 7-point scale while the two other projects – Latino Barometer and World Values Survey (WVS) – use a Likert-type 4-point scale. It is therefore impossible to estimate whether the between-project difference is due to the scale used or to other features of the projects. Since the global harmonized data set includes other projects that use the same scales and are conducted in other regions of the world, we can control for the scale characteristics and the region surveyed and estimate the variance explained by each aspect.

Table 20.1 presents the results obtained when we introduce separately and together in the analysis indicators of the source of data – LAPOP and World/European Values Survey (WVS-EVS) compared with the other sources – the scale length – 4, 5, 10, and 11-point scales compared with 7-point scales – and the context in which the different projects are conducted (see Durand et al. 2021 for an explanation of the country groupings). The results stem from a 4-level analysis with (i) the combination of country and source of data (project) at level 4 ($n = 364$), (ii) the individual surveys – a combination of country, source, and year – at level 3 ($n = 1327$), (iii) the respondents at level 2 ($n = 1,829,218$), and (iv) the measurement – answers to questions pertaining to trust – at level 1 ($n = 21,209,889$). The indicators are introduced at the country-source level. Results are controlled for the impact of predictors at lower

Table 20.1 The impact of source, scale length, and context on average institutional trust.

	Bivariate		Source + Scale		Source + Country groupings		Scale + Country groupings		Source + Scale + Country groupings	
	Coeff.	Std. Dev.	Coeff.	Std. Dev.	Coeff.	Std. Dev.	Coeff.	Std. Dev.	Coeff.	Std. Dev.
Survey protect (ref: other protects)										
Intercept	4.142	0.039 ***	3.767	0.122 ***	3.966	0.062 ***	4.138	0.087 ***	3.828	0.117 ***
LAPOP	0.009	0.102	0.382	0.151 *	0.392	0.103 ***			0.584	0.148 ***
WVS-EVS	−0.117	0.066	−0.318	0.069 ***	−0.042	0.055			−0.103	0.064
Variance at level 4	0.276									
% explained variance*	0.01%									
Scale length (ref 7-point scale)										
Intercept	4.015	0.075 ***								
4-point scale	0.182	0.082 *	0.575	0.130 ***			−0.150	0.077	0.234	0.121
5-point scale	0.037	0.117	0.289	0.150			0.034	0.107	0.304	0.126 *
10-point scale	−0.361	0.139 **	−0.108	0.167			−0.394	0.123 ***	−0.112	0.141
11-point scale	−0.179	0.151	0.071	0.176			−0.230	0.131	0.055	0.148
Variance at level 4	0.256		0.237							
% explained variance*	7.9%		14.7%							
Country-groupings (ref: Consolidated democracies)										
Intercept	3.990	0.061 ***								
West Asia and North Africa	0.078	0.100			0.125	0.098	0.085	0.101	0.069	0.099
Sub-Saharan Africa	0.539	0.096 ***			0.576	0.094 ***	0.542	0.099 ***	0.495	0.099 ***
Central and South America	−0.087	0.085			−0.248	0.092 **	−0.160	0.089	−0.313	0.094 ***
Rest of Asia	0.730	0.090 ***			0.764	0.088 ***	0.732	0.093 ***	0.681	0.094 ***
Postcommunist countries	−0.118	0.074			−0.081	0.073	−0.148	0.073 *	−0.094	0.073
Variance at level 4	0.178				0.166		0.166		0.157	
% explained variance[a]	36.3%				40.3%		40.2%		43.5%	
					Source 3.9%		Scale: 3.9%		Scale + Source: 7.2%	

*: $p < 0.05$; **: $p < 0.01$; ***: $p < 0.001$.

[a] Compared with 0.278 in the previous model including the main controls at levels 1 to 3.

levels: the institutional category is controlled for at the measurement level; age, gender, and item nonresponse are controlled at the respondent level; following the visualization of trends, time, time^2, and time^3 are introduced at the country-source-year level (see Durand et al. 2021 for a justification of these controls).

After controlling for the determinants at lower levels and before introducing any predictor at the country-source level, the variance at that level is 0.278, that is, 7.6% of the total variance of 3.684. When we introduce each group of variables separately (first column), the source of data does not explain any variance by itself. Four-point scales lead to higher estimates of trust (+0.182) than 7-point scales and 10-point scales, to lower estimates (−0.361). By itself, scale length explains close to 8% of the variance and the country groupings, 36.3%. The countries of Asia (+0.730) and Sub-Saharan Africa (+0.539) benefit from a substantially higher level of institutional trust than Consolidated Democracies, the reference category.

The second column shows that when we combine the source of data and scale length, most coefficients change drastically. Once scale length is controlled for, LAPOP appears to lead to estimates that are 0.381 higher than the other projects and WVS-EVS, 0.318 points lower. In addition, 4-point scales become the only type of scale that differs from the 7-point scales, and by far (+0.575). The combined indicators explain 14.7% of the variance.

When we introduce the source of data and country groupings together (third column), LAPOP becomes the only project that differs significantly (+0.392) from the other projects. Besides, the coefficient for Latin America (−0.248) – the region where LAPOP is mostly conducted – becomes significant and negative. The two series of indicators combined explain 40.2% of the variance, the source itself explaining close to four percentage points more variance than country groupings alone.

When scale length and country groupings are entered together, 10-point scales now differ from 7-point scales (−0.406), but the 4-point scales do not differ anymore. The coefficients for country groupings do not change substantially except for postcommunist countries where trust now appears significantly lower than in Consolidated Democracies. Scale length explains 3.9% more variance than country groupings alone.

Finally, the three series of indicators combined explain 43.5% of the variance and 7.2% is attributable to the survey projects. LAPOP now appears to lead to estimates that are 0.584 higher than the other projects, scale length is barely significant and institutional trust in South and Central America is 0.313 points lower than in Consolidated Democracies.

These analyses show that the source of data itself, the original scale length, and the context in which each survey project takes place are intertwined in such a way that controlling for only one indicator leads to erroneous conclusions. If the presence of LAPOP is not controlled for, we erroneously conclude to the absence of a difference in trust between South and Central America and Consolidated Democracies and if we do not control for scale length, we cannot observe that LAPOP estimates are systematically higher than those of other projects active in the same region. Therefore, when different sources use different scales and are active in different regions of the world, it is essential to control for the three aspects all at once.

20.3 Challenges in the Analysis of Combined Data Sets

We now present three issues related to analyzing harmonized data sets, that is (i) dealing with time trends, (ii) missing values, and (iii) weights.

20.3.1 Dealing with Time

Since all social phenomena are likely to change over time, introducing time of data collection in the analyses is essential. Visualization of trends is a first step that allows for a better assessment of change and of the possible impact of occurring events over the period studied. After examining the static impact of the source of data, we need also examine its possible dynamic impact. In short, do different sources of data portray similar trends in similar contexts?

Our data set includes data collected from 1995 to 2017. As a first step, a time-series file is produced with one line by country, source, year and institutional category surveyed ($n = 13,073$). Local regression (Cleveland and Devlin 1988; Fox 2000a, 2000b; Jacoby 2000; Loader 1999) performed on this file allows for estimating the trends weighted by the proximity of the data points. It is preferable to line graphs of means, which do not provide a representation of the dispersion nor a weighted estimate of trends.

Figure 20.1 shows the trends in average institutional trust in six country groupings. Each dot represents the average trust in one institutional category, for one country, year and source. The lines represent the weighted average trends for all the institutions and countries as estimated by local regression.

We focus on differences in trends for three country groupings. The graph for Central and South America shows a fish-like, cubic trend that seems related to the Left-Wing turn taken by many Latin American countries, starting in 1998 with the election of Hugo Chavez in Venezuela. The trend for West Asia and North Africa (WANA) portrays a decline in trust during the last part of the period right after what was called the Arab Spring. Finally, the trend for postcommunist countries appears almost flat and ends at a similar level as the trend for Consolidated Democracies.

The graphs presented in Figure 20.2, taken from Durand et al. (2021), allow for assessing whether the trends are similar according to the source of data. The first graph confirms that there is a systematic

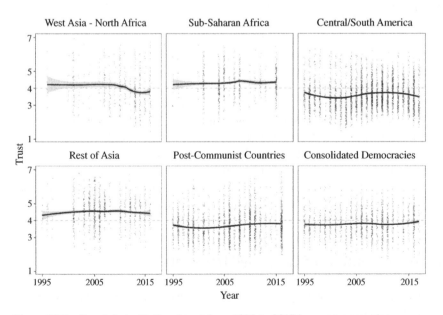

Figure 20.1 Trends in institutional trust from 1995 to 2017 by country groupings.

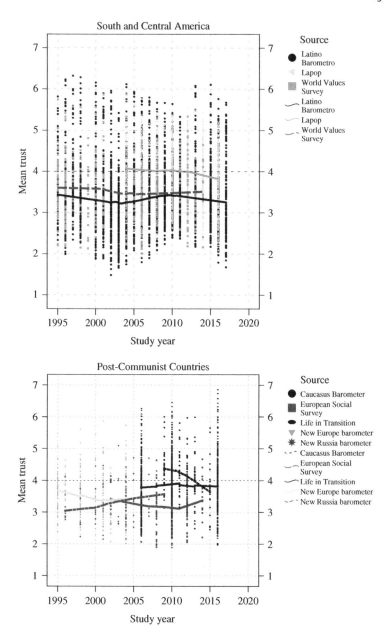

Figure 20.2 Trends in institutional trust from 1995 to 2017 by source of data in selected country groupings.

difference in the *level* of trust estimated by LAPOP compared with the two other projects conducted in Latin America. However, the dispersion of values and the form of the trends are quite similar. Therefore, we conclude that the fish-like trend is not due to the source of data and that we need to control for LAPOP's higher estimates of trust in subsequent analyses.

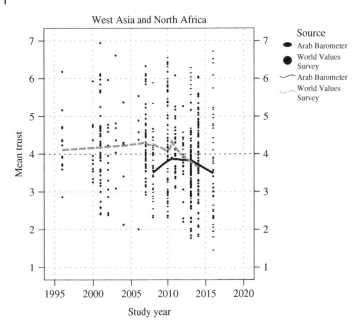

Figure 20.2 *(Continued)*

The second graph shows that the WVS and the Arab Barometer portray a similar decline in trust in WANA. The difference in trends when the Arab Barometer started its data collection in 2008 is due to the fact that different countries were surveyed by the two projects.

Finally, the last graph portrays some of the trends for the 11 different projects conducted in all or parts of the postcommunist countries of Eastern Europe and the former Soviet Union. The figure shows that the average trend for the region hides heterogeneity. During the first part of the period, the New Russia Barometer – the linear increasing trend on the left – shows an increase in trust while the New Europe Barometer shows a downward trend. On the right-hand side, the Caucasus Barometer portrays a sharp decline in trust for the countries of that region. The Life in Transition survey shows a stable trend and the ESS a lower average trend increasing at the end of the period. These trends vary partly because of the different countries surveyed. The graph also allows for observing that the different projects complement each other to improve the coverage of the region.

Guided by the data visualization, we estimate statistically the quadratic trend in the WANA region and the cubic trend in South and Central America. The first column of Table 20.2 is the same as the last column of Table 20.1, to which we add the information on the time components introduced at level 3. It shows that the linear and cubic components of time are significant, which means that the overall average trend in institutional trust during the period is a fish-like trend.

The second column shows the effect of the cross-level interactions of time components on the specific country groupings. It confirms that all else equal, the trend in trust in WANA is quadratic and the trend in South and Central America is cubic. Since the coefficients for the global time components drop

Table 20.2 Add in p time at the country grouping level.

	Model without time interactions			Model with time interactions		
	Coeff.	Std. Dev.		Coeff.	Std. Dev.	
Intercept	3.828	0.117	***	3.845	0.117	***
Survey level						
Time	0.008	0.004	*	−0.001	0.005	
Time^2 *100	0.033	0.024		0.017	0.033	
Time^3 *1000	−0.142	0.044	**	0.015	0.059	
Survey project (ref: other projects)						
LAPOP	0.584	0.148	***	0.578	0.147	***
WVS EVS	−0.103	0.064		−0.116	0.064	
Scale length (ref 7-point scale)						
Intercept						
4-point scale	0.234	0.121		0.240	0.120	*
5 point scale	0.304	0.126	*	0.279	0.126	*
10-point scale	−0.112	0.141		−0.134	0.141	
11-point scale	0.055	0.148		0.031	0.148	
Country-groupings (ref: Consolidated democracies)						
Intercept						
West Asia and North Africa	0.069	0.099		0.334	0.111	**
Time				−0.021	0.012	
Time^2				−0.004	0.001	**
Sub-Saharan Africa	0.495	0.099	***	0.471	0.099	***
Central and South America	−0.313	0.094	***	−0.365	0.096	***
Time				0.024	0.008	**
Time^2 *100				0.089	0.049	
Time^3 *1000				−0.350	0.089	***
Rest of Asia	0.681	0.094	***	0.661	0.094	***
Postcommunist countries	−0.094	0.073		−0.090	0.072	
Variance at level 4	0.157			0.155		
% explained variance[a]	**43.5%**			**44.2%**		
				Time interactions: 0.7%		

*: $p < 0.05$; ***: $p < 0.01$; ***: $p < 0.001$.
[a] Compared with 0.278 in the previous model including the main indicators at levels 1 and 2.

drastically and become nonsignificant, their significance in the first model may, therefore, be attributed uniquely to the specific regional trends. Though the interactions of the time components with these two regions explain less than 1% of the variance, it does explain it better since it is congruent with the trends presented graphically. The analysis now shows that in WANA, the situation is not that of a level of trust comparable to Consolidated Democracies, as estimated in the first model, but rather, that of a higher level of trust at the beginning of the period followed by a sharp drop in recent years.

One of the main advantages of combining time-series cross-sectional survey data is the possibility to examine and validate trends. Keeping all the data and using local regression and multilevel analysis allows for a complete and validated portrait of trends. Cross-level interactions allow for validating variations in trends by context, source of data, institutional groupings, etc.

20.3.2 Dealing with Missing Values

In harmonized data sets, since usually some questions are not asked in all surveys, and surveys are not conducted in all locations and years, missing values may be a major issue. Using multilevel analysis allows for keeping and analyzing all the available data without having to resort to replacement or imputation (Snijders and Bosker 2012; Hox 2010; Raudenbush et al. 2016). If we used only those surveys where all the needed information is present, it could bias the results and it would reduce power. However, if we need to impute data, harmonized data sets present interesting opportunities, since we can use all the information provided by all the units at all levels and periods in the imputation process.

Rubin's (1976) definition of three types of missingness seems to have been adopted by all authors (Snijders and Bosker 2012; Tabachnick and Fidell 2012). Missing values are categorized as either missing completely at random (MCAR), where all the information provided by the observed data is sufficient, missing at random (MAR) – where the presence of missing values is related to explanatory variables, – or not missing at random (NMAR), when missing values are related to the variable of interest but there is no information to help estimate their impact. In harmonized data sets, these different types of missing values may be present at different levels and their categorization at each level may not be that clear.

Depending on the type and level of missingness, solutions vary from dropping incomplete data – including dropping surveys, – to replacement with mean, regression estimates or EM (expectation maximization) imputation and finally, multiple imputation (Tabachnick and Fidell 2012; van Buuren 2018). When the proportion of missing values for a variable of interest is low (less than 5%), researchers tend to resort to classical solutions, that is, either dropping the observations for which data are missing, or replacing the missing information with the overall mean or group means when relevant (Tabachnick and Fidell 2012). When a variable presents a high level of nonrandom missing values, the best solution may be to include a valid category for missing values and group the other answers in categories (Tabachnick and Fidell 2012; see Polity IV categorization in Wutchiett and Durand 2021). However, according to Snijders and Bosker (2012), researchers must examine their data for missingness, since these classical solutions are no longer considered acceptable without some thinking.

20.3.2.1 Missing Values at the Respondent and Measurement Level

Since we use multilevel analysis and we include a measurement level, the unasked questions – missingness at the survey level – and the unanswered questions – classical item nonresponse – have been dropped from the file. According to Snijders and Bosker (2012), at the survey level, unasked questions are considered MAR, since country, source, and year are explanatory variables of missingness. For the respondent, unasked questions are characterized as MCAR, since missingness is not under his/her control.

However, when respondents do not answer some of the questions asked, missingness may not be random. In order to check whether this type of nonresponse biases the estimates, we computed a variable informing on the proportion of item nonresponse at the respondent level. In our harmonized data set, one quarter (26%) of the respondents did not answer at least one question pertaining to trust and 9% did not answer at least 25% of the questions they were asked. Therefore, item nonresponse is an issue.

We hypothesized that more cynical respondents or those living in a politically sensitive environment may be more reluctant to answer questions pertaining to trust. Therefore, item nonresponse would be negatively related to trust at the respondent level. The results show that, after controlling for the other predictors at the measurement and respondent levels, the relationship between nonresponse and trust goes in the opposite direction. An increase of one-tenth– from 0% to 10% nonresponse – results in an average level of trust that is 0.027 higher on the 7-point scale. However, item nonresponse explains at most 0.1% of the variance at the respondent level, but 0.3% of the variance at the survey level and 0.8% at the country-source level. Further checks show that the average proportion of nonresponse varies substantially by source, from 2% in the European Quality of Life project to 10% or more in the Consolidation of Democracy, Caucasus Barometer, and South Asia Barometer projects. We can thus conclude that item nonresponse is more a matter of each project's methods than a characteristic of individuals.

20.3.2.2 Missing Values at the Survey Level

At higher levels, the challenge is different. Researchers using harmonized data sets often resort to external explanatory variables of the phenomenon of interest, that is, indicators of the social, economic, and political characteristics of the countries in which the surveys are conducted. Missingness in that type of data is usually not random. Data are less available for new democracies, less wealthy, or less democratic countries. For the country-years under study in our data set, Table 20.3 shows that the Gini index (Solt 2020) is missing in more than 10% of the cases, while the other indicators, selected from the Quality of Government Database (2020) specifically for their high coverage level, are missing in around 3% of the cases. Performing the analyses with that type and level of missingness could bias the results. We resorted to multiple imputation[1] (van Buuren 2018) as suggested by Tabachnick and Fidell (2012), Snijders and Bosker (2012), and others. Imputation used the information provided by the nine variables and the main democracy indices provided by Freedom House, Polity IV, V_DEM, Global State of Democracy (n.d.), and the Economist Intelligence Unit index for the same years – 1995 to 2017 – and countries.

The imputed data show lower average economic characteristics, a higher proportion of young citizens, a lower proportion of older citizens, and a higher population growth, coupled with a larger variance. This is expected given the fact that missingness is related to socioeconomic development. Coherently, the main indices of democracy present lower averages and higher variance after imputation, confirming that missingness is related to the political, social, and economic context.

---

1 The imputation was performed by David Wutchiett using mice (van Buuren and Groothuis-Oudshoorn 2011) for the paper "In Democracy We Trust, Do We?" presented by Durand, Wutchiett, Peña Ibarra and Rezgui at the "Political Trust in Crisis Digital Conference" in October 2020. https://trustgov.net/events/2020/10/22/2-day-digital-conference-political-trust-in-crisis

Table 20.3 Socioeconomic indices before and after imputation.

	Before imputation			After imputation	
	% of missing values	**Mean**	**Std. Dev.**	**Mean**	**Std. Dev.**
Gini dispositional index	10.2	39.5	8.4	38.7	8.1
Log per capita gross product	3.0	8.4	1.2	8.1	1.6
UN Human Development Index	3.2	0.71	0.11	0.65	0.16
Population 0–14 yr old	2.9	26.8%	9.3%	29.7%	11.3%
Population 65+ yr old	2.9	8.9%	5.0%	7.7%	5.4%
Population growth	2.8	1.0%	1.2%	1.5%	1.4%
Urban population	2.7	63.0%	18.0%	62.7%	17.9%
Urban population growth	2.7	1.6%	1.6%	1.6%	1.6%

Statistics computed on country-source-year level file.

Table 20.4 shows the results of an analysis of the relationship between the selected socioeconomic variables and average trust, without and with imputed values. We use the same 4-level model with the same controls at the different levels indicated in Sections 20.2.2 and 20.2.3. Results show that the significant coefficients are larger and the standard errors of the estimates smaller for the imputed variables than before imputation. In summary, both analyses show that the demographic characteristics of the population that are related to development – population aged 65 years plus, population growth, and urban population growth – are associated with trust. An older population and a higher population

Table 20.4 Impact of socioeconomic indicators on Average Trust in Institutions.

	Before imputation			After imputation		
	Coeff.	**Std. error.**		**Coeff.**	**Std. error.**	
Gini dispositional	−0.005	0.004	ns	−0.006	0.004	ns
Ln National Product per capita	0.082	0.042	ns	0.073	0.037	ns
UN Human Development Index	0.843	0.607	ns	0.897	0.531	ns
Population 0–14	0.003	0.007	ns	−0.001	0.006	ns
Population 65+	−0.029	0.010	**	−0.042	0.009	***
Population growth	−0.064	0.040	ns	−0.070	0.035	*
Urban population	−0.002	0.002	ns	−0.003	0.002	ns
Urban population growth	0.107	0.028	***	0.111	0.026	***

ns: not significant; *: $p < 0.05$; **: $p < 0.01$; *** $p < 0.001$.

growth are associated with lower average trust; on the opposite, a higher urban population growth is associated with more trust.

In conclusion, although multilevel analysis takes care of a good part of the missingness problem, we still need to examine the level and type of nonresponse in order to make sure that it is not substantially related to other characteristics at any level of analysis. Indeed, Snijders and Bosker (2012) suggest introducing as many relevant explanatory variables as possible in order to make sure to control for the variables that may be related to missingness. At the country and year level, missingness like socio-economic characteristics is rarely completely random. Harmonized data sets provide much relevant information that may help achieve adequate multiple imputations.

20.3.3 Dealing with Weights

Although it is good practice to examine the survey weights, when reliable weights are provided and there is no relationship between the components used to compute the weights and the variable of interest, there is no need to use the weights. It would not affect the average estimates, but it would increase variance (Zielinski et al. 2017, chapter 19; Joye et al. 2019). When the weight components and the variable of interest, are related introducing the components as control in the analysis is usually good enough. This question is covered in Chapters 9 and 19.

A second question is whether we need to account for the varying population sizes in the countries surveyed. Weighting according to population size allows for estimating a global statistic for a region or even the world. Researchers rarely weight according to population size because they are rarely interested in estimating the distribution of the variable of interest in the world or even in a specific region (Joye et al. 2019; Zielinski et al. 2017). However, what if trust, for example, were related to population size?

If we compute weights for the whole world, the weight of China and India combined would be such that the estimates would reflect only these countries and the variance in weights would be very large. Joye et al. (2019) deal with this issue stating that "...a multilevel perspective takes into account the country's effect, and then we do not need population weights. This does not mean that weighting is not necessary in a multilevel perspective" (similarly Snijders and Bosker 2012).

A third question is whether we should control for the coverage, that is, the fact that the number of surveys conducted in each country and year varies. Both Joye et al. (2019) and Snijders and Bosker (2012) note that multilevel analysis takes care of the varying number of units at the higher level, that is, by country-source level in our case. They nonetheless suggest checking whether the variables that would be used for weighting are related to the variable of interest and decide whether weighting is appropriate based on these analyses. We apply this approach by introducing the log of the population size and the number of surveys conducted in each country – which ranges from 1 to 35 – as predictors of average institutional trust.

The results are reassuring. Table 20.5 shows that when we enter only these variables at the country-source level in the analysis (model 1b) and compare with the model without any control variable at any level (model 1a), the two coefficients are significant and explain 22.2% of the variance. Average institutional trust is higher in countries with larger population sizes in countries where less surveys have been conducted.

Table 20.5 Impact of population size and number of surveys conducted on institutional trust.

	Without any control		With controls at lower levels	
	Model 1a	Model 1b	Model 2a	Model 2b
Intercept	3.958 0.031 ***	3.449 0.272 ***	3.844 0.115 ***	4.324 0.273 ***
In (population)		0.053 0.017 **		−0.019 0.017
Number of surveys		−0.027 0.003 ***		−0.009 0.004 *
Variance				
Level 1: Measurement	1.051	1.051	2.262	2.262
Level 2: Respondents	2.432	2.431	1.061	1.061
Level 3: Survey (country-source-year)	0.088	0.08S	0.089	0.089
Level 4: Country-source	0.288	0.224	0.151	0.146
Explained variance		27.2%		3.3%

Note: Model 1a: model 0; Model 1b: impact of lpop and number of surveys only.
Model 2a: Variables at all levels are introduced; Model 2b: impact of LnPop and nb of surveys after control for all the variables.
ns: not significant; *: $p < 0.05$; **: $p < 0.01$; *** $p < 0.001$.

However, when we introduce all the controls at the different levels as in the models presented in Tables 20.1, 20.2, and 20.4, the log of the population size becomes nonsignificant and the number of surveys conducted, is barely significant. They do, however, increase the estimate of the intercept by half a point after control. Although the number of surveys conducted in each country is barely significant, it nonetheless means that in the countries where more surveys were conducted, trust tended to be lower, from 0.01 points to 0.35 (on a 7-point scale) for the country where the largest number of surveys have been conducted (Columbia). These variables explain 3.3% of the variance at the country-source level.

In conclusion, at the survey level, the fact that units have different population sizes or that some units are more surveyed than others should not be a concern, at least when using multilevel analysis. The impact of these variables may be controlled for, and this will make resorting to weights less needed.

20.4 Recommendations

The first recommendation pertains to keeping as much as possible all the data available on a specific topic and do it in such a way that the original data are easily available, and decisions may be modified all along the process including at the end. We have shown how it is possible to analyze these data adequately by resorting to multilevel analysis and the use of a within-respondent measurement level. The process presented may easily be applied to other topics: agreement on policies related to a similar object and various behaviors like political participation or belief in climate change, for example. It remains that the harmonized databases end up being rather complex and should be "handled with care."

Researchers should always be aware of the multiple occasions for multicollinearity issues in harmonized data sets. Properties of scales, clusters of methodologies, and contexts may be so intertwined that it is difficult to partial out the different effects.

The estimation of trends and the impact of events is often at the heart of our research questions. There may be much heterogeneity in trends. Using local regression allows for a reliable portrait of change over time. Following visualization, trends should be validated by introducing time components in the analysis. Multilevel analysis allows for modeling varying trends according to contexts, age, and gender groups or even institutions. Therefore, if allows for a better representation of the data.

Using multilevel analysis solves a number of issues regarding how to deal with missing values. It remains that computing a variable for item nonresponse allows for a specific analysis of the possible impact of missingness at the respondent level. In addition, missing values are common and rarely completely random in external country-year-level variables. They reduce the sample size and may introduce bias. Proceeding to data imputation helps correct for a possible impact of missing values at the country-year level.

Finally, unless we are looking for weighted marginals, weights at the survey level based on population size or coverage, are not very useful. However, introducing the variables on which the weights would be based may seriously modify the intercept. In the absence of adequate weighing, we should interpret the intercept with caution.

In conclusion, there are so many differences between survey projects, including the questions asked, the way they are asked, and where they are asked, that any analysis has to carefully assess how these different aspects interact with each other before reaching firm conclusions on specific effects.

References

van Buuren, S. (2018). *Flexible Imputation of Missing Data*. CRC Press.

van Buuren, S. and Groothuis-Oudshoorn, K. (2011). Mice: multivariate imputation by chained equations in R. *Journal of Statistical Software* 45 (3): 1–67.

Cleveland, W.S. and Devlin, S.J. (1988). Locally weighted regression: an approach to regression analysis by local fitting. *Journal of the American Statistical Association* 83 (403): 596–610.

Cole, L.M. and Cohn, E.S. (2016). Institutional trust across cultures: its definitions, conceptualizations, and antecedents across Eastern and Western European nations. In: *Interdisciplinary Perspectives on Trust* (ed. E. Shockley, T. Neal, L. PytlikZillig, and B. Bornstein). Cham: Springer https://doi.org/10.1007/978-3-319-22261-5_9.

Durand, C., Peña Ibarra, L.P., Rezgui, N. et al. (2020). Replication data for: institutional trust in the world. Université de Montréal Dataverse, V1. https://doi.org/10.5683/SP2/TGJV6G.

Durand, C., Peña Ibarra, L.P., Rezgui, N., and Wutchiett, D. (2021). How to combine and analyze all the data from diverse sources: a multilevel analysis of institutional trust in the World. Quality and Quantity, Special Issue on Data Harmonization. https://www.springerprofessional.de/en/how-to-combine-and-analyze-all-the-data-from-diverse-sources-a-m/18777770.

Economist Intelligent Unit (n.d.) https://www.eiu.com/n/campaigns/democracy-index-2020 (Retrieved 1 February 2022), data from 2006 to 2017.

Fox, J. (2000a). *Multiple and Generalized Nonparametric Regression*. Thousand Oaks: Sage.

Fox, J. (2000b). *Non Parametric Simple Regression*. Thousand Oaks: Sage.

Freedom House (n.d.) https://freedomhouse.org/report/freedom-world (Retrieved 1 February 2022), data from 1995 to 2017.

Global State of Democracy (n.d.) https://www.idea.int/data-tools/tools/global-state-democracy-indices (Retrieved 1 February 2022), data from 1995 to 2017.

Hox, J. (2010). *Multilevel Analysis. Techniques and Applications*, 2e. New York: Routledge.

Jacoby, W.G. (2000). Loess: a nonparametric, graphical tool for depicting relationships between variables. *Electoral Studies* 19: 577–613.

Joye, D., Sapin, M., and Wolf, C. (2019). Weights in comparative surveys. *Harmonization* 5 (2): 2–15.

Kolczynska, M. and Slomczynski, K. (2019). Item metadata as controls for ex post harmonization of international survey projects. In: *Advances in Comparative Survey Methodology* (ed. T.P. Johnson, B.-E. Pennell, I.A.L. Stoop, and B. Dorer), 1011–1034. Wiley.

Kwak, J. (2020). Inter-survey methodological variability in institutional trust from the survey data recycling project. *Harmonization* 6 (1): 18–27.

Kwak, J. and Slomczynski, K.M. (2019). Aggregating survey data on the National Level for Indexing Trust in Public Institutions: on the effects of lagged variables, data harmonization controls, and data quality. *Harmonization: Newsletter on Survey Data Harmonization in the Social Sciences* 5 (1): 5–13.

Loader, C. (1999). *Local Regression and Likelihood*. New York: Springer.

Mattes, R. and Moreno, A. (2017). Social and political trust in developing countries: sub-Saharan Africa and Latin America. In: *The Oxford Handbook of Social and Political Trust* (ed. E.M. Uslaner). https://doi.org/10.1093/oxfordhb/9780190274801.013.10.

Polity 4 (n.d.) https://www.systemicpeace.org/polity/polity4.htm (Retrieved 1 February 2022), data from 1995 to 2017.

PytlikZillig, L.M. and Kimbrough, C.D. (2016). Consensus on conceptualizations and definitions of trust: are we there yet? In: *Interdisciplinary Perspectives on Trust* (ed. E. Shockley, T. Neal, L. PytlikZillig, and B. Bornstein). Cham: Springer https://doi.org/10.1007/978-3-319-22261-5_2.

Quality of Government Data Base (2020). *Standard Dataset*. Sweden: University of Gothenburg https://www.gu.se/en/quality-government/qog-data.

Raudenbush, S.W. and Bryk, A.S. (2002). *Hierarchical Linear Models: Applications and Data Analysis Methods*. Thousand Oaks: Sage Publications.

Raudenbush, S.W., Bryk, A.S., Cheong, Y.F. et al. (2016). HLM7 Hierarchical Linear and Nonlinear Modeling User Manual: User Guide for Scientific Software International's (S.S.I.) Program, SSI, U.S.A., Skokie, Il, 366.

Rubin, D.B. (1976). Inference and missing data. *Biometrica* 63: 581–592.

Snijders, T. and Bosker, R. (2012). *Multilevel Analysis. An Introduction to Basic and Advanced Multilevel Modeling*, 2e. London: Sage Publications.

Solt, F. (2020). Measuring income inequality across countries and over time: the standardized world income inequality database. *Social Science Quarterly*. SWIID Version 9.1, May 2021.

Sullivan, L.E. (2009). Classical measurement theory. In: *The SAGE Glossary of the Social and Behavioral Sciences*, vol. 1, 76–76. SAGE Publications, Inc. https://doi.org/10.4135/9781412972024.n379.

Tabachnick, B.G. and Fidell, L.S. (2012). *Using Multivariate Statistics*. Boston: Pearson.

Tomescu-Dubrow, I. and Slomczynski, K.M. (2016). Harmonization of cross-National Survey Projects on political behavior: developing the analytical framework of survey data recycling. *International Journal of Sociology* 46 (1): 58–72.

V_Dem (n.d.) https://www.v-dem.net/fr.

Wutchiett, D. and Durand, C. (2022). Multilevel and time-series missing value imputation for combined survey and longitudinal context data. *Quality and Quantity: International Journal of Methodology*, Springer 56 (3): 1799–1828.

Zielinski, M.W., Powalko, P., and Kolczynska, M. (2017). The past, present, and future of statistical weights in international survey projects: implications for survey data harmonization. In: *Advances in Comparative Survey Methodology* (ed. T.P. Johnson, B.-E. Pennell, I.A.L. Stoop, and B. Dorer), 1035–1052. Wiley.

21

On the Future of Survey Data Harmonization

Kazimierz M. Slomczynski[1,2], Christof Wolf[3], Irina Tomescu-Dubrow[1,2,4], and J. Craig Jenkins[2,5]

[1]*Institute of Philosophy and Sociology, Polish Academy of Sciences, Warsaw, Poland*
[2]*Department of Sociology, The Ohio State University, Columbus, Ohio, USA*
[3]*GESIS Leibniz-Institute for the Social Sciences and University Mannheim, Germany*
[4]*Graduate School for Social Research, Polish Academy of Sciences, Warsaw, Poland*
[5]*The Mershon Center, The Ohio State University, Columbus, Ohio, USA*

This chapter focuses on the future of survey data harmonization. Rather than summarizing contributions presented in this volume, we highlight and discuss some lessons learned from practicing harmonization that we consider important for the future of empirical social science research. Our future orientation is justified by the speed and scope of changes that we are witnessing in terms of both the types of quantitative data available to researchers and the technologies to create and process such data. As one report by the American Association of Public Opinion Research, AAPOR, notes: *Public opinion research is entering a new era, one in which traditional survey research may play a less dominant role. The proliferation of new technologies, such as mobile devices and social media platforms, is changing the societal landscape across which public opinion researchers operate. The ways in which people both access and share information about opinions, attitudes, and behaviors have gone through perhaps a greater transformation in the last decade than at any previous point in history, and this trend appears likely to continue* (AAPOR 2014, p. 3).

Nonetheless, surveying public opinion by asking people questions about their attitudes and behaviors remains an important part of research in the social sciences, although the forms of contact with respondents change over time. The pandemic of 2020–2021 and policies restricting face-to-face contact had an obvious impact on studying public opinion. Restrictions on using PAPI and CATI forced researchers to use the Internet and smartphones for interviewing. It is important that even in these difficult times the studies of public opinion have been continued. Many of these studies, including international ones based on ex-ante harmonization, directly dealt with COVID and its social consequences; they were conducted by telephone or the Internet (Aizpurua 2020). Harmonizing such data would be easier by applying new technological tools to which we refer later in this chapter.

21.1 What We Have Learned from Contributions on Survey Data Harmonization in this Volume

From the variety of approaches to survey data harmonization that this volume presents, five issues emerge as particularly important for future research. The first issue is growing attention to data quality among secondary users, in addition to data producers and methodologists. Beginning with the concept of Total Survey Error, social scientists have been looking at survey data quality more carefully. Of course, data quality is most likely not a fixed characteristic but can only be assessed in relation to the user's purpose. In particular, a dataset may be optimal for one usage but suboptimal for another. Here, the concept of "fit for purpose" or "fitness for use" is especially appropriate.

Moreover, evaluation of the quality of data of a given survey, or set of surveys in the same project, at the time of data production, may differ substantially from evaluation of the same data quality years later in the process of ex-post harmonization. When harmonizing ex-post, researchers decide how to treat source data of unequal quality – shall "deficient" data be discarded, at the cost of losing substantively important information? The new approach presented in this volume goes beyond documenting source data properties to storing main survey metadata as methodological variables in the harmonized dataset. Because these indicators capture variation in aspects of survey quality, users can decide which subsets of the harmonized dataset fit their research needs best, and/or they can use these indicators in their analysis, thereby directly controlling for *quality aspects of the data.*

The second issue is related to the harmonization process (ex-ante and ex-post) *per se,* specifically documenting harmonization decisions. Researchers involved in ex-ante harmonization of cross-national surveys are aware that often even the best translation from the English master questionnaire into different languages does not exactly correspond to the original. For example, the meaning of trust in languages in which there is no distinction between trust and confidence is broader than in English since it captures both: trust and confidence.

The contributions in this volume pay attention to the documentation of all decisions dealing with the control of translations. Going a step further, one can postulate that ex-ante output harmonization would include metadata noting semantic differences between concepts in different languages of the cross-national research. In the case of ex-post harmonization, accounting for the differences in the wording of the questionnaire items is essential for further substantive analyses. To facilitate transparency and, at the same time, minimize information loss following ex-post harmonization, it is good practice to store harmonization process metadata in the form of variables as *harmonization controls.* Again, these variables, together with survey-quality indicators, can be used in substantive analyses and replication studies.

The third common lesson that emerges from the differentiated descriptions of projects included in this volume is the importance of *coordination among different stakeholders.* Both ex-ante and ex-post harmonization involve funding agencies, data producers, and users. Since users constitute the main stakeholders of the final product, that is, the harmonized dataset and its documentation, the problem is how to optimize users' access to and usability of the data. Data usability is a complex problem that researchers involved in harmonization recognize well. To aid usability, workshops for potential users often accompany the main activities of producers of harmonized datasets.

In the academic realm, funding agencies, data producers, archives, and data users are the main recognized stakeholders involved in the harmonization process so far. However, with the development and popularity of survey harmonization, specialization in this type of activity becomes necessary. The teams of researchers who perform data harmonization become stakeholders in their own rights and interests.

As a fourth lesson, we mention that both ex-ante and ex-post harmonization require *advanced methodology*, including statistical tools. In the case of ex-ante harmonization, in its output version, this pertains to checking the representativeness of samples and constructing weights. Statistical solutions for testing and accounting for measurement invariance, employing intergroup confirmatory factor analysis (FA), became a good practice at different stages of harmonization, including pilot studies and final stages of ex-post harmonization. International harmonization projects encompassing many countries and continuous in time are suitable for multilevel regression analysis in which respondents are nested in countries and periods.

For survey data harmonization, one crucial task is to analyze the comparability of the concepts understood as inter-survey equivalent entities. Recent works show that differential item functioning in the item response theory (IRT) and invariance in the FA provide converging information. All parameters in IRT and FA models reflect the characteristics of items in relation to the value of a latent variable (Bauer 2017). In the IRT framework, item thresholds indicate the likelihood of a person "endorsing" an item for the latent construct, while in the FA framework, item loadings show the strength of the relationship between an item and the latent construct. Different constraints, similar to the hierarchy of invariance types in FA may be considered: configural, metric, and scalar. The moderated nonlinear factor analysis (MNLFA) (Bauer 2017; Bauer and Hussong 2009; Curran et al. 2014) expands on traditional IRT. Another method of relaxing the cross-group equality is alignment optimization (Asparouhov and Muthén 2014), further developed by Pokropek et al. (2020). Both approaches allow for measurement differences among compared surveys, opening new opportunities for survey data harmonization.

The fifth lesson is that survey data harmonization develops interdisciplinarity on both substantive and methodological levels (e.g., Avazpor, Grundy and Zhu, 2019). Substantively, both ex-ante and ex-post survey data harmonization described in this volume involve scholars from different disciplines, with the composition of disciplines being project-specific. Interdisciplinary research often relies on data from multiple sources and disciplines, including survey data, to address complex research questions. However, integrating survey data collected across different studies or disciplines can be challenging due to variations in survey design, question wording, response categories, or data coding. Harmonizing survey data allows researchers to overcome these challenges and combine information from various sources to conduct interdisciplinary analyses and draw more robust conclusions.

By harmonizing survey data, researchers can compare findings across different studies, identify commonalities, and gain a broader perspective on the phenomenon under investigation. Integrated datasets can support interdisciplinary research by providing a more comprehensive understanding of complex issues, facilitating cross-disciplinary collaborations, and enabling researchers to explore relationships and patterns that may not be apparent within the confines of a single study. Overall, interdisciplinarity and survey data harmonization work together to enhance the depth and breadth of research. They promote collaboration, integration of knowledge, and the ability to tackle complex problems from multiple angles, ultimately leading to more holistic and impactful research outcomes.

21.2 New Opportunities and Challenges

We focus on the following opportunities and challenges: (i) reorientation of survey research in the era of new technology, (ii) advances in technical aspects of data management, (iii) harmonizing survey data with other types of data, (iv) developing a new methodology of harmonizing non-survey data, (v) emerging legal and ethical issues.

21.2.1 Reorientation of Survey Research in the Era of New Technology

The traditional model of conducting social surveys for academic research, i.e. nationwide face-to-face surveys, has come under pressure. Response rates of these surveys have been dropping constantly over the last few decades (de Leeuw and de Heer 2002; de Leeuw et al. 2018). This is also true for surveys applying rigorous methodology such as the European Social Survey (Beullens et al. 2018). At the same time, costs for conducting these surveys have severely increased (Wolf et al. 2021). Therefore, we are witnessing an intensive search for alternative, cheaper forms of data collection. These could be, for example, surveys in self-completion mode (Luijkx et al. 2021; Wolf et al. 2021) or the use of digital trace data (Golder and Macy 2014). Considering the spread of smartphones and the Internet, researchers are taking advantage of using new technology to generate data on people's behavior, values, and attitudes. Although there is no definitive answer to how the future of social science data collection will look, we can safely assume that the variety of data sources will grow. What are the consequences of this trend for survey data harmonization?

21.2.2 Advances in Technical Aspects of Data Management

The work presented in this volume shows that harmonizing existing survey data is a complicated endeavor engaging computer specialists who work with available tools and adjust them to the needs. The number of such tools on the market is growing very fast. Potentially, several computer programming tools can be used in ex-post survey data harmonization, where questionnaires, codebooks, computer records on respondent characteristics, and other parts of data documentation must be simultaneously examined. Data wrangling tools may be used to speed up data preparation for the harmonized database. A good data wrangler could allow researchers to interpret, clean, and transform data into a useful format and analyze the data to check for abnormalities. Among the most popular data wrangling tools are, for example, Scrapy (an open-source web framework built in Python), Talend (a suite of tools for various functions of data preparation activities), and Microsoft Power Query (built directly into MS Excel).

Then there are dedicated ETL tools: Extract, Transform, and Load (for a review, see Vassiliadis 2009; Mali and Bojewar 2015). In the first step of the ETL process, extracting, structured, and unstructured datasets are imported and consolidated into a single repository. Transformation of data involves (i) cleansing — resolving inconsistencies and missing values; (ii) standardization — applying formatting rules to the dataset; (iii) considering duplication — excluding or flagging redundant data; (iv) verification — removing unusable data and flagging anomalies; (v) sorting — organizing data according to the needs; (vi) other tasks specific to programs. Load organizes the data into a desirable format. There are several popular free/open-source ETL tools that can be checked on the Internet. Choosing the best ETL tool for survey data harmonization can be a daunting task, as each tool has its advantages and disadvantages.

Of course, not everything needed for survey data harmonization could fit the tools on the market. As researchers approach various datasets for harmonization, they will certainly come across challenges that will go beyond the solutions of the existing tools, and they will continue to learn new approaches to meet those challenges to construct new programming tools. Our point is that ex-post harmonization requires the collaboration of social scientists with IT specialists who know the existing data management tools well but who could also create new tools if needed. Such collaboration requires that both parties – social scientists and IT specialists – overcome differences in ways in which the problems are formulated and differences in the technical language characterizing each profession.

21.2.3 Harmonizing Survey Data with Other Types of Data

Some harmonization projects involve both survey and census data. A good example of coupling these two types of data is a comprehensive time series on educational attainment (and mean years of schooling), covering the period between 1970 and 2010 for 171 countries (Goujon et al. 2016). Besides census data, this project used international surveys (Labor Force Surveys, Demographic and Health Surveys, Multiple Indicator Cluster Surveys) and national surveys (for less developed countries). Harmonization included adjusting educational categories to a modified ISCED-1997 schema, with the assumption that the number of years of schooling can be assigned with the numbers 6 for primary, 3 for lower secondary, 3 for upper secondary, and 3 for college – that is similar to other projects involving census data or surveys.

Household survey data are used together with census data to produce maps of poverty, especially in developing countries. Almost all household surveys are too small to be representative at low levels of aggregation, representing small geographic units, such as cities, towns, or villages. Census data cover these units well but usually lack information on income. Hentschel et al. (2000) demonstrate how sample survey data can be combined with census data to yield predicted poverty rates for the population covered by the census. In some cases, household surveys are combined with rich census data, allowing researchers to compare the quality of data from two sources (Alderman et al. 2002).

On a larger scale, the issues of combining census and survey data are subject of concern for IPUMS. Since the release of the Integrated Public Use Microdata Series in 1993, IPUMS has prepared interoperable and accessible census and survey data. Presently, IPUMS presents microdata from international censuses and surveys on health, time use, employment, and other topics. Record-level integration to create interoperable datasets with comprehensive metadata and documentation facilitates analyses (Kugler and Fitch 2018). The Demographic and Health Surveys (DHS), including 101 samples from 23 African and Asian countries from 1980 to the present, can be analyzed in combination with census data on an aggregate level.

Harmonized survey data can be integrated with Big Data. The increased availability of various forms of Big Data could supplant survey data in some settings. As is evidenced by the AAPOR Task Report on Big Data in Survey Research (Japec et al. 2015), the structure of data organization makes this integration feasible. One of the examples of data integration is given in Turner and Zielinski (2020), where the Google search data and harmonized survey data are jointly analyzed. Another example is the ex-ante harmonized cross-national panel in which survey answers are matched with information about respondents' online behavior (Torcal et al. 2023).

21.3 Developing a New Methodology of Harmonizing Non-Survey Data

The problems of data harmonization are not limited to survey research. They apply to a variety of data. For example, the Terra Populus project aims at assembling a globe-spanning and temporally extensive collection of environmental data in various formats, providing workflows for an integrated database (Kugler et al. 2015). Data harmonization is an important part of the Genomic Data Commons, where incoming genetic data are reviewed by a team of bioinformaticians who determine how to proceed with harmonization based on the data type, quality, and available computational resources (Zhang et al. 2021). Avillach et al. (2013) describe the procedure used for harmonizing the extraction from eight European databases of medical events of interest deemed to be important in pharmacology safety. In one of the papers (Firnkorn et al. 2015), the authors provide a generic approach for the harmonization of data from three sites on lung cancer phenotype. In a transportation project, the programing tools, called data harmonizers, grab data from car sensors and other mechanical devices, clean and integrate all information to be stored in the harmonized database (Figueiras et al. 2018).

Harmonization of big data coming from the Internet is a separate large topic. Since these data are usually characterized by the 3V model (large Volume, great Velocity, and Variety of data sources), pre-analysis data processing includes combining, cleaning, and integrating available information (Bhadani and Jothimani 2016). According to experts' assessments of dealing with this type of data, the handcrafted work — "data wrangling," "data munging," and "data janitor work" — consumes from 50% to 80% of the scientists' time before the data can be explored (Lohr 2014). Although these mundane activities are rarely discussed in the context of data harmonization, in reality they involve it. However, some papers are directly devoted to harmonization of big data (Patel and Sharma 2017), while others focus on specific aspects such as heterogeneous textual data sets (Kumar et al. 2021), or image harmonization (Sunkavalli et al. 2010; Lu et al. 2021).

Researchers involved in the harmonization of Big Data can learn a lot from emerging standards of survey data harmonization. We have in mind here the way in which data quality is assessed and the comparability of indicators is described and justified. And, simultaneously, researchers involved in survey data harmonization can learn from the accumulated knowledge of harmonizing other types of data, especially administrative data, where the same information from different sources must be reconciled. The development of the general methodology of data harmonization will require interdisciplinary collaboration on the international level.

21.3.1 Emerging Legal and Ethical Issues

Researchers have few incentives or mechanisms to share or interlink cleaned data sets, as Haak et al. (2012) point out. These authors write: *Access to these data is limited by a patchwork of laws, regulations, and practices that are unevenly applied and interpreted. A Web-based infrastructure for data sharing and analysis could help. Data exchange standards are a first step* (p. 196). The problem is that creating these standards is difficult, exactly because of legal and ethical issues (Dove et al. 2016).

Legally, the main issue deals with the ownership rights to the data. In the case of survey data harmonization, researchers can use the data that are in the public domain but usually have no right to redistribute even these parts of original data that they harmonized. In other words, they are allowed to share harmonized target variables (T) but not original source variables (S). If the team of harmonization provides the code for translating S into T, researchers aiming at replication could go to the original data set on the owner's platform. However, on this platform, a new version of the original data may be displayed, and the exact replication could not be possible. Sometimes, obtaining the version used for harmonization is cumbersome since the data distributors may have difficulty identifying all changes introduced in subsequent versions of publicly available datasets. It would be much easier if harmonized variables could be presented together with the original ones.

From an ethical point of view, even if the collection of data was approved by ethics committees in different countries, this does not mean that these committees worked with the same guidance. Should the team working on the harmonization of survey data from different countries accept the decisions of the national ethics committees or independently assess these data with respect to issues of potential harm that the use of these data may cause? There is a need to establish ethical guidelines for data harmonization from an international perspective. The problems are like those in establishing open standards for operating platforms for managing and sharing genomic and clinical data (Knoppers 2014).

21.4   Globalization of Science and Harmonizing Scientific Practice

Science, including social science, has reached a new stage. According to the popularized view, science developed from being observational (first paradigm), through theory-driven hypotheses testing (second paradigm), computational (third paradigm), to the beginnings of the data-intensive stage (fourth paradigm). The consequences for survey research and survey data harmonization are far-reaching: *The fourth paradigm, especially as applied to the intersection of Big Data and survey research, would require cross-discipline sharing of data, code, and knowledge. . . We are, despite the challenges, getting closer to the fourth paradigm. Social scientists, however, must become completely "data aware," that is, they must recognize that the data they collect or acquire, manage and munge, use and reuse, disseminate, and share must be properly stewarded – and, more importantly, that plans for the future of their data must be built at the beginning of a research study. We can see increasing evidence that these data-aware protocols are being taught to the students of today, who will, in turn, become the scientists of tomorrow.* (Hill et al. 2020, p. 729)

In the fourth paradigm, data harmonization is spreading to various disciplines, and the international scientific community should take this process into account in planning various activities. We end this chapter with a note that even conducting science is not free from proposals for harmonization, as evidenced by the following quote: *The globalization of science and its growing economic importance underline the need to establish and harmonize codes of good scientific practice. . .* (Bosch 2010, p. 252)

References

AAPOR (2014). Mobile Technologies for Conducting, Augmenting and Potentially Replacing Surveys: Report of the AAPOR Task Force on Emerging Technologies in Public Opinion Research https://www.aapor.org/getattachment/EducationResources/Reports/REVISED_Mobile_Technology_Report_Final_revised10June14.pdf.aspx

Aizpurua, E. (2020). Interview the expert: the societal experts action network (SEAN) COVID-19 survey archive, with gary langer. *Survey Practice* 13 (1): https://doi.org/10.29115/SP-2020-0006.

Alderman, H., Babita, M., Demombynes, G. et al. (2002). How low can you go? Combining census and survey data for mapping poverty in South Africa. *Journal of African Economies* 11 (2): 169–200. https://doi.org/10.1093/jae/11.2.169.

Asparouhov, T. and Muthén, B. (2014). Multiple-group factor analysis alignment. *Structural Equation Modeling: A Multidisciplinary Journal* 21: 495–508.

Avazpour, I., Grundy, J., and Zhu, L. (2019). Engineering complex data integration, harmonization and visualization systems. *Journal of Industrial Information Integration* 16: 100103. https://www.sciencedirect.com/science/article/pii/S2452414X18301511.

Avillach, P., Coloma, P.M., Gini, R. et al. (2013). Harmonization process for the identification of medical events in eight European healthcare databases: the experience from the EU-ADR project. *Journal of the American Medical Informatics Association* 20 (1): 184–192. https://doi.org/10.1136/amiajnl-2012-000933.

Bauer, D.J. (2017). A more general model for testing measurement invariance and differential item functioning. *Psychological Methods* 22 (3): 507–526. https://doi.org/10.1037/met0000077.

Bauer, D.J. and Hussong, A.M. (2009). Psychometric approaches fordeveloping commensurate measures across independent studies: traditional and new models. *Psychological Methods* 14: 101–125. http://dx.doi.org/10.1037/a0015583.

Beullens, K., Loosveldt, G., Vandenplas, C., and Stoop, I. (2018). Response rates in the European Social Survey: Increasing, decreasing, or a matter of fieldwork efforts? Survey Insights: Methods from the Field. https://surveyinsights.org/?p=9673.

Bhadani, A. and Jothimani, D. (2016). Big data: challenges, opportunities and realities. In: *Effective Big Data Management and Opportunities for Implementation* (ed. M.K. Singh and D.G. Kumar), 1–24. Pennsylvania, USA: IGI Global.

Bosch, X. (2010). Safeguarding good scientific practice in Europe: The increasingly global reach of science requires the harmonization of standards. *EMBO Reports* 11 (4): 252–257.

Curran, P.J., McGinley, J.S., Bauer, D.J. et al. (2014). A moderated nonlinear factor model for the development of commensurate measures in integrative data analysis. *Multivariate Behavioral Research* 49: 214–231. http://dx.doi.org/10.1080/00273171.2014.889594.

Dove, E.S., Townend, D., Meslin, E.M. et al. (2016). Ethics review for international data-intensive research: Ad hoc approaches mix and match existing components. *Science* 351 (6280): 1399–1400.

Figueiras, P., Guerreiro, G., Silva, R. et al. (2018). Data processing and harmonization for intelligent transportation systems: an application scenario on highway traffic flows. In: *Learning Systems: From Theory to Practice* (ed. V. Sgurev, V. Piuri, and V. Jotsov), 281–301. Cham: Springer.

Firnkorn, D., Ganzinger, M., Muley, T. et al. (2015). A generic data harmonization process for cross-linked research and network interaction. *Methods of Information in Medicine* 54 (5): 455.

Golder, S.A. and Macy, M.W. (2014). Digital footprints: opportunities and challenges for online social research. *Annual Review of Sociology* 40: 129–152. https://doi.org/10.1146/annurev-soc-071913-043145.

Goujon, A., Samir, K.C., Speringer, M. et al. (2016). A harmonized dataset on global educational attainment between 1970 and 2006 – An analytical window into recent trends and future prospects in human capital development. *Journal of Demographic Economics* 82 (3): 315–363. https://doi.org/10.1017/dem.2016.10.

Haak, L.L., Baker, D., Ginther, D.K. et al. (2012). Standards and infrastructure for innovation data exchange. *Science* 338 (6104): 196–197. https://doi.org/10.1126/science.1221840.

Hentschel, J., Olson, J., Lanjouw, O.J. et al. (2000). Combining census and survey data to trace the spatial dimensions of poverty: a case study of ecuador. *The World Bank Economic Review* 14 (1): 147–165. https://doi.org/10.1093/wber/14.1.147.

Hill, C.A., Biemer, P.P., Buskirk, T.D. et al. (ed.) (2020). *Big Data Meets Survey Science: A Collection of Innovative Methods*. Wiley.

Japec, L., Kreater, F., Berg, M. et al. (2015). Big data in survey research: AAPOR task force report. *Public Opinion Quarterly* 79 (4): 839–880. https://doi.org/10.1093/poq/nfv039.

Knoppers, B.M.I. (2014). International ethics harmonization and the global alliance for genomics and health. *Genome Medicine* 6 (13): 1–3. https://doi.org/10.1186/gm530.

Kugler, T. and Fitch, C. (2018). Interoperable and accessible census and survey data from IPUMS. *Scintific Data* 5: 180007. https://doi.org/10.1038/sdata.2018.7.

Kugler, T., Van Riper, D.C., Manson, S.M. et al. (2015). Terra populus: workflows for integrating and harmonizing geospatial population and environmental data. *Journal of Map & Geography Libraries* 11 (2): 180–206. https://doi.org/10.1080/15420353.2015.1036484.

Kumar, G., Basr, S., Imam, A.A. et al. (2021). Data harmonization for heterogeneous datasets: a systematic literature review. *Applied Sciences* 11 (17): 8275. https://doi.org/10.3390/app11178275.

de Leeuw, E. and de Heer, W. (2002). Trends in household survey nonresponse: a longitudinal and international comparison. In: *Survey Nonresponse* (ed. R.M. Groves, D.A. Dillman, J.L. Eltinge, and R.J.A. Little), 41–54. New York: Wiley.

de Leeuw, E., Hox, J., and Luiten, A. (2018). International nonresponse. Trends across countries and years: an analysis of 36 years of Labour Force Survey data. Survey Insights: Methods from the Field. https://surveyinsights.org/?p=10452. 10.13094/SMIF-2018-00008.

Lohr, S. (2014). For big-data scientists, 'janitor work' is key hurdle to insights. *New York Times* 17, B4.

Lu, X., Huang, Sh., Niu, L. et al. (2021). YouTube: Video Harmonization Dataset. arXiv preprint https://arxiv.org/pdf/2109.08809.pdf.

Luijkx, R., Jónsdóttir, G.A., Gummer, T. et al. (2021). The European values study 2017: on the way to the future using mixed-modes. *European Sociological Review* 37 (2): 330–347. https://doi.org/10.1093/esr/jcaa049.

Mali, M. and Bojewar, S. (2015). A survey of ETL tools. *International Journal of Computer Techniques* 2 (5): 20–27. http://www.ijctjournal.org/Volume2/Issue5/IJCT-V2I5P3.pdf.

Patel, J.A. and Sharma, P. (2017). Big Data harmonization – challenges and applications. *International Journal on Recent and Innovation Trends in Computing and Communication* 5 (6): 206–208. https://ijritcc.org/index.php/ijritcc/article/view/748.

Pokropek, A., Lüdtke, O., and Robitzsch, A. (2020). An extension of the invariance alignment method for scale linking. *Psychological Test and Assessment Modeling* 62 (2): 305–334.

Sunkavalli, K., Johnson, M.K., Matusik, W., and Pfister, H.H. (2010). Multi-scale image harmonization. *ACM Transactions on Graphics* 29 (4): 1–10.

Torcal, M., Carty, M., Comellas J. M., et al. (2023). The dynamics of political and affective polarisation: Datasets for Spain, Portugal, Italy, Argentina, and Chile (2019-2022). Data in Brief 48 id: 109219. https://doi.org/10.1016/j.dib.2023.109219

Turner, A. and Zieliński, M. (2020). To what extent do aggregate measures of google searches relate to individual responses to survey items? On harmonizing data from different sources. *Harmonization: Newsletter on Survey Data Harmonization in the Social Sciences* 6 (2): 2–15.

Vassiliadis, P. (2009). A survey of extract–transform–load technology. *International Journal of Data Warehousing and Mining* 5 (3): 1–27: https://doi.org/10.4018/jdwm.2009070101

Wolf, C., Christmann, P., Gummer, T. et al. (2021). Conducting General Social Surveys as self-administered mixed-mode surveys. *Public Opinion Quarterly 85* (2).

Zhang, Z., Hernandez, K., Savage, J. et al. (2021). Uniform genomic data analysis in the NCI Genomic Data Commons. *Nature Communications* 12: 1226. https://doi.org/10.1038/s41467-021-21254-9.

Index

Note: *Italicized* and **bold** page numbers refer to figures and tables, respectively.

a

AAPOR *see* American Association of Public Opinion Research (AAPOR)
AB *see* Afrobarometer (AB); Asia Barometer (AB)
AB Bible 59
Accessible Resources for Integrated Epigenomics Studies (ARIES) 200
ACS *see* African Charter on Statistics (ACS)
activity harmonization 292, 299
AfDB *see* African Development Bank (AfDB)
African Charter on Statistics (ACS) 134, 135, **136**
African Development Bank (AfDB) 128, 135, 138, 140
African Information Highway (AIH) portal system 135
African Peer Review Mechanism (APRM) 138
African Program for the Accelerated Improvement of Civil Registration and Vital Statistics (APAI-CRVS) 132
African quality assurance framework (AQAF) 134, 135
African Statistics System (AfSS) 134, 135
African Statistics Yearbook (ASY) 133
African Trade Statistics Yearbook (ATSY) 133, 139
African Union Commission (AUC) 128, 129, 131, 133, 134, 140
Afrobarometer (AB) 57
  applied harmonization methods
    data management 65

    documentation 65–66
    fieldwork and data collection 62
    questionnaire 62–64
    sampling 60–61
    training 61–62
    translation 64
  challenges to harmonization 67–70
  core principles 58–60
  country selection, harmonization and 66
  recommendations 70–71
  software tools and harmonization 66–67
AfSS *see* African Statistics System (AfSS)
AIH portal system *see* African Information Highway (AIH) portal system
ALSPAC cohort *see* Avon Longitudinal Study of Parents and Children (ALSPAC) cohort
American Association of Public Opinion Research (AAPOR) 8–9, 367, 371
American Time Use Survey (ATUS) 288, 289, 290, 291, 299
APAI-CRVS *see* African Program for the Accelerated Improvement of Civil Registration and Vital Statistics (APAI-CRVS)
APRM *see* African Peer Review Mechanism (APRM)
AQAF *see* African quality assurance framework (AQAF)
Arab Barometer 356
Arab Spring 354

Survey Data Harmonization in the Social Sciences, First Edition.
Edited by Irina Tomescu-Dubrow, Christof Wolf, Kazimierz M. Slomczynski, and J. Craig Jenkins.
© 2024 John Wiley & Sons Inc. Published 2024 by John Wiley & Sons Inc.

archives 27, 32–33
ARIES *see* Accessible Resources for Integrated Epigenomics Studies (ARIES)
ASEP *see* Austria Socio-Economic Panel (ASEP)
Asia Barometer (AB) 114, 359
"ask-the-same-question" rule 53
ASYCUDA 140
ASY *see* African Statistics Yearbook (ASY)
ATSY *see* African Trade Statistics Yearbook (ATSY)
ATUS *see* American Time Use Survey (ATUS)
AUC *see* African Union Commission (AUC)
Austria Socio-Economic Panel (ASEP) 180
Avon Longitudinal Study of Parents and Children (ALSPAC) cohort 192

b
barometers 23
baseline health and risk factor questionnaire (BL-HRFQ) 233, 234, 238
BHPS *see* British Household Panel Study (BHPS)
bias correlations 313
Big Data 9, 32, 33, 142, 371, 372
BL-HRFQ *see* baseline health and risk factor questionnaire (BL-HRFQ)
BLS *see* Bureau of Labor Statistics (BLS)
Bolshevik revolution 287
British Household Panel Study (BHPS) 177
British Occupational Based Social Class 192
Bureau of Labor Statistics (BLS) 76, 79

c
CAME *see* correction for attenuation due to measurement error (CAME)
CanPath 230, **232**, 233, 234, 235, 236, **236**, 237, 238, **238**, 239–243
CAPI *see* computer-assisted personal interview (CAPI)
CATI *see* computer-assisted telephone interview (CATI)
Caucasus Barometer 356, 359
CAWI *see* computer-assisted web-interview (CAWI)
CDD-Ghana *see* Center for Democratic Development in Ghana (CDD-Ghana)
CDU *see* Christian Democratic Union (CDU)
censuses 139, 142, 207–210, 217
Center for Democratic Development in Ghana (CDD-Ghana) 59

Centre for Time Use Research (CTUR) 288, 300
CESSDA *see* Consortium of European Social Science Data Archives (CESSDA)
CFA *see* confirmatory factor analysis (CFA)
CFPS *see* China Family Panel Study (CFPS)
CGSS *see* Chinese General Social Survey (CGSS)
challenges to survey harmonization 9
  data dissemination 13
  data preparation 12
  data processing, quality controls, and adjustments 12
  fieldwork 11–12
  instruments and their adaptation 10–11
  population representation 10
  preparation for interviewing 11
Chapel Hill Expert Surveys (CHES) 104
CHES *see* Chapel Hill Expert Surveys (CHES)
China Family Panel Study (CFPS) 177
Chinese General Social Survey (CGSS) 107, 108, 111, 113, 114, 115, 118, 121, 122
Christian Democratic Party (KDNP) 99
Christian Democratic Union (CDU) 103
civil registration and vital statistics (CRVS) 129, 132
Classical Invariance Analysis 326–327
CLOSER *see* Cohort and Longitudinal Studies Enhancement Resources (CLOSER)
CNEF project *see* Cross-National Equivalent File (CNEF) project
coding consistency 296
Cohort and Longitudinal Studies Enhancement Resources (CLOSER) 189
  applied harmonization methods 191
    body size/anthropometric data 193–194
    harmonization methods: divergence and convergence 195–196
    mental health 194–195
    occupational social class 191–193
  challenges to harmonization 198–199
  documentation and quality assurance 196–198
  harmonization-related work packages *190*
  recommendations 200–202
  software tools 199–200

collaborative partnerships   68

combined data sets   353

   missing values   358

      at respondent and measurement level   359

      at survey level   359–361

   time, dealing with   354–358

   weights, dealing with   361–362

comparability and measurement invariance   323

   Classical Invariance Analysis   326–327

   invariance, approximate   327–328

   latent variable framework   324–325

   measurement equivalence, empirical assessment of   325–326

   multiple indicators, beyond   329

   partial invariance   327

   partial invariance, approximate   328–329

Comparative Manifesto Project (MARPOR)   101, 104

Comparative Study of Electoral Systems (CSES)   89

   CSES Integrated Module Dataset (IMD)   90, 101

      demographic variables in   101–102

      harmonizing party data in   102–104

   data products, as of April 2022   **90**

   Election Study detailed in a CSES Codebook   *95*

   Election Study Note concept in CSES Codebook   *94*

   electoral alliances and their coding in CSES Codebook   *100*

   ex-ante input harmonization   91

      macro data   95–97

      module questionnaire   92–95

   ex-ante output harmonization   97

      demographic variables in CSES modules   97–98

      derivative variables   99–101

      harmonizing party data in modules   98–99

   harmonization principles and technical infrastructure   91

   harmonizing party codes across time in CSES IMD   **103**

   Macro Report   96

   Planning Committee (PC)   90

   relational data structure   *98*

   taking stock and new frontiers in harmonization   104

Comparative Survey Design and Implementation (CSDI)   9

comparative surveys   333

   design weights   335–337

   population weights   341–343

   post-stratification weights   337–341

composite coding   213, 215, 217

computer-assisted personal interview (CAPI)   43, 79–80, 81

computer-assisted telephone interview (CATI)   79, 367

computer-assisted web-interview (CAWI)   43, 295

configural invariance   326

confirmatory factor analysis (CFA)   54, 310–311

congeneric reliability   312, 319

Consortium of European Social Science Data Archives (CESSDA)   27

constant learning   59

constrained flexibility   60

construct validity   309–310

content validity   309

context-sensitive standardization   58

control variables   314

convergent validity   310

correction for attenuation due to measurement error (CAME)   312

Corruption Perception Index (CPI) data   129

countries, harmonizing survey data across and within   7–8

country selection, harmonization and   66

country-specific questions (CSQs)   63, 65

Covid-19 pandemic   69, 79, 189

CPI data *see* Corruption Perception Index (CPI) data

criterion validity   309–310

Cronbach's Alpha   312

cross-cultural comparability   316

   construct match   316–317

      harmonizing units across cultures and instruments   318

      harmonizing units of localized versions of the same instrument   318

      reliability   317

      translation and cognitive probing   317

      units of measurement   318

   of multi-item instruments   318–319

Cross-National Equivalent File (CNEF) project   1, 169
    applied harmonization methods   170–176
    challenges to harmonization   183–185
    country data sources, current and planned   176
    current partners
        British Household Panel Study (BHPS)   177
        China Family Panel Study (CFPS)   177
        Household Income and Labor Dynamics in
            Australia (HILDA) Survey   176
        Italian Lives (ITA.LI)   178
        Japan Household Panel Study (JHPS)   178
        Korea Labor and Income Panel Study
            (KLIPS)   179
        Panel Study of Income Dynamics
            (PSID)   **173**, 179–180
        Russia Longitudinal Monitoring Survey
            (RLMS-HSE)   178–179
        Socio-Economic Panel (SOEP)   177–178
        Survey of Labor and Income Dynamics
            (SLID)   176–177
        Swedish pseudo-panel   179
        Swiss Household Panel (SHP)   179
        Understanding Society, UKHLS   178
    documentation and quality assurance   181–183
    planned partners   180
        Austria Socio-Economic Panel (ASEP)   180
        Israel Longitudinal Study (ILS)   180
        Longitudinal and International Study of Adults
            (LISA)   180
        Mexican Family Life Surveys (MxFLS)   180–181
        National Income Dynamics Study (NIDS)   181
        Panel Study of Family Dynamics (PSFD)   181
    recommendations for researchers interested in
        harmonizing panel survey data   185–186
cross-national secondary analysis,
        harmonization for   147
    challenges to harmonization   156–161
    documentation and quality assurance   155–156
    recommendations
        for researchers interested in harmonizing survey
            data ex post   162–163
        for SDR2 users   163–164
    software tools of the SDR project   161–162
    survey data recycling (SDR) project   149–155

survey data recycling database v.2.0 (SDR2)
        database   **149**, 150–155, **160**
crosswalk coding. 151   154
Crosswalk Tables (CWTs)   154, 155, 160
CRVS *see* civil registration and vital statistics (CRVS)
CSDI *see* Comparative Survey Design and
        Implementation (CSDI)
CSES *see* Comparative Study of Electoral
        Systems (CSES)
CSQs *see* country-specific questions (CSQs)
CTUR *see* Centre for Time Use Research (CTUR)
CWTs *see* Crosswalk Tables (CWTs)

d

data collection event (DCE)   230, 232, 233, 234,
        235, 237
data dictionaries   151, 154, 155, 161. 162, 210–211,
        212–213, 216, 221, 234, 235
Data Documentation Initiative (DDI) standards   51
Data Documentation Initiative-Lifecycle (DDI-L)   197
data janitor work   372
data management   65
    advances in technical aspects of   370–371
data munging   372
data producers   1, 25, 26–27, 32–33, 127, 368
DataSchema variable   235–236, **236**, 237–238, 241, 243
data science   9, 150
data sets   65, 67, 250, 253
    challenges in the combination of   347
        assessing the equivalence of survey
            projects   351–353
        no censorship inclusive approach   348–349
        using multilevel analysis and introducing a
            measurement level   349–351
    combined data sets, challenges in the
        analysis of   353
        missing values   358–362
        time, dealing with   354–358
        weights, dealing with   361–362
    country data sets   65
    harmonized data set   250, 253, 255, 347
    HaSpaD data set   253, 256, 258, 259, 260, 263
    pairfam data set   253
    source data sets   263

data users   27–28, 33, 54, 121, 123, 209, 244, 259, 281, 312
data wrangling   370, 372
DCE *see* data collection event (DCE)
DDI standards *see* Data Documentation Initiative (DDI) standards
DDI-L *see* Data Documentation Initiative-Lifecycle (DDI-L)
Demographic and Health Surveys (DHS)   371
demographic variables   42, 45, 97, 101–102, 337
designed data   32, 33
design weights   335–337, 343
Detailed Source Variables Reports (DVRs)   151, 154, 155
DHS *see* Demographic and Health Surveys (DHS)
differential item functioning (DIF)   324, 369
DIF *see* differential item functioning (DIF)
disposable household income   271, 272, *272*
divergent validity   310
DVRs *see* Detailed Source Variables Reports (DVRs)

e
EABS *see* East Asia Barometer Survey (EABS)
EA map *see* enumeration area (EA) map
EASSDA *see* East Asian Social Survey Data Archive (EASSDA)
EASS project *see* East Asian Social Survey (EASS) project
East Asia Barometer Survey (EABS)   114
East Asian Social Survey (EASS) project. 10   107–111
   challenges to harmonization   118
      data collection phase, difficulty in synchronizing   121–122
      translating "fair" and restriction by copyright   118–121
   harmonization process of   111
      answer choices and scales, harmonization of   114–115
      module theme, establishing   111–112
      questions and answer choices, translation of   115
      standard background variables, harmonization of   113–114
      subtopics and questions, selecting   112–113

integrated data
   documentation of   117–118
   steps to harmonize   115–117
recommendations   122–123
software tools   122
East Asian Social Survey Data Archive (EASSDA)   108, 118
East Asia Value Survey (EAVS)   114
EAVS *see* East Asia Value Survey (EAVS)
EB *see* Eurobarometer (EB)
ECOWAS   139
EDC *see* electronic data capture (EDC)
EFA *see* exploratory factor analysis (EFA)
e-GDDS *see* Enhanced General Data Dissemination System (e-GDDS)
Election Study Notes (ESNs)   93, 95, 96, 97
electronic data capture (EDC)   61, 62, 65, 67, 70, 71
employer history roster   81
end-product of survey harmonization   14–15
Enhanced General Data Dissemination System (e-GDDS)   135
enumeration area (EA) map   58, 60, 61
equivalence   6, 54, 174, 327
ERETES   140
ESCEG *see* European Standard Classification of Cultural and Ethnic Groups (ESCEG)
ESM *see* experience sampling method (ESM)
ESNs *see* Election Study Notes (ESNs)
ESOMAR *see* European Society of Opinion and Marketing Research (ESOMAR)
ESS *see* European Social Survey (ESS)
ethical and legal issues   15–16, 233
ETL tools *see* Extract, Transform, and Load (ETL) tools
Eurobarometer (EB)   4, 23
European Social Survey (ESS)   1, 4, 10, 11, 23, 102, 335, 351, 356
European Society of Opinion and Marketing Research (ESOMAR)   9, 13, 15
European Standard Classification of Cultural and Ethnic Groups (ESCEG)   102
European Values Study (EVS)   7, 9, 41, 43, 46
EUROSTAT recommendations   296
EUROTRACE User Group   140
EVS *see* European Values Study (EVS)

ex-ante and *ex-post* harmonization, exploring interplay between   101
  demographic variables in CSES IMD   101–102
  harmonizing party data in CSES IMD   102–104
ex-ante harmonization   7, 13, 14, 138, 318, 367, 368, 369
ex-ante harmonization of official Statistics in Africa   125
  civil registration and vital statistics (CRVS), development of   132
  common software tools used   139–140
  ex-ante harmonization, challenges with   138
  ex-post harmonization, challenges with   139
  ex-post harmonization, examples of   132
    African Statistics Yearbook (ASY)   135
    African Trade Statistics Yearbook (ATSY)   133
    International Comparison Program for Africa (ICP-Africa)   135
    KeyStats   135
    labor market indicators, guideline for producing   132
  governance, peace and security (GPS) statistics initiative   131, *131*
  National Strategies for the Development of Statistics (NSDS)   137–138
  quality assurance framework   134–136
  recommendations   140–142
ex-ante input harmonization   1–2, 3–4, 9, 11, 91
  macro data   95–97
  module questionnaire   92–95
ex-ante output harmonization   3–4, 9, 12, 44, 97, 368
  demographic variables in CSES modules   97–98
  derivative variables   99–101
  harmonizing party data in modules   98–99
experience sampling method (ESM)   294
exploratory factor analysis (EFA)   328
ex post harmonization   2–4, 7, 15–16, 24–32, 34, 42, 43, 59, 62, 147, *270*, 275, 279, 280, 281, 282, 289, 294, 307, 368–369, 371
  construct match   308
    consequences of a mismatch   309
    construct and criterion validity   309–310
    improving construct comparability   311
    multi-item instruments, techniques for   310–311
    qualitative research methods   309

cross-cultural comparability   316
  of multi-item instruments   318–319
  reliability   317
  translation and cognitive probing   317
  units of measurement   318
measurement, units of   312
  instrument characteristics, controlling for   314
  unit comparability, improving   313–314
  unit differences, consequences of   313
measurement and reality   307–308
reliability differences   311
  assessment   312
  consequences of   311–312
  improving reliability comparability   312
repeated measurements, harmonizing units based on   315
  from the same population   315–316
ex-post harmonization of time use data   285
  applied harmonization methods   289
    harmonizing the matrix of the diary   289–291
    time use data   294
    variable harmonization   291–293
    variables   293–294
  challenges to harmonization   297–299
  documentation   294–295
  quality checks   296–297
  recommendations   301–302
  software tools   300–301
ex-post harmonization   3, 24–32, 34, 42–43, 101, 103, 129
  challenges with   139
  examples of   132–133
Extract, Transform, and Load (ETL) tools   370–371

f

factor analysis (FA)   310, 313, 319, 369
FAIR data principles   15, 243
FA *see* factor analysis (FA)
Fidesz Party   99
fieldwork and data collection   62
fieldwork supervision   62
future of survey data harmonization   367
  globalization of science and harmonizing scientific practice   373

new opportunities and challenges   370
   data management, advances in technical aspects of   370–371
   harmonizing survey data with other types of data   371
   reorientation of survey research in the era of new technology   370
   non-survey data, developing a new methodology of harmonizing   372
   emerging legal and ethical issues   372–373

g

GDPR *see* General Data Protection Regulation (GDPR)
General Data Protection Regulation (GDPR)   15
General Public License (GPL)   242
General Social Survey (GSS)   8, 107
General Target Variable Report (GVR)   152, 153, 155, 156, 160
Generations and Gender Survey (GGS)   253, 254, 255
Genomic Data Commons   372
geographic information system (GIS) boundary files   220
German Life History Studies (GLHS)   253–254
GGS *see* Generations and Gender Survey (GGS)
GIS boundary files *see* geographic information system (GIS) boundary files
GLHS *see* German Life History Studies (GLHS)
Google Scholar   53, 169
governance, peace and security (GPS) statistics initiative   129, 131, *131*, 138, 141
GPL *see* General Public License (GPL)
GPS statistics initiative *see* governance, peace and security (GPS) statistics initiative
GSS *see* General Social Survey (GSS)
GVR *see* General Target Variable Report (GVR)

h

harmonization controls   149, 153, **154**, 157, 158, 159, 163. 368
harmonization cost   29
harmonization fit   29, 30
harmonization principles and technical infrastructure   91
harmonization quality   29–30
harmonized consumer price index (HCPI)   129, 141

Harmonized European Time Use Survey (HETUS) data collection   287–288, 290, 291, 292, 299
Harmonizing and Synthesizing Partnership Histories from Different Research Data Infrastructures (HaSpaD) project   249
   biographical data, harmonizing
      methodological recommendations   262–263
      procedural recommendations   263
      technical recommendations   263
   challenges to harmonization   258
      harmonized complex survey data, analyzing   258–259
      sporadically and systematically missing data   259–260
   cumulative HaSpaD data set   263–264
   data search strategy and data access   250–252
   documentation   255–256
   processing and harmonizing data   253
      additional variables, harmonizing   254–255
      partnership biography data, harmonizing   253–254
   quality assurance
      harmonized HaSpaD data set, benchmarking the   256–258
      process-related quality assurance   256
   software tools   260–262
   survey programs and sub-studies used in   **251–252**
harmonizing data formats   290
HaSpaD-Harmonization Wizard   250, 256, 261–262, *262*, 263
HaSpaD project *see* Harmonizing and Synthesizing Partnership Histories from Different Research Data Infrastructures (HaSpaD) project
HCPI *see* harmonized consumer price index (HCPI)
HETUS data collection *see* Harmonized European Time Use Survey (HETUS) data collection
hierarchical coding.   **47**, 292, 293, 297, 299
HILDA Survey *see* Household Income and Labor Dynamics in Australia (HILDA) Survey
Hope and Glory (film)   75, 86
Household Income and Labor Dynamics in Australia (HILDA) Survey   174, 176
household survey data   371

i

ICATUS *see* International Classification of Time Use Activities (ICATUS)

ICLS *see* International Conference of Labor Statisticians (ICLS)

ICP *see* international comparison program (ICP)

ICPSR *see* Inter-university Consortium for Political and Social Research (ICPSR)

ICP-Africa *see* International Comparison Program for Africa (ICP-Africa)

ILS *see* Israel Longitudinal Study (ILS)

income and wealth data   269
 applied harmonization methods   271–275
 challenges to harmonization   278–281
 documentation   278
 quality assurance   275
  harmonization   276
  selection of source datasets   276
  validation – "green light" check   276–278
 software tools   281–282

individual participant data (IPD)   229–230, 237, 241, 250

input harmonization   3, 11, 24, 43, 44, 48, 49, 50, 52–53, 108, 113

instrument characteristics, controlling for   314

international actors and funding agencies   26

International Classification of Time Use Activities (ICATUS)   288, 299

international comparison program (ICP)   129

International Comparison Program for Africa (ICP-Africa)   133, 135, 138

International Conference of Labor Statisticians (ICLS)   132

International Physical Activity Questionnaire (IPAQ)   235

International Social Survey Program (ISSP)   1, 4, 7–8, 12, 16, 23, 98, 107, 108, 111–115, **119**, *120*, 121, 161, 341

International Standard Classification of Education (ISCED)   3, 13, 27, 28, 217, 274, 339
 ISCED-97-scale   254, 371
 ISCED-2011   46, 51

international survey programs   23, 317, 318

Internet   31, 43, 50, 367, 370, 371, 372

Inter-university Consortium for Political and Social Research (ICPSR)   108, 117, 150

invariance, approximate   327–328, 330

IPAQ *see* International Physical Activity Questionnaire (IPAQ)

IPD *see* individual participant data (IPD)

IPUMS DCP   221

IPUMS International   1, 207, 371
 applied harmonization methods   210–215
 challenges to harmonization   217–220
 Data Conversion Program (DCP)   215
 documentation and quality assurance   215–217
 project history   208
  web dissemination system, evolution of   210
 recommendations   223–225
 software tools   221
  data harmonization   221–222
  data reformatting   221
  dissemination system   222
  metadata tools   221
 team organization and project management   222–223

IPUMS Time Use   300, 301

IRT-Models   319

IRT *see* item response theory (IRT)

ISCED *see* International Standard Classification of Education (ISCED)

ISCED-97-scale   254

ISCED-2011 unified coding scheme   45–46, 51, 52

ISCO coding scheme   292

Israel Longitudinal Study (ILS)   180

ISSP *see* International Social Survey Program (ISSP)

ISTAT *see* Italian National Institute of Statistics (ISTAT)

ITA.LI *see* Italian Lives (ITA.LI)

Italian Lives (ITA.LI)   178

Italian National Institute of Statistics (ISTAT)   178

item response theory (IRT)   315, 369

j

Japanese General Social Survey (JGSS) International Symposium   107, 108, 111–118, 120–122

Japan Household Panel Study (JHPS)   178

JDSurvey software   51–52

JGSS International Symposium *see* Japanese General
  Social Survey (JGSS) International Symposium
JHPS *see* Japan Household Panel Study (JHPS)

k

Kaplan–Meier estimates   257
KeyStats   133, 135, 139
KGSS *see* Korean General Social Survey (KGSS)
KLIPS *see* Korea Labor and Income Panel
  Study (KLIPS)
Korea Labor and Income Panel Study (KLIPS)   179
Korean General Social Survey (KGSS)   107, 108, 111,
  112, 114, 118, 121, 122

l

Labour Force Survey (LFS)   339
LAPOP *see* Latin America Public Opinion
  Project (LAPOP)
large-scale survey programs   30
Latent Class Analysis   310, 319
latent variable framework   324–325
Latin America Public Opinion Project (LAPOP)   351,
  353, 356
Latino Barometer   351
LAT partnerships *see* living apart together (LAT)
  partnerships
LFS *see* Labour Force Survey (LFS)
Life in Transition survey   356
Likert-type 4-point scale   351
linear stretching   314
LISA *see* Longitudinal and International Study of
  Adults (LISA)
LIS project *see* Luxembourg Income Study (LIS) project
Lived Poverty Index   65
living apart together (LAT) partnerships   254
Longitudinal and International Study of Adults
  (LISA)   180
Longitudinal Study of Australian Children (LSAC)   82
LSAC *see* Longitudinal Study of Australian
  Children (LSAC)
Luxembourg Income Study (LIS) project   1, 170, 269
  applied harmonization methods   271–275
  challenges to harmonization   278–281
  documentation   278
  ex post harmonization at   *270*

quality assurance   275, *276*
  harmonization   276
  selection of source datasets   276
  validation – "green light" check   276–278
  software tools   281–282
Luxembourg Wealth Study (LWS) *Database*   269
LWS *Database see Luxembourg Wealth Study* (LWS)
  Database

m

macro data harmonization   91, 96
Maelstrom Research   229
  applied harmonization methods   230
    assembling study information and selecting final
      participating studies   234
    DataSchema variable   235–236, **236**, 243
    harmonized datasets, producing   236–237
    initiating activities and organizing the operational
      framework   233–234
  challenges to harmonization   240–241
  documentation and quality assurance   238–240
  recommendations   243–244
  for rigorous retrospective harmonization   **231–232**
  software tools   241–242
MarkDoc document   260–261, *261*
MARPOR project   101, 104
MAR *see* missing at random (MAR)
MASTER file   154, 155
Mature Women cohort   74
maximum support   60
MCAR *see* missing completely at random (MCAR)
MDGs *see* Millennium Development Goals (MDGs)
measurement, units of   307, 308, 312
  instrument characteristics, controlling for   314
  repeated measurements, harmonizing units
    based on   315
    obtained from the same population   315–316
  unit comparability, improving   313–314
  unit differences, consequences of   313
measurement difference   313, 369
measurement equivalence   53, 194, 195, 198,
  324, 325–326
measurement invariance   54, 318–319, 324,
  326, 329–330
memory effect   8

Metadata Information System (METIS)   274

methodological indicators   149, 151, 153, 157, 159, 162, 163

methodological variables   153, 161, 314, 368

METIS *see Metadata Information System* (METIS)

metric equivalence   326

Mexican Family Life Surveys (MxFLS)   180–181

MG-BSEM *see* Multiple-Group Bayesian Structural Equation Modeling (MG-BSEM)

MG-CFA modeling *see* Multiple-Group Confirmatory Factor Analysis (MG-CFA) modeling

MGCFA *see* multigroup confirmatory factor analysis (MGCFA)

Mica-Opal software toolkit   242

Millennium Development Goals (MDGs)   128

MINDMAP project   230, **232**, 233–234, 237, 238, 239–241, 243, 244

MI *see* modification indices criteria (MI); multiple imputation (MI)

missing at random (MAR)   358

missing completely at random (MCAR)   358, 359

missing values   **160**, 217, 271, 277, 281, 293, 295, 302, 358

 at respondent and measurement level   359

 at survey level   359–361

MNLFA *see* moderated nonlinear factor analysis (MNLFA)

moderated nonlinear factor analysis (MNLFA)   369

modification indices criteria (MI)   260, 327

MTMM *see* multitrait–multimethod approach (MTMM)

MTUS *see* Multinational Time Use Study (MTUS)

multigroup confirmatory factor analysis (MGCFA)   54, 310, 318–319

multi-item instruments   308, 310, 312, 315, 316, 318–319

multilevel modeling approaches   314

Multinational Time Use Study (MTUS)   1, 288, 293

Multiple-Group Bayesian Structural Equation Modeling (MG-BSEM)   328

Multiple-Group Confirmatory Factor Analysis (MG-CFA) modeling   310–311, 325, *325*, 327–328, 329

multiple imputation (MI)   260, 315, 359, 361

multitrait–multimethod approach (MTMM)   312, 314

MxFLS *see* Mexican Family Life Surveys (MxFLS)

n

National Income Dynamics Study (NIDS)   181

National Institute of Child Health and Human Development (NICHD)   186

National Institutes on Aging (NIA)   170

National Longitudinal Surveys (NLS) program   73–74, 76, 80, 82–87, *85*

National Longitudinal Surveys of Youth (NLSY)   73

 applied harmonization   76–80

 challenges to harmonization   80–82,

 cross-cohort design   75–76

 documentation and quality assurance   82–84

 NLSY79 cohort   73–87

 NLSY97 cohort   75–87

 recommendations and some concluding thoughts   86–87

 software tools   84–86

National Quality Assurance Framework   134

National Statistics Offices (NSOs)   60, 61, 128, 131, 133, 137, 138, 139, 140, 142

National Statistics Systems (NSSs)   129, 134, 135, 138, 139, 141

National Strategies for the Development of Statistics (NSDS)   128–129, 137–138

NEAT design   315, 316, 319

NEAT research design   319

NEC design *see* nonequivalent groups with covariates (NEC) design

need to harmonize social survey data   5–6

net worth   271, 272–273, *273*, 278

NIA *see* National Institutes on Aging (NIA)

NICHD *see* National Institute of Child Health and Human Development (NICHD)

NIDS *see* National Income Dynamics Study (NIDS)

NLS program *see* National Longitudinal Surveys (NLS) program

NLSY *see* National Longitudinal Surveys of Youth (NLSY)

NMAR *see* not missing at random (NMAR)

nomological validity   310

nonequivalent groups with covariates (NEC) design   316

non-survey data, new methodology of
    harmonizing   372–373
not missing at random (NMAR)   358
NSDS *see* National Strategies for the Development of
    Statistics (NSDS)
NSOs *see* National Statistics Offices (NSOs)
NSSs *see* National Statistics Systems (NSSs)

o

observed score equating   316, 318
OECD *see* Organization for Economic Cooperation and
    Development (OECD)
operational comparability   272, 273, 279, 280
organic data   22, 31–34
Organization for Economic Cooperation and
    Development (OECD)   174, 271, 274

p

Panel Study of Family Dynamics (PSFD)   181
Panel Study of Income Dynamics (PSID)   170, 172,
    173, 179–180, 182, 184, 186
Pan-African institutions   129
paper-and-pencil interview (PAPI)   43, 79, 295
PAPI *see* paper-and-pencil interview (PAPI)
ParlGov *see* Parliaments and Governments database
    (ParlGov)
Parliaments and Governments database
    (ParlGov)   104
parquet   221
partial invariance   327
    approximate   328–329
PI *see* principal investigators (PI)
PLUG-COUNTRY file   154–155
PLUG-SURVEY   154–155
population weights   341–343, 361
post-stratification weights   258, 337–341
PPP *see* purchasing power parity (PPP)
PPPS *see* probability proportionate to population
    size (PPPS)
pretest databases   309
primary sampling units (PSUs)   49, 60
principal investigators (PI)   3, 42, 107, 111, 112, 256
probability proportionate to population size (PPPS)   60
producers   1, 24, 32

PSFD *see* Panel Study of Family Dynamics (PSFD)
PSID *see* Panel Study of Income Dynamics (PSID)
PSUs *see* primary sampling units (PSUs)
psychometric equivalence   6
psychometric multi-item instruments   315
psychometry   315
purchasing power parity (PPP)   12, 133, 277

q

qualitative cognitive interviews   317
qualitative pretests   309
quality assurance framework   134–136
quality assurance of harmonization products   239
quality-cost trade-off and implications for
    harmonization   68–69
quality of the input and end-product of survey
    harmonization   14–15
questionnaire translation   44–45, 48, 51

r

random group design   316
RECs *see* Regional Economies Communities (RECs)
Regional Economies Communities (RECs)   128–129,
    133, 135, 139
Regional Strategies for Development of Statistics
    (RSDS)   128–129
Registrar General's Social Class   192
reliability differences   308, 311
    assessment   312
    consequences of   311–312
    improving reliability comparability   312
repeated measurements, harmonizing units
    based on   315
    obtained from the same population   315–316
Research Institute of the McGill University Health
    Centre (RIMUHC)   230
RIMUHC *see* Research Institute of the McGill
    University Health Centre (RIMUHC)
RLMS-HSE *see* Russia Longitudinal Monitoring Survey
    (RLMS-HSE)
RSDS *see* Regional Strategies for Development of
    Statistics (RSDS)
Russia Longitudinal Monitoring Survey
    (RLMS-HSE)   178–179

S

sampling start point (SSP) 61

SAT *see* Senior Advisory Team (SAT)

SBVs *see* standard background variables (SBVs)

Scientific Social Surveys and Research: An Introduction to the Background, Content, Methods, and Analysis of Social Studies (Young and Schmid) 8

SDDS Plus *see* Special Data Dissemination Standard Plus (SDDS Plus)

SDF *see* Study Description Form (SDF)

SDMX *see* Statistical Data and Metadata exchange (SDMX)

SDR2 *see* survey data recycling database v.2.0 (SDR2)

SDR Database *see* Survey Data Recycling (SDR) Database

SDR Portal 149, 154, 155, 157, 161–162, 163

SDR project *see* survey data recycling (SDR) project

*SDR*Querier 161

semantic harmonization techniques 314

Senior Advisory Team (SAT) 68

SHaSA *see* Strategy for the harmonization of Statistics in Africa (SHaSA)

SHP *see* Swiss Household Panel (SHP)

single-item instruments 308, 312, 316, 318, 319

SLID *see* Survey of Labor and Income Dynamics (SLID)

SMC-FCS *see* substantive model compatible fully conditional specification (SMC-FCS)

SMQ *see* Study Monitoring Questionnaire (SMQ)

social transfers in kind (STIK) 272

sociodemographic variables 13, 43, 250, 338

Socio-Economic Panel (SOEP) 170, 173, **173**, 177, 253, 257

British Household Panel Study (BHPS) 177

Understanding Society, UKHLS 178

SOEP *see* Socio-Economic Panel (SOEP)

source data quality controls 153, 154

sources of knowledge for survey data harmonization 8–9

Southern African Democracy Barometer 59

Special Data Dissemination Standard Plus (SDDS Plus) 135

Specialized Technical Groups (STGs) 129, **130**

SPSS software packages 67, 122

SSP *see* sampling start point (SSP)

stakeholders 26

archives 27

in the collection of organic data

archives 32–33

harmonization of organic data 33

producers 32

users 33

data producers 26–27

data users 27–28

international actors and funding agencies 26

standard background variables (SBVs) 108, 113, 116, 122

standardization 13, 22, 53, 70, 157, 197, 243, 273, 274, 291

statistical analysis, using harmonized data in 347

combined data sets, challenges in the analysis of 353

missing values, dealing with 358–361

time, dealing with 354–358

weights, dealing with 361–362

data sets, challenges in the combination of 347

assessing the equivalence of survey projects 351–353

no censorship inclusive approach 348–349

using multilevel analysis and introducing a measurement level 349–351

recommendations 362–363

Statistical Data and Metadata exchange (SDMX) 135, 140

STGs *see* Specialized Technical Groups (STGs)

STIK *see* social transfers in kind (STIK)

Strategy for the harmonization of Statistics in Africa (SHaSA) 127, 128

civil registration and vital statistics (CRVS), development of 132

common software tools used 139–140

ex-ante harmonization, challenges with 138

ex-post harmonization, challenges with 139

ex-post harmonization, examples of 132

African Statistics Yearbook (ASY) 135

African Trade Statistics Yearbook (ATSY) 133

International Comparison Program for Africa (ICP-Africa) 135

KeyStats   135
  labor market indicators, guideline for
      producing   132
  governance, peace and security (GPS) statistics
      initiative   131, *131*
  National Strategies for the Development of Statistics
      (NSDS)   137–138
  quality assurance framework   134–136
  recommendations   140–142
Study Description Form (SDF)   117
Study Monitoring Questionnaire (SMQ)   117
subjective well-being (SWB)   174
substantive model compatible fully conditional
      specification (SMC-FCS)   260
Survey Data Recycling (SDR) Database   1
survey data recycling (SDR) project   149–155
  challenges to harmonization   156–161
  documentation and quality assurance   155–156
  recommendations
      for researchers interested in harmonizing survey
          data ex post   162–163
      for SDR2 users   163–164
  SDR Harmonization Workflow   *150*
  software tools of the SDR project
      161–162
survey data recycling database v.2.0
      (SDR2)   **149**, 150–155
  codes for missing values in   **160**
  data quality indicators in   **166–167**
survey harmonization and standardization
      processes   13
Survey of Health, Ageing, and Retirement in Europe   1
Survey of Labor and Income Dynamics (SLID)   176–177
survey research process, data harmonization on   1, 21,
      53, 211, 372
  designed data and organic data   31–32
  early conceptions of standardization and
      harmonization   22
  growing impact of data harmonization   23–25
  harmonization cost   29
  harmonization fit   30
  harmonization quality   29–30
  international survey programs, foundational
      work of   23
  moving forward   30–31

  stakeholders   26
      archives   27
      data producers   26–27
      data users   27–28
      international actors and funding agencies   26
  stakeholders in the collection of organic data
      archives   32–33
      data users   33
      harmonization of organic data   33
      producers   32
SWB *see* subjective well-being (SWB)
Swedish pseudo-panel   179
Swiss Household Panel (SHP)   179
synchronization meetings   64

t
Taiwan Social Change Survey (TSCS)   107, 108, 111,
      112, 114, 115, 118, 121, 122
technical information form (TIF)   65–66
test–retest reliability   312
3V model   372
TIF *see* technical information form (TIF)
time use data, *ex-post* harmonization of   285
  applied harmonization methods   289
      harmonizing the matrix of the diary   289–291
      variable harmonization   291–293
      variables   293–294
  challenges to harmonization   297–299
  documentation   294–295
  quality checks   296–297
  recommendations   301–302
  software tools   300–301
TIMXLS   140
Topical Guide segment   82
Total Survey Error (TSE) model   24, 45, 50, 153,
      334, 368
translation   44–45, 64
  and cognitive probing   317
  questionnaire   44–45, 48, 51
  of questions and answer choices   115
Translation, Review, Adjudication, Pretest,
      Documentation (TRAPD) model   11, 317
TSCS *see* Taiwan Social Change Survey (TSCS)
TSE model *see* Total Survey Error (TSE) model
12-Item Short-Form Health Survey (SF-12)   118

u

UKHLS *see* UK Household Longitudinal
        Study (UKHLS)
UK Household Longitudinal Study (UKHLS)   177, 178
UK National Institutes on Aging (NIA)   170
Understanding Society, UKHLS   177, 178
UNECA *see* United Nations Economic Commission for
        Africa (UNECA)
unit comparability, improving   313–314
unit differences, consequences of   313
United Nations Economic Commission for Africa
        (UNECA)   128
United Nations Statistics Division (UNSD)   134
UNSD *see* United Nations Statistics Division (UNSD)

v

Variable documentation   210, 215
Variable harmonization   91, 213, 256, 291, 295

w

wall materials variable (WALL)   220
WALL *see* wall materials variable (WALL)
WANA *see* West Asia and North Africa (WANA)
WAPOR *see* World Association of Public Opinion
        Research (WAPOR)
WDIs *see* World Development Indicators (WDIs)
web dissemination system   209, 210, 211, 221

West Asia and North Africa (WANA)   354, 356, 358
within-country harmonization   8
work history arrays   76–78, 79, 80–81, 82, 83, 85
World Association of Public Opinion Research
        (WAPOR)   8, 14, 15
World Development Indicators (WDIs)   277
World/European Values Survey (WVS-EVS)   351, 353
World Values Survey (WVS)   1, 4, 16, 23, 41–46, **42, 46**,
        47, 48–54, 114, 115, 351, 356
World Values Survey Association (WVSA)   43, 45,
        48, 52, 53
world value survey, harmonization in   41
    applied harmonization methods   42–48
    challenges to   49–51
    documentation and quality assurance   48–49
    ex-ante output harmonization methods   44
    ex-post harmonization   42
    input harmonization   44
    recommendations   52–54
    software tools   51–52
WVSA *see* World Values Survey Association (WVSA)
WVS *see* World Values Survey (WVS)
WVS-EVS *see* World/European Values Survey
        (WVS-EVS)

y

Young Men cohort   74, 75
Young Women cohort   74, 75

Printed and bound by CPI Group (UK) Ltd, Croydon, CR0 4YY

16/04/2025

14658371-0002